Sustainable Civil Infrastructures

About this Series

Sustainable Infrastructure impacts our well-being and day-to-day lives. The infrastructures we are building today will shape our lives tomorrow. The complex and diverse nature of the impacts due to weather extremes on transportation and civil infrastructures can be seen in our roadways, bridges, and buildings. Extreme summer temperatures, droughts, flash floods, and rising numbers of freeze-thaw cycles pose challenges for civil infrastructure and can endanger public safety. We constantly hear how civil infrastructures need constant attention, preservation, and upgrading. Such improvements and developments would obviously benefit from our desired book series that provide sustainable engineering materials and designs. The economic impact is huge and much research has been conducted worldwide. The future holds many opportunities, not only for researchers in a given country, but also for the worldwide field engineers who apply and implement these technologies. We believe that no approach can succeed if it does not unite the efforts of various engineering disciplines from all over the world under one umbrella to offer a beacon of modern solutions to the global infrastructure. Experts from the various engineering disciplines around the globe will participate in this series, including: Geotechnical, Geological, Geoscience, Petroleum, Structural, Transportation, Bridge, Infrastructure, Energy, Architectural, Chemical and Materials, and other related Engineering disciplines.

More information about this series at http://www.springer.com/series/15140

Hany Shehata · Youssef Rashed
Editors

Numerical Analysis of Nonlinear Coupled Problems

Proceedings of the 1st GeoMEast International Congress and Exhibition, Egypt 2017 on Sustainable Civil Infrastructures

Springer

2017 GEOMEAST

SSIGE
SOIL-STRUCTURE INTERACTION GROUP IN EGYPT
المجموعة المصرية للهندسة الجيوتقنية والبنية التحتية

Editors
Hany Shehata
Soil-Structure Interaction Group in Egypt (SSIGE)
Cairo
Egypt

Youssef Rashed
Cairo University
Cairo
Egypt

ISSN 2366-3405 ISSN 2366-3413 (electronic)
Sustainable Civil Infrastructures
ISBN 978-3-319-61904-0 ISBN 978-3-319-61905-7 (eBook)
DOI 10.1007/978-3-319-61905-7

Library of Congress Control Number: 2017946471

Printed on acid-free paper

This Springer imprint is published by Springer Nature
The registered company is Springer International Publishing AG
The registered company address is: Gewerbestrasse 11, 6330 Cham, Switzerland

Preface

Toward building sustainable and longer civil infrastructures, the engineering community around the globe continues undertaking research and development to improve existing design, modeling, and analytical capability. Such initiatives are also the core mission of the Soil-Structure Interaction Group in Egypt (SSIGE) to contribute to the ongoing research toward sustainable infrastructure. This conference series "GeoMEast International Congress and Exhibition" is one of these initiatives.

Ancient peoples built their structures to withstand the test of time. If we think in the same way, our current projects will be a heritage for future generations. In this context, an urgent need has quickly motivated the SSIGE and its friends around the globe to start a new congress series that can bring together researchers and practitioners to pursue "Sustainable Civil Infrastructures." The GeoMEast 2017 is a unique forum in the Middle East and Africa that transfers from the innovation in research into the practical wisdom to serve directly the practitioners of the industry.

More than eight hundred abstracts were received for the first edition of this conference series "GeoMEast 2017" in response to the Call for Papers. The abstracts were reviewed by the Organizing and Scientific Committees. All papers were reviewed following the same procedure and at the same technical standards of practice of the TRB, ASCE, ICE, ISSMGE, IGS, IAEG, DFI, ISAP, ISCP, ITA, ISHMII, PDCA, IUGS, ICC, and other professional organizations who have supported the technical program of the GeoMEast 2017. All papers received a minimum of two full reviews coordinated by various Session Chairs and supervised by the volumes editors through the Editorial Manager of the SUCI "Sustainable Civil Infrastructure" book series. As a result, 15 volumes have been formed of the final +320 accepted papers. The authors of the accepted papers have addressed all the comments of the reviewers to the satisfaction of the Session Chairs, the volumes editors, and the proceedings editor. It is hoped that readers of this proceedings of the GeoMEast 2017 will be stimulated and inspired by the wide range of papers written by a distinguished group of national and international authors.

Publication of this quality of technical papers would not have been possible without the dedication and professionalism of the anonymous papers reviewers. The names of these reviewers appear in the acknowledgment that follows. For any additional reviewers whose names were inadvertently missed, we offer our sincere apologies.

We are thankful to Dr. Hany Farouk Shehata, Dr. Nabil Khelifi, Dr. Khalid M. ElZahaby, Dr. Mohamed F. Shehata, and to all the distinguished volumes editors of the proceedings of the GeoMEast 2017. Appreciation is extended to the authors and Session Chairs for their significant contributions. Thanks are also extended to Springer for their coordination and enthusiastic support to this conference. The editors acknowledge the assistance of Ms. Janet Sterritt-Brunner, Mr. Arulmurugan Venkatasalam in the final production of the 15 edited volumes "Proceedings of GeoMEast 2017".

Contents

Deciding for Remediation of the Seepage Barrier System in Foundation of an Earthfill Dam Based on the Monitoring Data and Numerical Analysis

Hamed Farshbaf Aghajani(✉)

Department of Civil Engineering,
Faculty of Engineering, Azarbaijan Shahid Madani University,
Kilometer 35 of Tabriz/Azarshahr Road, P.O. Box 53714-161, Tabriz, Iran
h.farshbaf@azaruniv.ac.ir

Abstract. Today, dams play vital infrastructure role in modern civilizations. For preventing the unfortunate failure during dam construction and operation, the frequent surveillance of dam safety should be considered. In the Gheisaragh dam, due to prevalence of solution in the foundation, the remedial cutoff wall was constructed. Regarding to some practical limitations and more critical condition, the cutoff wall is sketched in the middle and the right abutment of the foundation and extending over the left abutment is postponed to future requirements.

This paper aims to evaluate seepage behavior of the dam foundation in the left abutment and if any, makes a decision for remediation requirements. Hence, the long-term data of piezometers and observation wells installed in the left abutment foundation are investigated. Besides, the anticipated seepage pattern of dam foundation is determined from 2-D finite element transient seepage analysis and compared with measurements.

The results indicate that in both numerical and monitoring analyses, the water level in piezometers is considerably affected by the reservoir fluctuation after a short delay due to the short length of seepage paths in the foundation. However, after drawing down the reservoir, significant falling is observed in the piezometric water level with more intensity in comparison with the numerical analysis. Besides, the ground water level in the downstream of left abutment is considerably dropped and lies in hydrostatic condition. These consequences are inferred that the ground of the dam in left abutment has less ability to retain the water and thus, some open channel with higher permeability exist on the foundation. Therefore, the covering the left abutment of the dam foundation with the cutoff wall should be considered.

1 Introduction

Today, dams play a vital role in human civilizations in supplying the water and energy. Regarding to giant dimensions of dam body and reservoir, any defect in dam body can lead to unfortunate events. One of the main factors threading the dam safety is the solution of the evaporite rock type existed in dam foundation (Foster et al. 2000).

H. Shehata and Y. Rashed (eds.), *Numerical Analysis of Nonlinear Coupled Problems*,
Sustainable Civil Infrastructures, DOI 10.1007/978-3-319-61905-7_1

In worldwide, many evidences are found that the solution of bedrock makes the unfortunate events and even lead to failure of the dam (Nusier et al. 2002; Turkmen 2003; Rice and Duncan 2010; Taluki et al. 2015). For preventing the aforementioned events, regular surveillance and controlling the dam safety via analyzing the instrumentation data and numerical simulation are recommended by many dam experts (Johansson 1997; ASCE 2000).

The Gheisaragh dam was functioned since 2007. Regarding to some geological specifications of dam foundation, this dam has been faced with solution problem leading to extra settlement. For remediation of these problems, a cutoff wall was constructed at right and middle part of the dam foundation and extending the water barrier system toward the left abutment was postponed to the future requirements. This paper aims to investigate the critical seepage behavior in dam foundation at left abutment and answer to this question that whether any remedial works are required in left abutment. To this end, the water seepage pattern in dam foundation is determined by analyzing the long term instrumentation data and comparing with numerical analysis.

2 Reviewing Previous Events of Dam

The Gheisaragh dam is an earth fill type dam located in north-west of Iran with a distance of 50 km from Tabriz city and went under operation since 2007. The height and crest length of dam body are 16 and 988 m, respectively. The total volume of dam reservoir is about 2.7 MCM.

As the initial plan for water sealing of dam foundation, the trench of clay core with a depth of 5 m below the ground surface covers the upper pervious layer of foundation. However, one year after the reservoir operation, some unfortunate evidences of sinkholes in downstream valley and extra settlement in dam crest were reported. Besides, the volume of water collected in toe drainage of the dam was unexpectedly increased. At the same time, the chemical analysis of water exerted from downstream sinkholes indicated the high content of gypsum content in water. All of these concerns were summarized that the dam foundation is suspected to solution process. For stopping the solution of dam foundation, a cement-bentonite cutoff wall to a depth of 20 m was constructed in the upstream heel of the dam. Furthermore to complete the seepage barrier system, the upstream face of the dam is covered by an impervious concrete shell and connected to the cutoff wall. Regarding to more evidences of solution in the right and middle parts of the dam foundation and some limitations on construction season, the cutoff wall was constructed in these parts of the foundation and extending the cutoff wall toward the left bank of dam foundation was postponed to future requirements. The detailed specifications of dam rehabilitation works can be found in other publications (Moradi et al. 2009). The layout of dam body and extension of rehabilitation works (cutoff wall and impervious upstream shell) are illustrated in Fig. 1.

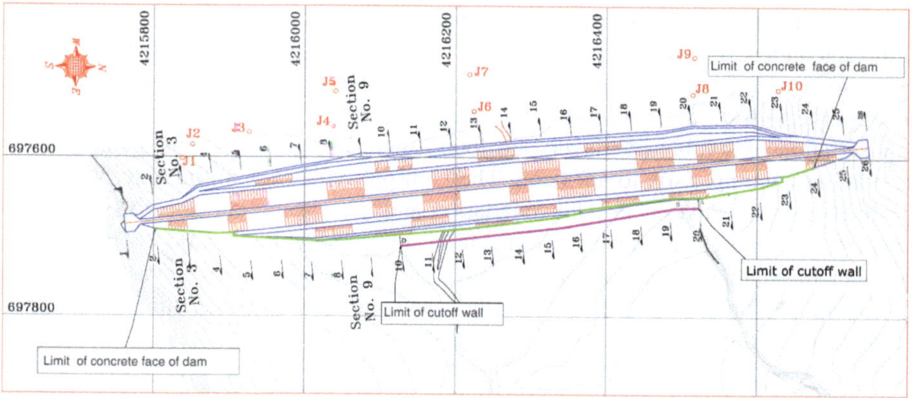

Fig. 1. The general plan of dam body and extension of remediation works (the coordinates are in UTM system)

3 Geological Illustration of Dam Foundation

By conducting a comprehensive site investigation program, the geological condition of dam foundation was thoroughly explored. The longitudinal geological profile of the dam foundation is presented in Fig. 2. The upper layer of foundation is comprised of alluvial soils with fine grained texture which some gypsum and lime beds and streaks are occasionally disturbed inside the alluvium. The average thickness of the shallow alluvial layer is about 5 m.

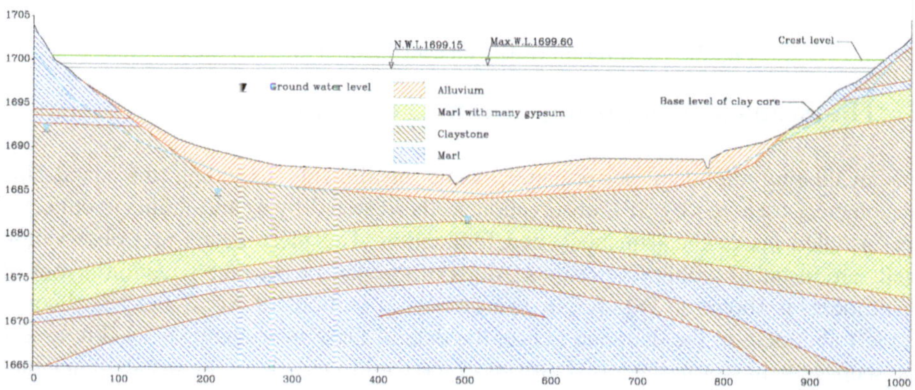

Fig. 2. The longitudinal geological section of dam (The ground water level was captured when boreholes were bored)

Beneath the alluvial layer, the dam foundation is comprised of marl rocks. According to the BS classification system (BS1377-1 1990), the bedrock is categorized as the firm rocks. However, most of rock samples are exhibited as the moderately cemented soil with the average plastic limit about 18. Before the aforementioned event of solution was occurred, the permeability of the bedrock was measured about 10^{-8} m/s in accordance with Lugeon test (Lugeon 1933) and hence categorized as impervious rock.

In some situations, the gypsum strikes with different thickness and degree of purity are observed inside the marl bedrock which are shown in Fig. 3. The intensity and concentration of soluble interlayers are found more in right bank and middle part of foundation in comparison to left bank. The high purity of gypsum interlayer increases the likelihood of solution in dam foundation and even some evidence of solution is detected in the right abutment of foundation after reservoir operation. The solution event in the dam foundation consequently leads to sinkholes in downstream, extra settlement of dam body and changes the chemical quality of exerting water from downstream springs.

Fig. 3. The gypsum strike embedded within the marl rocks

4 The Seepage Monitoring in Dam Foundation

For evaluating the dam performance, some instruments including the stand-pipe piezometers (denoted by SP), observation wells (denoted by J) and drainage collector are installed inside the dam body and foundation. The piezometers and observation wells are installed within five cross sections of the dam, where the situation layout of instruments is shown in Fig. 1. The instruments in sections of No. 3 and No. 9 cover the left abutment of the dam and hence, the data of these instruments is used to assess the seepage behavior in the left abutment of the dam foundation. The installation detail and situations of the instruments in sections of No. 3 and No. 9 are shown in Fig. 4. In No. 3 section, piezometers of SP-3(1) and SP-3(2) are installed in foundation at various depths. In section of No. 9, the piezometers of SP-9(1) and SP-9(3) are installed in the foundation and clay core and piezometers of SP-9(2) and SP-9(4) are located in the downstream filter and downstream foundation, respectively.

The water level in all piezometers and observation wells that are in proper condition have been regularly measured from the earliest date of reservoir operation (year of 2009) up to the present time (year of 2014). However, other instruments are unfortunately damaged or clogged during reservoir operation and hence, cannot be functioned.

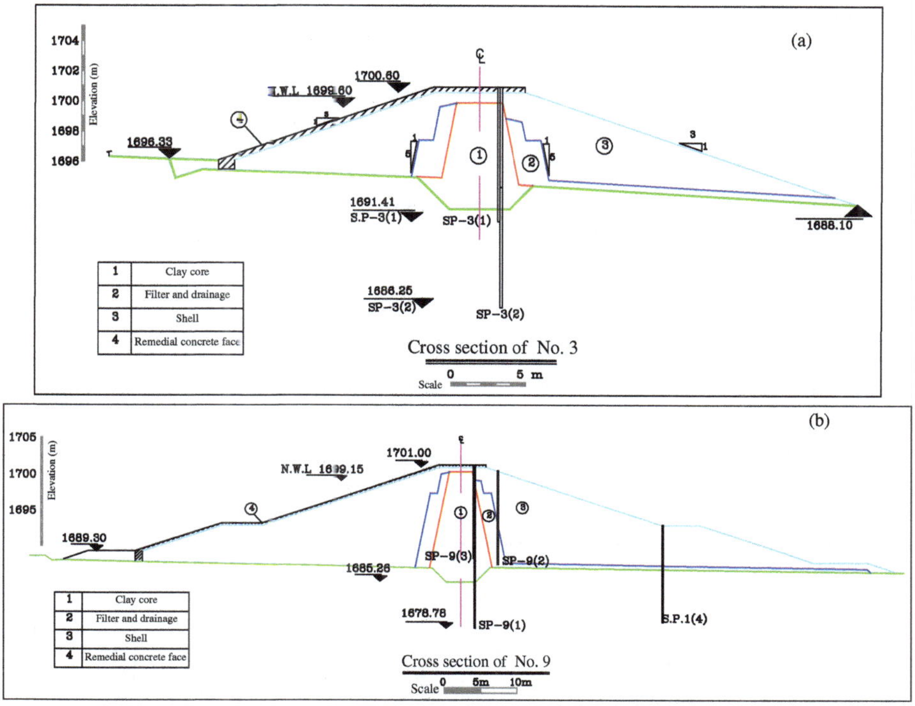

Fig. 4. The installation location of instruments in dam body: (a) Cross sections of No. 3; (b) Cross section of No. 9

The graph of long-term monitoring of water level in piezometers of SP-3(1), SP-3(2) and SP-9(1) together with the graph of reservoir elevation are presented in Fig. 5. As seen, the variation of water levels in the dam foundation completely conforms to fluctuation of reservoir and rising and lowering the reservoir elevation instantly influence the piezometric level in the foundation. However, the effect of reservoir fluctuation on the variation of water level in the SP-9(1) piezometer in No. 9 section is more distinctive than the SP-3(2) piezometer in No. 3 section. When the reservoir goes under operation and reservoir elevation fall below the ground surface of No. 3 section (i.e. elevation of 1696.33 m), the water level in piezometers of No. 3 section goes to the steady condition and are not influenced by reservoir fluctuation. Furthermore, during the full state of the reservoir when the reservoir elevation is located top of the ground surface of Section No. 3, the difference of total energy head between the piezometers of Section No. 3 (SP-3(1) and SP-3(2)) and reservoir head is about 3.7 m. At the same time, the energy head drop in piezometer of SP-9(1) measured as 6.7 m. It means that more energy dissipation occurs in dam foundation at Section No. 9 in comparison with Section No. 3.

From monitoring graph, it can be seen that even though the different embedment depth of both piezometers of SP-3(1) and SP-3(2), the identical water level in both piezometers indicates that the water have equal total head in both situations and hence

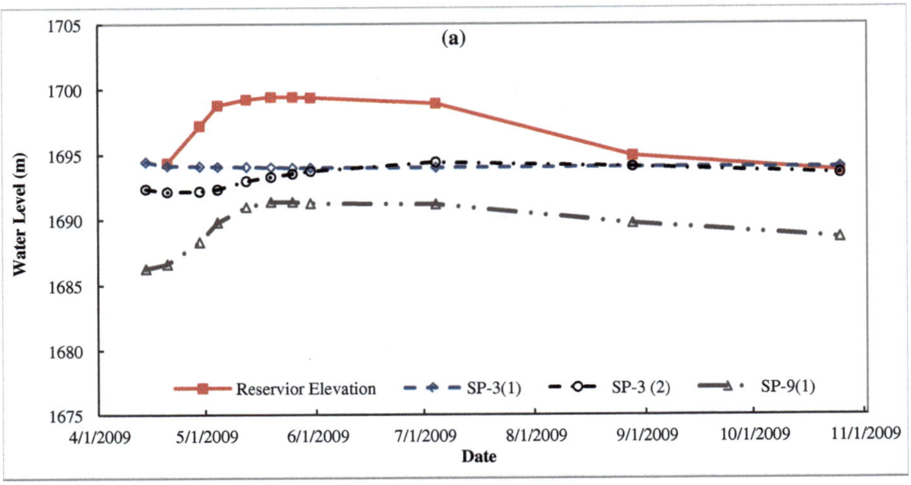

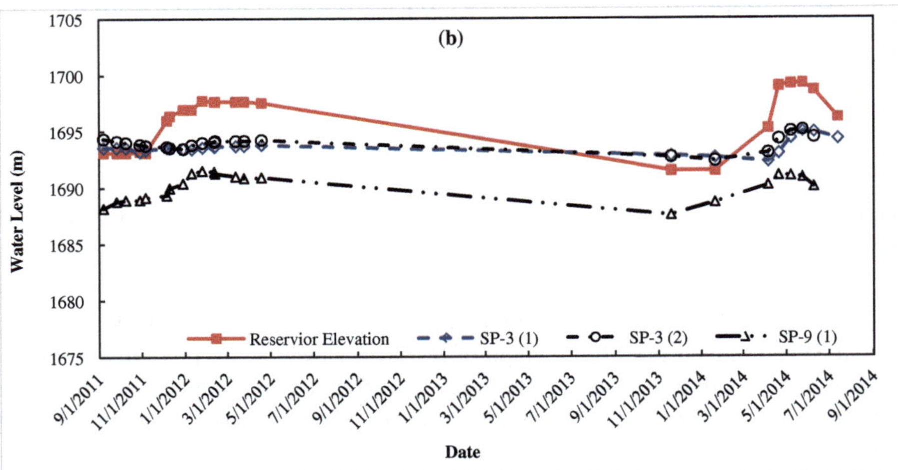

Fig. 5. The graph of reservoir elevation and water level in the dam foundation measured from piezometers of SP-3(1) and SP-3(2) in No. 3 section and piezometer of SP-9(1) in No. 9 section: (a) during year of 2009, (b)from 2011 to 2014

both piezometers are located on the identical equipotential line. In other words, regarding to the absence of the seepage barrier system in left abutment foundation, the water is directly flowed beneath the clay core trench along a direct path which is perpendicular to the equipotential line connecting between piezometers of SP-3(1) and SP-3(2). In addition, when the reservoir elevation is drawn down below the ground surface of No. 3 section (i.e. elevation of 1696.33 m) and reservoir surface limit is approached to ground of No. 9 section, the water level in both piezometers of No. 3 section lie in equivalent to reservoir elevation. This consequent indicates that after the reservoir drawn down, the piezometric level in the whole of the left abutment of dam foundation is consequently dropped until the hydrostatic equilibrium condition is established in the left abutment of dam foundation.

The water flowed through right and left banks of dam body and foundation are separately measured in drainage system which is embedded in the toe of the dam along the section of No. 9. The long-term graph of drainage flux measured in both right and left banks during 2011 to 2014 years is presented in Fig. 6. At earlier times, more water is seeped from left bank in comparison with the right bank. However, in later date, the seepage water in right bank is considerably enhanced and overtakes the left bank drainage water.

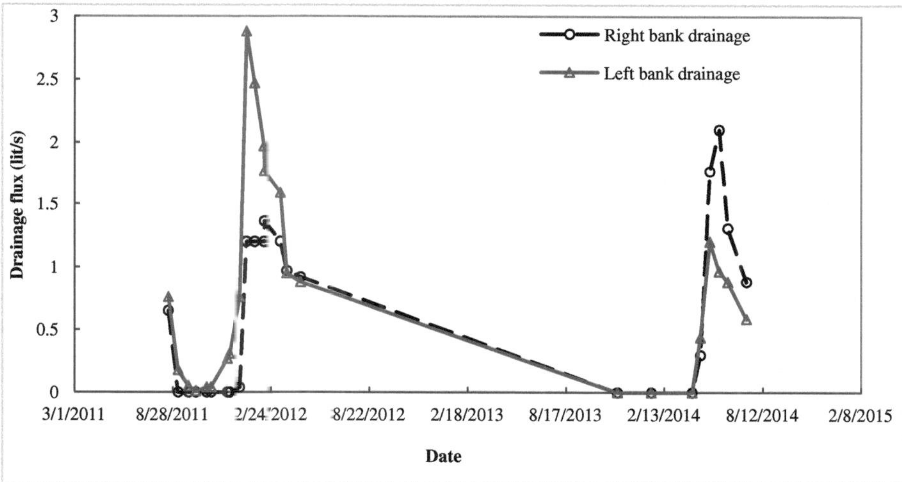

Fig. 6. The graph of the measured flux of water collected in the drainage collector system at the toe of dam body

5 Numerical Analysis of Seepage

The anticipated seepage behavior in the left abutment of dam body and foundation is determined by performing finite element seepage analysis via GeoStudio software (GEOSLOPE International 2004). Because of the long length of dam body, the supposition of 2-dimensional flow of water across the dam can be valid and hence, the seepage analysis is performed on the 2-dimensional cross sections of the dam where the instruments are installed. The extension and geometry of regions in numerical model completely coincide to the geological profile of foundation and dam body which is shown in Fig. 7. Correspondingly, the permeability of dam body and foundation layers are determined based on the in-situ permeability tests for foundation material (measured before dam construction) and laboratory permeability test for dam body materials and presented in Fig. 7.

For considering the fluctuation of the dam reservoir in numerical analysis, the time-depended boundary condition is assigned to the upstream face of dam body

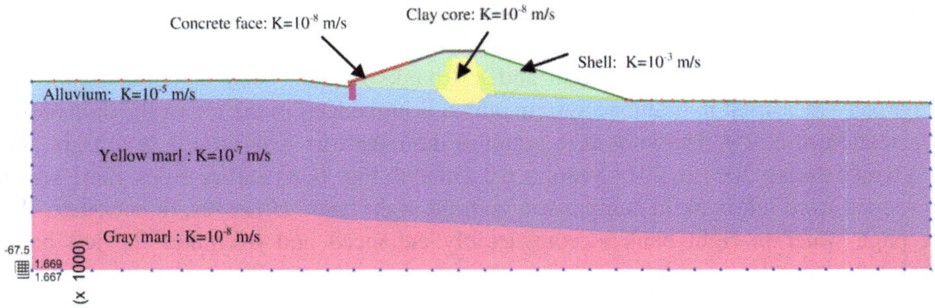

Fig. 7. The geometry of No. 3 section used in numerical analysis

conforming to the real condition of reservoir elevation and thus, the seepage analysis is performed as transient type.

In Figs. 8a and b, the equipotential lines in dam body and foundation along the cross section of No. 3 are presented for two upper and lower limit conditions of the reservoir. When the reservoir is filled up to a maximum elevation (1699.3 m), the drop of total head in dam body is mostly occurred in the remedial concrete face. Moreover, the equipotential line belongs to total head contour of 1695 m is located beneath the dam center line where both piezometers of SP-3(1) and SP-3(2) are positioned along this equipotential line. This means that from the numerical analysis aspect, the water in both piezometers should be located in equivalent level. This consequent has formerly

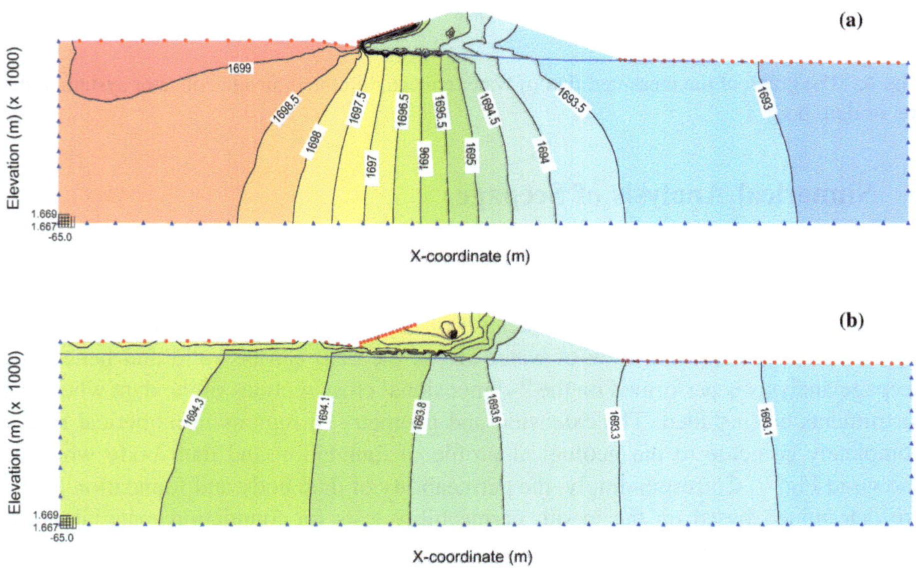

Fig. 8. The equipotential lines in dam body and foundation of No. 3 section, (a) when the reservoir is filled up to the maximum elevation (1699.3 m), (b) the reservoir is drawn down to the minimum elevation (1691.5 m)

been demonstrated in monitoring graph of both piezometers and hence, the findings numerical analysis is confirmed.

The graph of total head of water at the identical locations of the instruments of SP-3 (1) and SP-9(1) is computed from the numerical analysis and compared with the measured water level in these piezometers and presented in Fig. 9. As seen, general conformity exists between the predicted and measured water level in these piezometers and both measured and predicted water level is completely influenced with reservoir fluctuation. However, when the reservoir elevation is drawn down, significant difference is observed between predicted and measured water level in such that the actual water level in piezometer of SP-3(1) is located about 0.9 m lower than the predicted level. The difference between the predicted and measured water level in piezometer of SP-9(1) in No. 9 section is about 1.4 m. In fact, lowering the reservoir level causes more dropping in piezometer water level which exceeds the expectation value from numerical analysis.

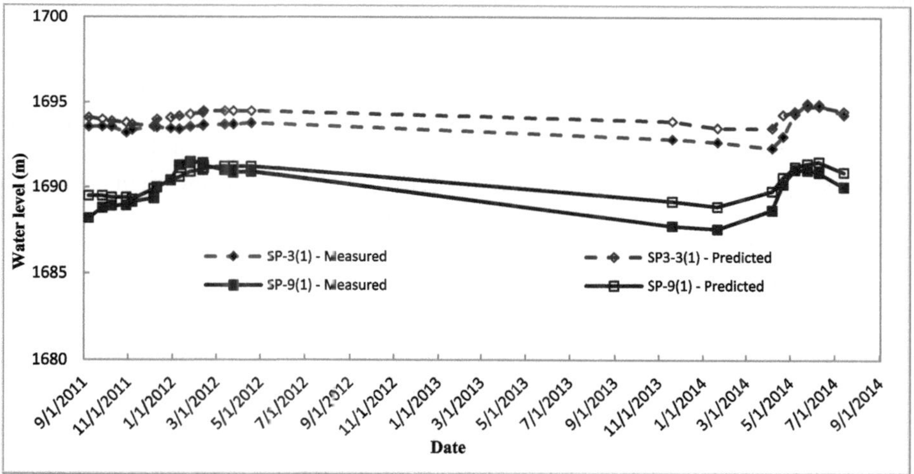

Fig. 9. Comparison between predicted and measured water level in the dam foundation at piezometer locations of SP-3(1) in section of No. 3 and SP-9(1) in section of No. 9

6 Discussion on the Seepage Pattern in the Left Abutment

By reviewing and comparing the monitoring data and result of numerical analysis, some consequents are emerged which specify the seepage behavior of the dam foundation in left abutment. The first is that as water is flowed through the dam foundation, less energy of water is dissipated and the difference of total energy head between the reservoir and beneath the dam core is insignificant. This phenomenon is more distinctive in the section of No. 3 rather than the section of No. 9. Furthermore, any little fluctuation in reservoir elevation is instantaneously transmitted to the piezometric regime of foundation and the piezometers are influenced by the reservoir variations. These phenomena are captured from both monitoring and numerical analysis and would be expected and common for this dam due to the nonexistence of the water tightening system in the left abutment of the dam foundation and thus, water travels through a short and direct pass in dam foundation with a less energy drop.

Nevertheless, the most prominent feature of seepage pattern in the left abutment is the unexpected falling of piezometric level during the reservoir drawn down. When the reservoir elevation is lowered below the ground surface, the piezometric regime is continuously dropped until the hydrostatic equilibrium condition is established in the ground out of reservoir limits. Even though the drop of water level in the ground due to reservoir lowering is detected in both measurements and numerical analysis, however, this phenomenon is observed with more intensity than expectations from numerical analysis.

The dropping and then attaining the hydrostatic equilibrium in piezometric regime is observed not only in the foundation beneath the dam body, but also in the left bank of the downstream valley of the dam. For more clarification, the graph of ground water level (GWL) measured in the observation wells during various dates is presented in Fig. 10. For well illustration, the topography graph of the top surface of the wells is shown in this figure. The observation wells are installed in the downstream of the dam which the layout of wells is presented in Fig. 1 by notation of J and assigned number. The wells of J1, J2 and J4 are located at the left bank of the downstream valley of the dam. The wells of J1 and J2 are arranged in downstream of No. 3 section and J4 well covers the downstream of No. 9 section. As seen from the graph of Fig. 10, during all stages of reservoir operation, the ground water level in the left bank of the downstream valley is significantly located in lower depth in comparison to the right bank. For instance, even though the ground water level in J1 well is ranged between depth of 8.7 and 10.7 m below the ground surface, the GWL in J10 well (along the section of No. 23) is located at the depth of 4.5 m.

By comparing the GWL in the wells of J1 and J10 which correspond to the left and right bank of downstream valley respectively, one can realize that the range of GWL

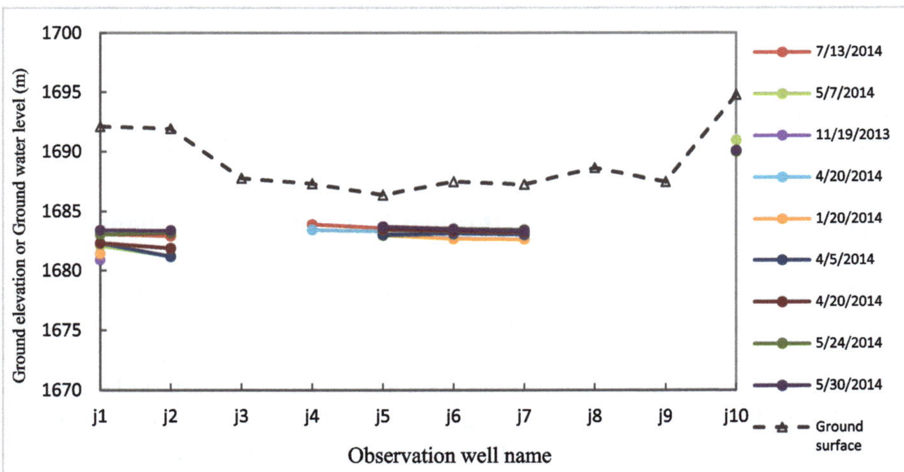

Fig. 10. The longitudinal profile of ground water level measured in downstream observation wells during various dates together with top ground elevation of wells

variation in the left bank is significantly greater than the right bank. Also, variation of GWL in the left bank of the downstream valley completely depends on the reservoir level fluctuation. In fact, when the reservoir level is drawn down to the lowest elevation, the lowest level of ground water in wells of left bank is recorded and by rising the reservoir elevation, the GWL in left bank is increased until the hydrostatic condition is reached. In contrast, according to the proper functioning of the water tightening system in the dam foundation at the right abutment, the collaboration between GWL at the right bank of downstream valley and reservoir is completely broken and thus, the fluctuation of GWL in the right abutment becomes minor.

The other characteristic feature emerged for ground water regime at the left bank is the insignificant hydraulic gradient of GWL along the longitude direction. As seen in GWL profile shown in Fig. 10, the ground water level observed in all wells of left bank is almost located in equivalent level with wells located at dam center (i.e. wells of J5, J6 and J7) and at some dates, hydrostatic equilibrium condition is established between the dam center and the left bank of the downstream valley. In contrast to the left bank, the ground water level in the right bank of the downstream valley is located about 6.5 m higher than the dam center and hydraulic gradient permanently exists for ground water level in right bank. This gradient cause to flow the ground water toward the valley center and thus, extra water is concentrated in the right abutment drainage collector system.

Dropping the piezometric level in the dam foundation and GWL in the downstream valley at the left abutment due to reservoir drawn down and then, attaining the hydrostatic equilibrium condition imply the less ability of dam foundation materials to retain the water. In other hand, even though the in-situ permeability test results indicate low permeability of foundation material, however, unexpected falling of water level infers that some open channels are locally existed inside the foundation strata's which provide conductive paths for water to flow through the dam. These local channels are gradually established by solution of gypsum strikes and beds embedded inside the alluvial and marl layers. As thought earlier, the more intensity of the solution process in the right and middle part of the foundation led to aforementioned events and thus the remedial works were focused on the right abutment foundation to rehabilitate and seal the foundation. However, even though the less distribution of evaporite interlayer and strikes are found in left abutment foundation, developing the solution of small strikes of gypsum cause to enhance the size and number of open channels in foundation and hence provided paths for flowing the water through the dam foundation may be increased. For avoiding to further solution in foundation, sealing the dam foundation at left abutment and breaking the flow paths of water inside the foundation layers are necessary. Therefore, extending the cutoff wall toward the left abutment for remediating and tightening the dam foundation is recommended.

7 Conclusions

Regarding to high costs and practical difficulty of any remedial works in dams, the requirement of remedial work for treating the foundation and body of dams should be thoroughly examined based on the up-to-date surveillance of dam condition. In this paper, the monitoring of piezometric regime in the left abutment of the dam foundation indicates the considerable falling in piezometric water level and attaining the hydrostatic equilibrium condition after drawing down of reservoir elevation. These phenomena are related to existing the open channel and voids in the impervious layer of foundation that are established by solution of evaporite gypsum beds and strikes within marl rocks.

In order to decide the critical depth and optimal method required for treating the dam foundation at left abutment, conducting the comprehensive complementary site investigation program in the left bank of foundation should be considered. This program can be included the boreholes boring within the left abutment to recognize the extension depth of solution in foundation, in-situ permeability test to identify the current value of rock permeability in both intact and defected zone of the foundation and non-destructive geophysical investigation to recognize the horizontal and vertical extension of defected zone.

References

ASCE Task Committee: Guidelines for Instrumentation and Measurements for Monitoring Dam Performance. ASCE Task Committee on Instrumentation and Dam Performance. ASCE Publications, Dey 11, 1378 AP (2000)

BS1377-1: Methods of Test for Soils for Civil Engineering Purposes. British Standard Institution, BS 1377-3 (1990)

GEOSLOPE International: GeoStudio software (2004). http://www.geo-slope.com/

Foster, M., Fell, R., Spannagle, M.: The statistics of embankment dam failures and accidents. Can. Geotech. J. **37**(5), 1000–1024 (2000)

Johansson, S.: Seepage monitoring in embankment dams. Ph.D. thesis, Institutionen för anläggning och miljö, Swenden (1997). ISBN: 91-7170-792-1

Lugeon, M.: Barrage et Géologie, Paris, France (1933)

Moradi, G., Abbasnejad, A.R., Farshbaf Aghajani, H.: The investigation of excess seepage of Gheisaragh dam and treatment method. In: Proceedings of 2nd International Conference on Long Term Behaviour of Dams (LTBD 2009), Graz, Austria, 12–13 October 2009. ISBN: 978-3-85125-070-1

Nusier, O.K., Alawneh, A., Malkawi, A.: Remedial measures to control seepage problems in the Kafrein dam, Jordan. Bull. Eng. Geol. Environ. **61**(2), 145–152 (2002)

Rice, J., Duncan, J.: Findings of case histories on the long-term performance of seepage barriers in dams. J. Geotech. Geoenviron. Eng. **136**(1), 2–15 (2010)

Talouki, H.H., Lashkaripour, G.R., Ghafoori, M., Saba, A.: Assessment and presentation of a treatment method to seepage problems of the alluvial foundation of Ghordanloo dam, NE Iran. J. Geol. Soc. India **85**(3), 377–384 (2015)

Turkmen, S.: Treatment of the seepage problems at the Kalecik dam. Eng. Geol. **68**(3–4), 159–169 (2003)

Nonlinear Seismic Response of Concrete Gravity Dams

Djamel Ouzandja[1](✉), Boualem Tiliouine[2], and Toufiq Ouzandja[3]

[1] Civil Engineering Department, Faculty of Technology, University of Msila, M'Sila, Algeria
dj_ouzandja@yahoo.com

[2] Civil Engineering Department, Ecole Nationale Polytechnique, Algiers, Algeria
boualem.tiliouine@g.enp.edu.dz

[3] Civil Engineering Department, Faculty of Technology, University of Bejaia, Béjaïa, Algeria

Abstract. This paper aims to present the nonlinear seismic response of concrete gravity dams considering dam-foundation interaction. For illustrative purpose, the Oued Fodda concrete gravity Dam, located in Chlef (northwestern Algeria), is selected as an example. Linear and nonlinear analyses are performed using ANSYS. The Druker-Prager model is used for dam concrete and foundation rock in nonlinear analysis. The hydrodynamic pressure of the reservoir water is modeled as added mass using the Westergaard approach. The maximum displacements and principal stresses are shown by the height of the dam. The results obtained from linear and non-linear analyses are compared with each other.

Keywords: Concrete gravity dams · Dynamic dam-foundation interaction · Nonlinear dynamic analysis · Drucker-Prager model · Finite element method

1 Introduction

Concrete gravity dams represent complex constructive systems of strategic importance. They are particularly used for electricity generation, water supply, flood control, irrigation, recreation, and other purposes. They are a fundamental part of the society's infrastructure system.

There are several factors affecting the dynamic response of concrete gravity dams to earthquake ground motions. Some of them are the interaction of the dam with the foundation rock and water in reservoir, and non-linear behavior of dam itself and its fondation. The dam-foundation interaction problem was investigated by Chopra and Chakrabarti [1], Leger and Boughoufalah [2], Nuss et al. [3], Bayraktar et al. [4], Lemos and Gomes [5], Moussaoui and Tiliouine [6], Saleh and Madabhushi [7], Lebon et al. [8], Saouma et al. [9], Hariri-Ardebili and Mirzabozorg [10], Burman et al. [11], Hariri-Ardebili and Mirzabozorg [12], and Ouzandja and Tiliouine [13].

This study investigates the nonlinear seismic response of a concrete gravity dam. The Oued Fodda Dam is considered in the numerical analyses. For this purpose,

H. Shehata and Y. Rashed (eds.), *Numerical Analysis of Nonlinear Coupled Problems*, Sustainable Civil Infrastructures, DOI 10.1007/978-3-319-61905-7_2

two-dimensional dam-foundation finite element model is used. The hydrodynamic pressure of the reservoir water is modeled utilizing the added mass concept [14]. The Druker-Prager model [15] is used for dam concrete and foundation rock in nonlinear analysis. All numerical analyses are performed using ANSYS [16]. The results obtained from linear and nonlinear analyses are compared with each other.

2 Numerical Model of Oued Fodda Concrete Gravity Dam

2.1 Material Properties

The Oued Fodda concrete gravity dam is located approximately 20 km of Oued Fodda (Chlef), in northwestern Algeria, and founded over a massive limestone known as "Koudiat Larouah". The reservoir is mainly used for irrigation purposes. The capacity of the dam is 125.5 hm^3. The maximum height "H" and base width of the dam are 101 m and 67.5 m, respectively. The dam crest is 5 m wide and the maximum height of the reservoir water is considered as 96.4 m. The dimensions of the dam-foundation system are shown in Fig. 1.

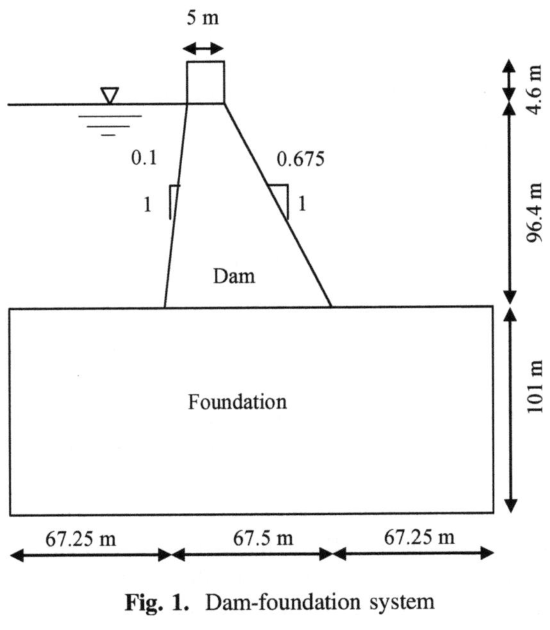

Fig. 1. Dam-foundation system

The material properties of Oued Fodda dam including its foundation are reported in Table 1 below. The Drucker-Prager model [15] is used in nonlinear analysis for concrete of the dam and foundation soil. The cohesion and the angle of internal friction of the dam body and foundation rock are assumed as to be 2.50 Mpa and 35°, respectively. The concrete of the dam has tensile strength of 1.6 MPa and compressive strength of 20 MPa.

Table 1. Material properties of Oued Fodda concrete gravity dam

Material	Material properties		
	Modulus of elasticity (MPa)	Poisson's ratio	Mass density (kg/m^3)
Dam	24600	0.20	2640
Foundation	20000	0.33	2000

2.2 Finite Element Model of Dam-Foundation System

The dam-foundation system is investigated using the two-dimensional finite element model shown in Fig. 2. The dynamic effect of the reservoir during the analysis is modeled by the Westergaard approach [14] based on the added mass concept. This finite element model is created using software ANSYS [16].

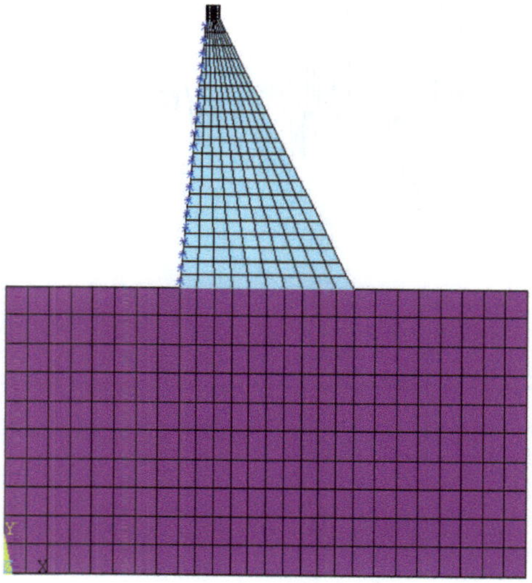

Fig. 2. Finite element discretization of dam-foundation system

A two-dimensional (2D) finite element model with 500 plane solid elements (PLANE 82) is used to model the dam and the foundation. A finite element model with 20 structural masses (Mass21) is used to model the reservoir water. The solid elements used in the analysis have eight nodes and 2 × 2 integration points. Solid element matrices are computed using the Gauss numerical integration technique [17].

3 Numerical Results and Discussion

This study investigates the seismic response of Oued Fodda concrete gravity dam considering barrage-foundation interation. For this purpose, the horizontal component of the 1980 El Asnam earthquake acceleration scaled by factor of 2.5 is utilized in analyses (Fig. 3). In 1980, El Asnam Province has already been shaken by the strong earthquake (M7). Unfortunately, we only have a record of a replica of this earthquake with peak ground acceleration (PGA) 0.132 g. Consequently, we chose the record of replica earthquake with a scaling factor of 2.5 to obtain an earthquake acceleration record with PGA 0.33 g, nearly equal to PGA of record of the strong earthquake (M7) which occurred in 1980. The linear and nonlinear time history analyses are performed using ANSYS [16]. The maximum horizontal displacements in upstream direction and the maximum principal stresses in the dam along its height are presented.

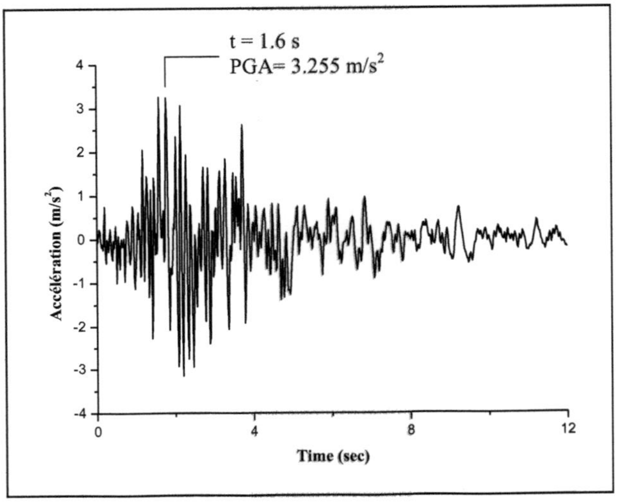

Fig. 3. Time history of horizontal acceleration for 1980 El Asnam earthquake record scaled by factor of 2.5

3.1 Displacements

The Fig. 4 represents the maximum horizontal displacements of dam in upstream direction obtained from linear and nonlinear transient analyses.

The Fig. 5 shows the time history of horizontal displacement at the crest of dam in linear and nonlinear analyses. The maximum horizontal crest displacements are equal to 4.10 and 3.54 cm, respectively, in linear and nonlinear analyses. However, it should be recognized that the use of nonlinear material models at the foundation and dam could increase or decrease the displacements depending on ground motion characteristics, surrounding foundation properties and the dam mechanical properties [18, 19].

3.2 Stresses

The Figs. 6 and 7 represent the principal tensile and compression stress distributions along the dam height.

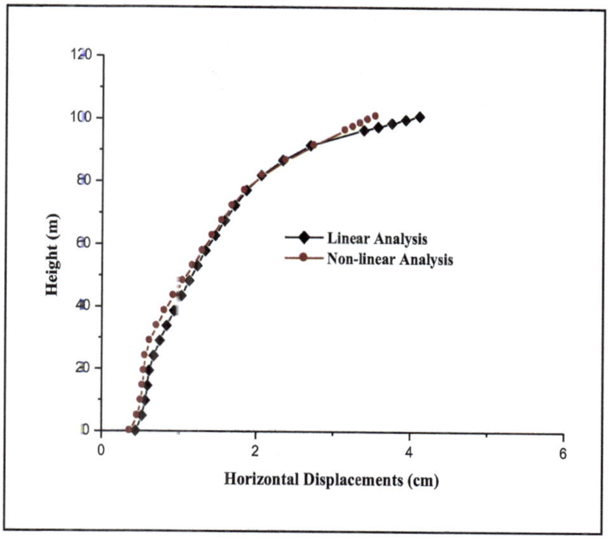

Fig. 4. Maximum horizontal displacements in upstream direction according to linear and nonlinear analyses

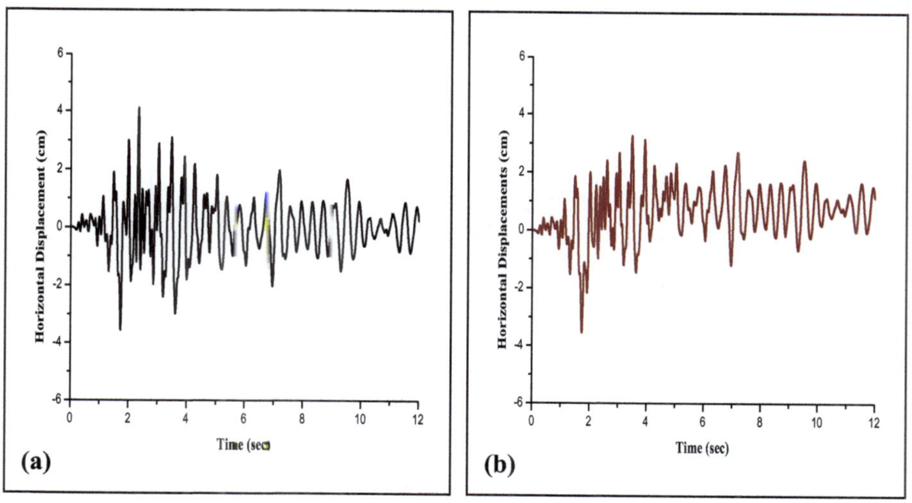

Fig. 5. Time history of horizontal displacement at crest of dam: (a) linear analysis and (b) nonlinear analysis

The maximum values of principal tensile and stresses were observed to be 7683.90 and 2347.09 kN/m^2, respectively, in linear and nonlinear analyses. The maximum values of principal compression stress were observed to be −6376.90 and −2513.83 kN/m^2, respectively, in linear and nonlinear analyses. Therefore, in nonlinear analysis, a decrease of 70 and 60%, respectively, in the magnitude of principal tensile and compression stresses were noticed.

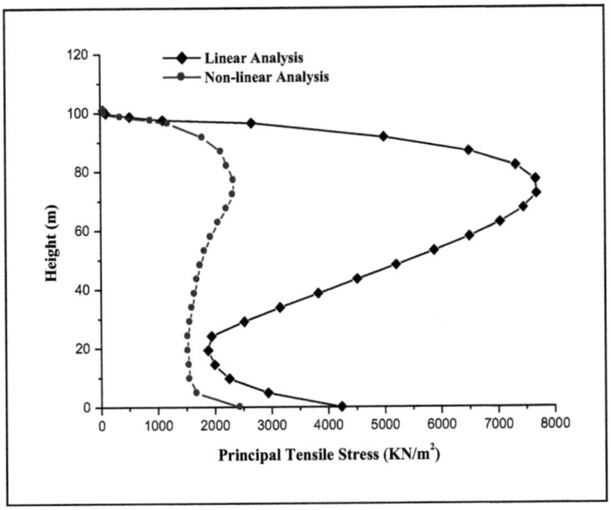

Fig. 6. Maximum principal tensile stresses along dam height according to linear and nonlinear analyses

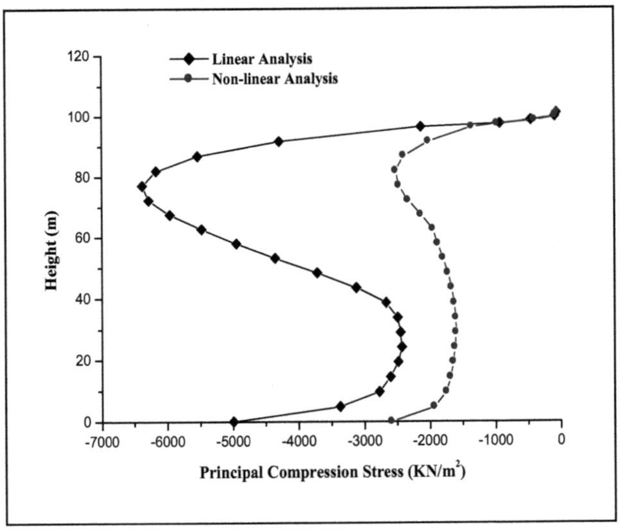

Fig. 7. Maximum principal compression stresses along dam height according to linear and nonlinear analyses.

When comparing linear and nonlinear analyses, a substantial decrease in the distribution of principal tensile and compression stresses due to material nonlinearity effects can be observed from Figs. 6 and 7. Table 2 below shows the maximum principal stress values at heel of the dam using both linear and nonlinear analyses.

It is observed that material nonlinearity of the foundation rock and dam decreases the principal stresses at heel of the dam when compared to the case when foundation

Table 2. Maximum principal stress values at heel of the dam in linear and nonlinear analyses

Principal stresses	Linear analysis	Nonlinear analysis
Principal tensile stress (kN/m^2)	4229.046	2443.51
Principal compression stress (kN/m^2)	−4998.51	−2581.37

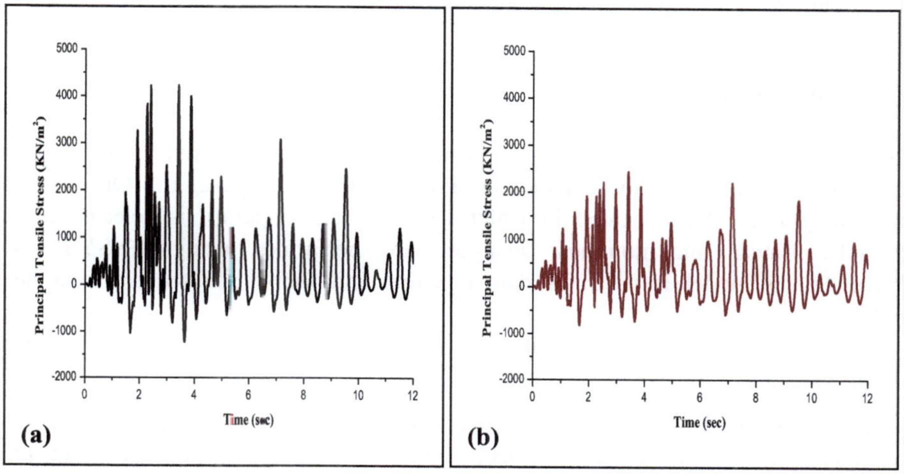

Fig. 8. Time history of principal tensile stress at heel of dam: (a) linear analysis and (b) nonlinear analysis

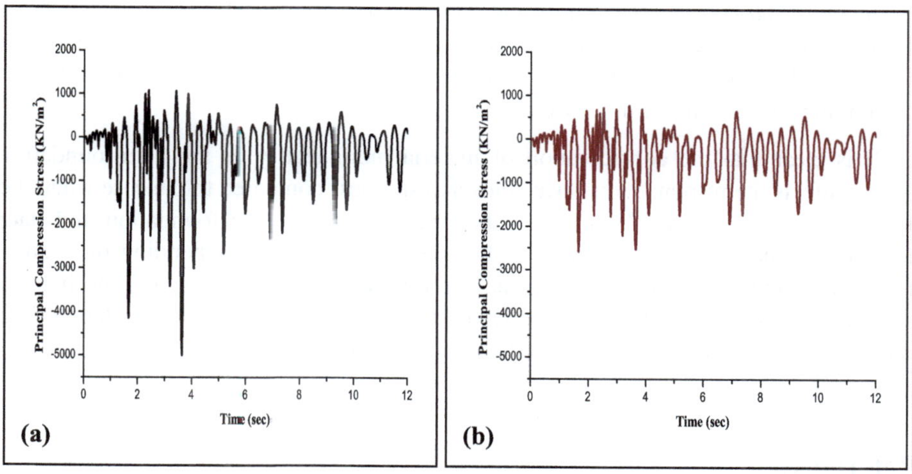

Fig. 9. Time history of principal compressive stress at heel of dam: (a) linear analysis and (b) nonlinear analysis

and dam material are considered to be linear. Therefore, foundation and dam nonlinearity could decrease the stress values in the dam body depending upon the ground motion characteristics [18, 19].

The Figs. 8 and 9 show the time history of principal tensile and compression stresses at heel of dam in linear and non-linear analyses. In linear analysis, the principal tensile and compression stresses at heel are 4229.05 and −4998.51 kN/m^2, respectively, while these reduce to 2443.51 and −2581.37 kN/m^2, respectively, in nonlinear analysis. Therefore, in nonlinear analysis, a decrease of 42 and 48%, respectively, in the magnitude of principal tensile and compression stresses were noticed.

4 Conclusions

This paper presents the non-linear seismic response of concrete gravity dams consedering dam-foundation interaction. For illustrative purpose, the Oued Fodda concrete gravity Dam, located in Chlef (northwestern Algeria), is selected as an example. Linear and nonlinear analyses are performed using ANSYS. The Druker-Prager model is used for dam concrete and foundation rock in nonlinear analysis.

From the numerical results obtained in the study, the following conclusions can be drawn:

1. Linear analysis of the dam shows high tensile stresses at the heel and top of the dam. Therefore, the upper and heel regions of the dam are the most severely stressed zones, hence one may expect the appearance of cracks around these parts.
2. When the material nonlinearity of dam-foundation system is considered, the horizontal displacements and principal stresses decrease compared to the linear case.
3. Nonlinear analysis of dam-fondation system leads to low values of displacement at the crest and stresses at the heel and near the neck region of dam.
4. In nonlinear analysis, the tensile stresses at the heel reduce but no such reduction is observed at the top of the dam. Upper and lower parts of the dam are still susceptible to cracking in this case.

It is clear that the consideration of material nonlinearity of the dam-foundation system affects the response (displacements and stresses) compared to the case when the dam-foundation system is assumed to be linear. However, the foundation and dam nonlinearity may increase or decrease the response parameters depending on ground motion characteristics, surrounding foundation properties and the structure mechanical properties. Thus, it is important to carry out the nonlinear analysis of foundation-dam interaction system to achieve more reliable results of dam behavior.

References

1. Chopra, A.K., Chakrabarti, P.: Earthquake analysis of concrete gravity dams including dam-water-foundation rock interaction. Earthquake Eng. Struct. Dynam. **9**, 363–383 (1981)
2. Leger, P., Boughoufalah, M.: Earthquake input mechanisms for time domain analysis of dam foundation systems. Eng. Struct. **11**, 37–46 (1989)

3. Nuss, L.K., Munoz, R.L., Jackmauh, F.J., Chopra, A.K.: Influence of dam-foundation interaction in seismic safety evaluation of two arch dams. In: Proceedings 12th World Conference on Earthquake Engineering, Auckland, New Zealand (2000)
4. Bayraktar, A., Hancer, E., Akkose, M.: Influence of base-rock characteristics on the stochastic dynamic response of dam-reservoir-foundation systems. J. Eng. Struct. **27**, 1498–1508 (2005)
5. Lemos, J.V., Gomes, J.P.: Modeling seismic failure scenarios of concrete dam foundations. In: Sousa, L.R., Fernandes, M.M., Vargas Jr., E.A., Azevedo (eds.) Applications of Computational Mechanics in Geotechnical Engineering, pp. 341–349. Taylor and Francis, London (2007)
6. Moussaoui, S.E., Tiliouine, B.: Analyse du comportement dynamique des barrages poids compte tenu de l'interaction fluide-sol-structures. In: 1st International Conference on Sustainable Built Environment Infrastructures in Developing Countries, ENSET, Oran, Algeria (2009)
7. Saleh, S., Madabhushi, S.P.G.: Response of concrete dams on rigid and soil foundations under earthquake loading. Earthquake Tsunami **4**(3), 251–268 (2010)
8. Lebon, G., Saouma, V., Uchita, Y.: 3D rock-dam seismic interaction. Dam Eng. **21**(2), 101–130 (2010)
9. Saouma, V., Miura, F., Lebon, G., Yagome, Y.: A simplified 3D model for soil-structure interaction with radiation damping and free field input. Bull. Earthquake Eng. **9**(5), 1387–1402 (2011)
10. Hariri-Ardebili, M.A., Mirzabozorg, H.: Effects of near-fault ground motions in seismic performance evaluation of a symmetry arch dam. Soil Mech. Found. Eng. **49**(5), 192–199 (2012)
11. Burman, A., Nayak, P., Agrawal, P., Maity, D.: Coupled gravity dam-foundation analysis using a simplified direct method of soil-structure interaction. Soil Dynam. Earthquake Eng. **34**, 62–68 (2012). doi:10.1016/j.soildyn.2011.10.008
12. Hariri-Ardebili, M.A., Mirzabozorg, H.: A comparative study of the seismic stability of coupled arch dam-foundation-reservoir systems using infinite elements and viscous boundary models. Int. J. Struct. Stab. Dyn. **13**(6), 1350032-1–1350032-24 (2013)
13. Ouzandja, D., Tiliouine, B.: Effects of dam-foundation contact conditions on seismic performance of concrete gravity dams. Arab. J. Sci. Eng. **40**(11), 3047–3056 (2015). doi:10.1007/s13369-015-1770-2
14. Westergaard, H.M.: Water pressures on dams during earthquake. Trans. ASCE **98**, 418–433 (1933)
15. Drucker, D.C., Prager, W.: Soil mechanics and plastic analysis of limit design. Q. Appl. Math. **10**(2), 157–165 (1952)
16. ANSYS: Theory User's Manual. Swanson Analysis Systems Inc., Houston (2009)
17. Wilson, E.L., Khalvati, M.: Finite elements for the dynamic analysis of fluid-solid systems. Int. J. Numer. Methods Eng. **19**(11), 1657–1668 (1983)
18. Leger, P., Katsouli, M.: Seismic stability of concrete gravity dams. Earthquake Eng. Struct. Dynam. **18**(6), 889–902 (1989). doi:10.1002/eqe.4290180611
19. Halabian, A.M., Naggar, E.: Effect of nonlinear soil-structure interaction on seismic response of tall slender structures. Soil Dyn. Earthquake Eng. **22**(8), 639–658 (2002). doi:10.1016/S0267-7261(02)00061-1

Applied GIS to the Monitoring of Building Work Case Study: Construction of 2000 Houses in Ghadames-Lybia

Samir Medhioub[1(✉)], Mohamed Baklouti[2], and Slah Bouraoui[1]

[1] High Institute of Technological Studies of Sfax, Sfax, Tunisia
samir.medhioub@gmail.com, s.bouraoui@gmail.com
[2] Sfax, Tunisia
baklouti.mohamed@gmail.com

Abstract. Chaâbane and Co Tunisian company has been contracted by "El Jihez Iskene and El Marafek Libya" to achieve 2,000 villa's in Ghadames and Derj (Nalout). The agreed implementation period was 30 months. This project covers an area of approximately 260 ha. The extent important project and the remarkable number of housing building and the diversity of construction activities (Reinforced concrete foundation, reinforced concrete in elevation, masonry, cladding, etc.) deserve the establishment of an efficient management of work through the control of the fixed time and the reserved budget. To do so, a process of planning, organization and management as well the affected resources must be put in place. This process is to break down the project managing in simple tasks or activities, to provide forecasts for these tasks and then to monitor their progress. The conventional planning tools (Primavera, Msproject, ...), generally adapted by project managers, allow a representation in time of the tasks to be performed, specifying the executions' duration and conditions of each one. However, this representation does not allow a spatial or geographic monitoring of the performed tasks within the scope of project intervention. From then, the application of Geographic Information System (GIS) complements the above monitoring tools for civil engineering projects occupying large areas on the one hand and having different types of scattered structures on the other. In fact, GIS can identify the collected data on task progress, can structure and present them in maps form in order to graphically and geographically visualize the overall progress and to conveniently extract useful summaries to the decision making.

1 Introduction

Chaabane Company (CC) from Tunisia has been contracted with the Housing Infrastructure Board (H.I.B) of Libya for the construction of 2000 housing units in Shaabiat Nalut (1440 houses in Ghadames city and 560 houses in Derj) with various services equipments building (Fig. 1).

This study is intended to provide HIB an easy and comprehensible tool for the monitoring of the works progress, regarding to the huge and complicated data to manage as well as the spatial extent of the project on 260 ha. In fact, the traditional

H. Shehata and Y. Rashed (eds.), *Numerical Analysis of Nonlinear Coupled Problems*, Sustainable Civil Infrastructures, DOI 10.1007/978-3-319-61905-7_3

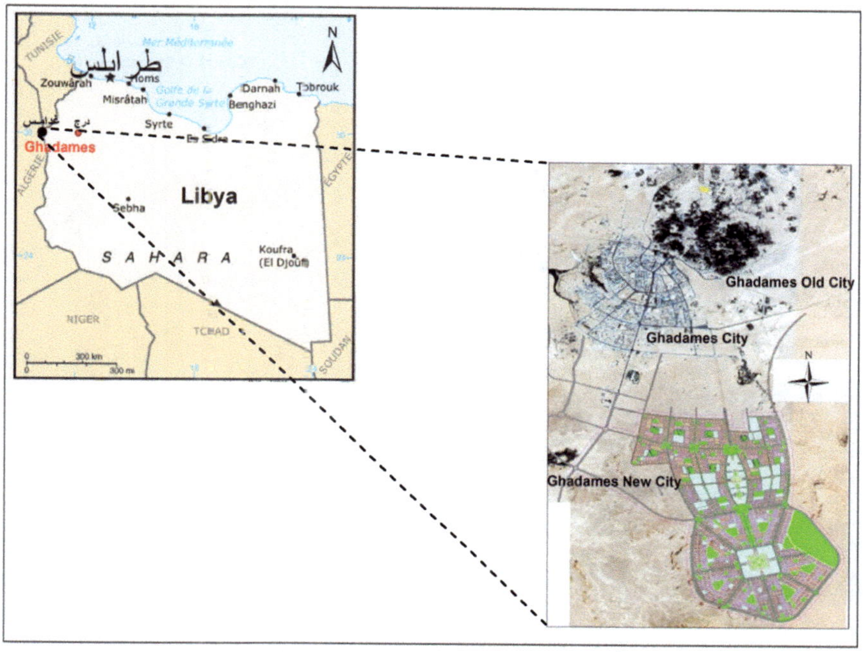

Fig. 1. Site location

approach for scheduling and progress monitoring techniques likes bar charts, CPM, PERT etc. are still being used by the project managers for planning. These are a serious disadvantage in the decision making purpose, as they fail to provide the necessary spatial aspects and data (Gopal et al. 2011). A type of bar chart, a Gantt charts show the start and finish dates of the different required tasks of a project. It shows as well the ratio of progress for each works task or phase. However, these softwares are unable to provide, a spatial or geographic map for monitoring works task for each house. Therefore, GIS allows project managers and different people involved in the project with different backgrounds to get the accurate information of the project and monitoring of activities. The project manager and client can use the visualization aspects at any stage of the project to monitor the activities and cost flow (Gopal et al. 2011). Therefore GIS is the best tool to reply to the latter need by ensuring the easy identification of the data collected, making presentations in many thematic maps. Showing in the construction space (Zhang et al. 2001) (graphic form) the global work progress and extracting all the kind of information that we are looking for with a simple attributed requests.

2 Brief Presentation of the Project

The 2000 houses are divided on five types A, B, C, D and D'. Figures 2 and 3 show the distribution of these houses on the project areas.

Number of housing units
عدد الوحدات السكنية

Models of houses	Ghadames غدامس	نماذج الوحدات
Type A	270	نموذج أ
Type B	100	نموذج ب
Type C	128	نموذج ج
Type D	822	نموذج د
'Type D	120	نموذج د'
Total	**1440**	الأجمالي

Fig. 2. Distribution of houses in Ghadames

Number of housing units
عدد الوحدات السكنية

Models of houses	Derj درج	نماذج الوحدات السكنية
Type A	130	نموذج أ
Type B	18	نموذج ب
Type C	94	نموذج ج
Type D	238	نموذج د
'Type D	80	نموذج د'
Total	**560**	الأجمالي

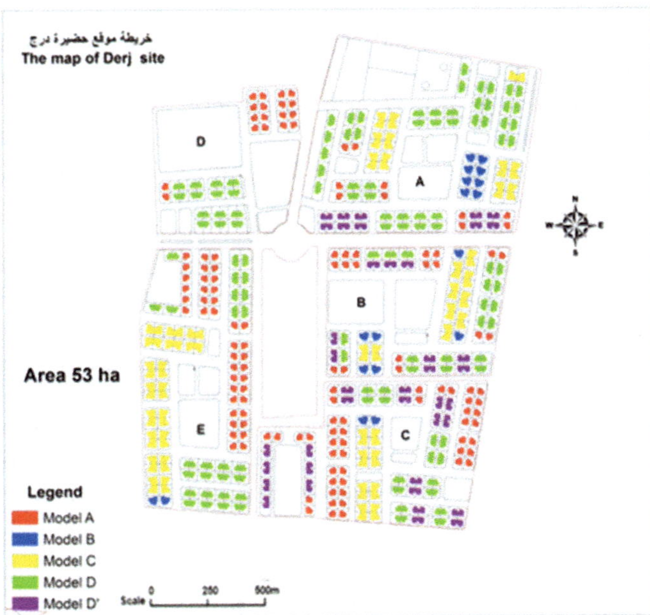

Fig. 3. Distribution of houses in Derj

The contract between Chaabane Company and HIB is a percent complete contract (Table 1). As shown in this table, the project was divided into 18 main activities based on which Chaabane Company establishes his monthly invoice according to the work progress. Therefore, the HIB was interested to see not only the total ratio of work progress for each activity but the work progress for each activity at each house. In other terms, the deal is to convert all collected work progress data in a map in order to provide HIB a clear view of what is being built when and where.

Table 1. Percentage contract value for the main activities

Number of contract execution phases	Description	Percentage of contract value
1	Studies and designs works	2.5%
2	Works of preparation and paving	1%
3	Excavation works	2%
4	Refilling works	2%
5	Foundation works including concrete for bases and pillars	10%
6	Works of concrete structures	24%
7	Masonry works and thermal insulation	9%
8	Tiles and surfaces insulating layer	3%
9	Internal and external walls plastering	7%
10	Wall coating works	6%
11	Floors coating works	8%
12	Windows, PVC items and doors	9%
13	Internal painting	2%
14	External painting	3%
15	Sewage and water extensions	2.5%
16	Electrical extensions	3.5%
17	Electrical accessories	2.5%
18	Sanitary kits and accessories	3%
Total		100%

3 Applied GIS to the Monitoring of Building Work Progress

A tool that has flourished within civil engineering in recent years is Geographical Information System (GIS). A GIS is a computer system for processing (assembling, storing, manipulating, and displaying) geographically related information. GIS is a special class of information system which has four components involving a computer system, GIS software, human expert and data. GIS activities may be grouped into spatial and attribute data management, data display, data exploration, data analysis and modeling (Rajesh et al. 2012).

GIS has been recognized in a majority of the civil engineering disciplines as a beneficial technology. This fact is illustrated by the growing number of articles finding their way into civil publications devoted to GIS (Oloufa et al. 1994).

The application of GIS in this project was conducted, in three principles steps:

Step 1: Modeling a geographic database (GDB);
Step 2: Modeling of an alphanumeric database (ADB);
Step 3: Analysis and interpretation using the principal functions of GIS.

The first step of this work is the modeling of a Geographic Database (GDB) which represents the technical and the practical part of the work. At this level, we create a GDB model which is based on various information layers about housing units, master plan, foundation and structure works, internal and external masonry and cement coating works, tiling works of floors and walls, electrical extensions works, installation of water proofing system, etc. Figure 4 shows the detailed methodology of the creation of the GDB.

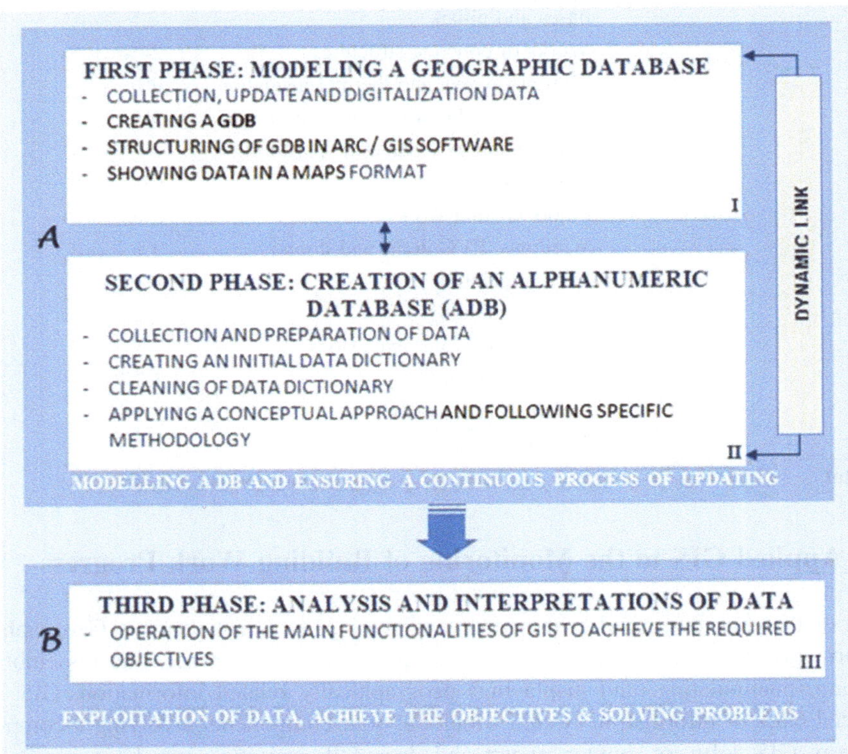

Fig. 4. Methodology adopted for a geographic database modeling

The second step is the creation of an Alphanumeric Database (ADB) to be linked to the GDB. The notion of modeling an ADB by Merise[1] method offers a breakdown of the project in phases ranging from general to specific (systemic approach) and advocates the use of models in different phases of analysis, design and implementation the system. In other terms, the task is to transform the real world into a conceptual model understandable by the machine. So, we started with collection and preparation of data concerning all the details of the project such as:

- codes and numbers of different types of housing;
- houses location for both Ghadames and Derj cities;
- identification of subcontractors and all teams (name, specialty, type of contract, etc.);
- identification of different phases of implementation according to the contract financial terms agreed between CC company and the client;
- distribution of housing units for the subcontractors and CC teams according to their specialties and the stage of construction.

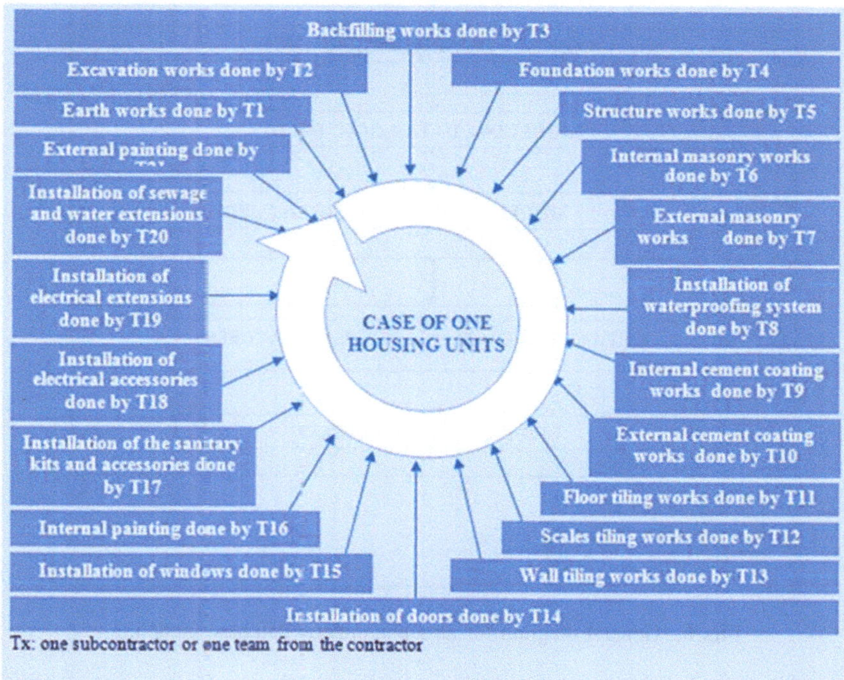

Fig. 5. Intervening parties for one construction house

[1] **MERISE:** A method for analyzing information. It's a software method for the construction of an information system using the Model Entity/Relation.

This process makes Chaabane and Co company in front of so many and various complicated situations in which it needs to ensure a good and fast data management. Figure 5 shows the huge daily data that should be collected for one house under construction. We can see that there are 21 teams and subcontractors that they participate in the construction of one house.

These brut dictionary data should be arranged, managed and cleaned. After that, we apply a conceptual approach and a specific methodology (Fig. 6) in order to create the alphanumeric database by using Microsoft Excel (Fig. 7).

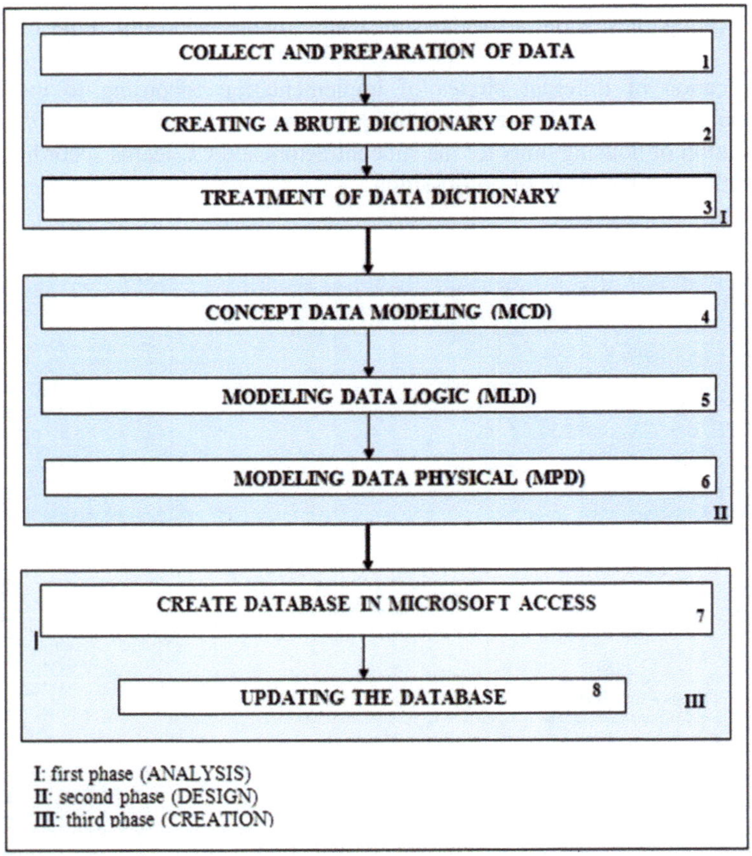

Fig. 6. General diagram for an alphanumeric database modeling

The applied general methodology needs to ensure a continuous updating of the database (DB), therefore, the adopted methodology in this project was a closed loop.

The third and the last step is to use the principal functions of GIS in order to ensure and to perform the management, scheduling as well as the monitoring of the construction works. So many tasks were done:

- making a monthly, weekly and daily control of the works progress;
- making a permanent control of the worker's performance (Fig. 8);

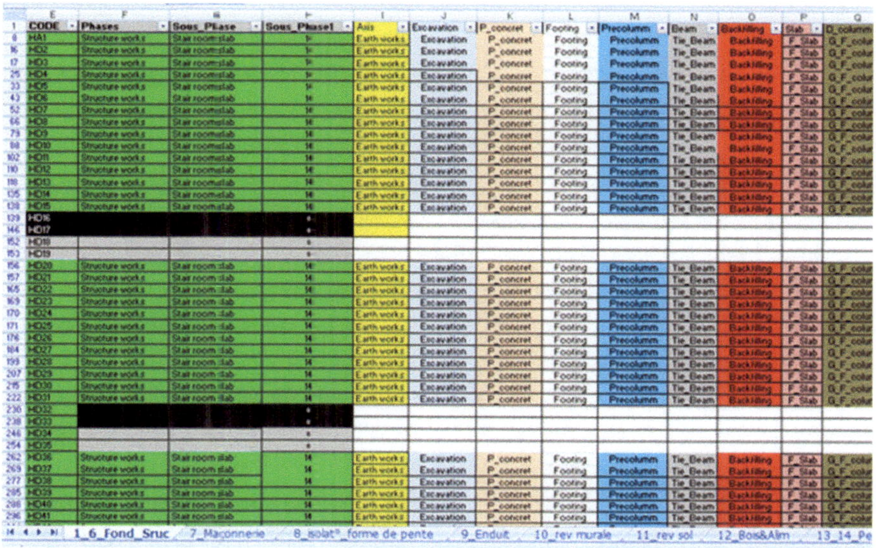

	E	F	G	H	I	J	K	L	M	N	O	P	Q
1	CODE	Phases	Sous_PLase	Sous_Phase1	Axis	Excavation	P_concret	Footing	Precolumm	Beam	Backfilling	Slab	D_column
8	HA1	Structure works	Stair room slab	14	Earth works	Excavation	P_concret	Footing	Precolumm	Tie_Beam	Backfilling	F_Slab	G_F_colu
16	HD2	Structure works	Stair room slab	14	Earth works	Excavation	P_concret	Footing	Precolumm	Tie_Beam	Backfilling	F_Slab	G_F_colu
17	HD3	Structure works	Stair room slab	14	Earth works	Excavation	P_concret	Footing	Precolumm	Tie_Beam	Backfilling	F_Slab	G_F_colu
25	HD4	Structure works	Stair room slab	14	Earth works	Excavation	P_concret	Footing	Precolumm	Tie_Beam	Backfilling	F_Slab	G_F_colu
33	HD5	Structure works	Stair room slab	14	Earth works	Excavation	P_concret	Footing	Precolumm	Tie_Beam	Backfilling	F_Slab	G_F_colu
43	HD6	Structure works	Stair room slab	14	Earth works	Excavation	P_concret	Footing	Precolumm	Tie_Beam	Backfilling	F_Slab	G_F_colu
52	HD7	Structure works	Stair room slab	14	Earth works	Excavation	P_concret	Footing	Precolumm	Tie_Beam	Backfilling	F_Slab	G_F_colu
66	HD8	Structure works	Stair room slab	14	Earth works	Excavation	P_concret	Footing	Precolumm	Tie_Beam	Backfilling	F_Slab	G_F_colu
79	HD9	Structure works	Stair room slab	14	Earth works	Excavation	P_concret	Footing	Precolumm	Tie_Beam	Backfilling	F_Slab	G_F_colu
88	HD10	Structure works	Stair room slab	14	Earth works	Excavation	P_concret	Footing	Precolumm	Tie_Beam	Backfilling	F_Slab	G_F_colu
102	HD11	Structure works	Stair room slab	14	Earth works	Excavation	P_concret	Footing	Precolumm	Tie_Beam	Backfilling	F_Slab	G_F_colu
110	HD12	Structure works	Stair room slab	14	Earth works	Excavation	P_concret	Footing	Precolumm	Tie_Beam	Backfilling	F_Slab	G_F_colu
118	HD13	Structure works	Stair room slab	14	Earth works	Excavation	P_concret	Footing	Precolumm	Tie_Beam	Backfilling	F_Slab	G_F_colu
135	HD14	Structure works	Stair room slab	14	Earth works	Excavation	P_concret	Footing	Precolumm	Tie_Beam	Backfilling	F_Slab	G_F_colu
138	HD15	Structure works	Stair room slab	14	Earth works	Excavation	P_concret	Footing	Precolumm	Tie_Beam	Backfilling	F_Slab	G_F_colu
139	HD16			-									
146	HD17			-									
152	HD18			-									
153	HD19			-									
156	HD20	Structure works	Stair room slab	14	Earth works	Excavation	P_concret	Footing	Precolumm	Tie_Beam	Backfilling	F_Slab	G_F_colu
157	HD21	Structure works	Stair room slab	14	Earth works	Excavation	P_concret	Footing	Precolumm	Tie_Beam	Backfilling	F_Slab	G_F_colu
165	HD22	Structure works	Stair room slab	14	Earth works	Excavation	P_concret	Footing	Precolumm	Tie_Beam	Backfilling	F_Slab	G_F_colu
169	HD23	Structure works	Stair room slab	14	Earth works	Excavation	P_concret	Footing	Precolumm	Tie_Beam	Backfilling	F_Slab	G_F_colu
170	HD24	Structure works	Stair room slab	14	Earth works	Excavation	P_concret	Footing	Precolumm	Tie_Beam	Backfilling	F_Slab	G_F_colu
171	HD25	Structure works	Stair room slab	14	Earth works	Excavation	P_concret	Footing	Precolumm	Tie_Beam	Backfilling	F_Slab	G_F_colu
176	HD26	Structure works	Stair room slab	14	Earth works	Excavation	P_concret	Footing	Precolumm	Tie_Beam	Backfilling	F_Slab	G_F_colu
184	HD27	Structure works	Stair room slab	14	Earth works	Excavation	P_concret	Footing	Precolumm	Tie_Beam	Backfilling	F_Slab	G_F_colu
199	HD28	Structure works	Stair room slab	14	Earth works	Excavation	P_concret	Footing	Precolumm	Tie_Beam	Backfilling	F_Slab	G_F_colu
207	HD29	Structure works	Stair room slab	14	Earth works	Excavation	P_concret	Footing	Precolumm	Tie_Beam	Backfilling	F_Slab	G_F_colu
215	HD30	Structure works	Stair room slab	14	Earth works	Excavation	P_concret	Footing	Precolumm	Tie_Beam	Backfilling	F_Slab	G_F_colu
222	HD31	Structure works	Stair room slab	14	Earth works	Excavation	P_concret	Footing	Precolumm	Tie_Beam	Backfilling	F_Slab	G_F_colu
230	HD32			-									
238	HD33			-									
246	HD34			-									
254	HD35			-									
262	HD36	Structure works	Stair room slab	14	Earth works	Excavation	P_concret	Footing	Precolumm	Tie_Beam	Backfilling	F_Slab	G_F_colu
269	HD37	Structure works	Stair room slab	14	Earth works	Excavation	P_concret	Footing	Precolumm	Tie_Beam	Backfilling	F_Slab	G_F_colu
277	HD38	Structure works	Stair room slab	14	Earth works	Excavation	P_concret	Footing	Precolumm	Tie_Beam	Backfilling	F_Slab	G_F_colu
285	HD39	Structure works	Stair room slab	14	Earth works	Excavation	P_concret	Footing	Precolumm	Tie_Beam	Backfilling	F_Slab	G_F_colu
288	HD40	Structure works	Stair room slab	14	Earth works	Excavation	P_concret	Footing	Precolumm	Tie_Beam	Backfilling	F_Slab	G_F_colu
296	HD41	Structure works	Stair room slab	14	Earth works	Excavation	P_concret	Footing	Precolumm	Tie_Beam	Backfilling	F_Slab	G_F_colu

1_6_Fond_Sruc / 7_Maçonnerie / 8_isolat°_forme de pente / 9_Enduit / 10_rev murale / 11_rev sol / 12_Bois&Alm / 13_14_Pe

Fig. 7. Extract of an alphanumeric database

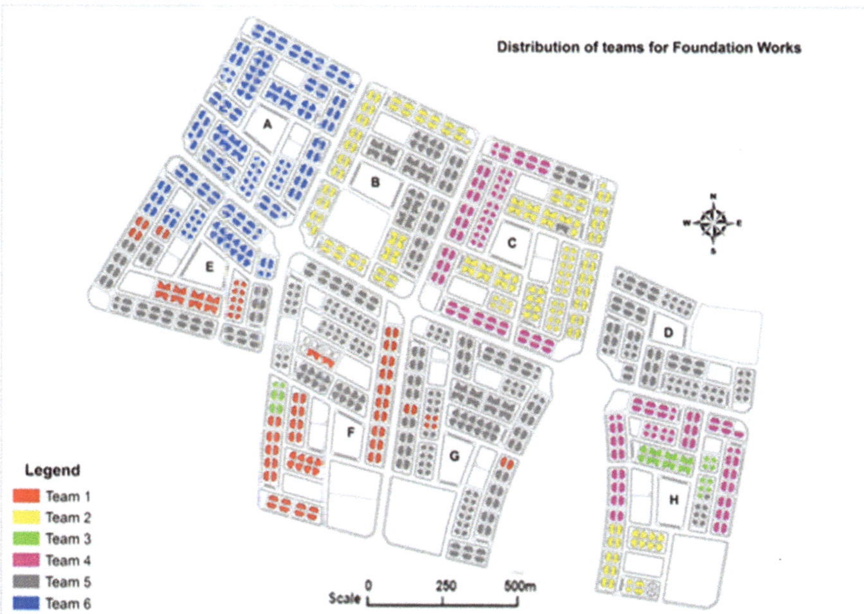

Fig. 8. Distribution teams for fondation works

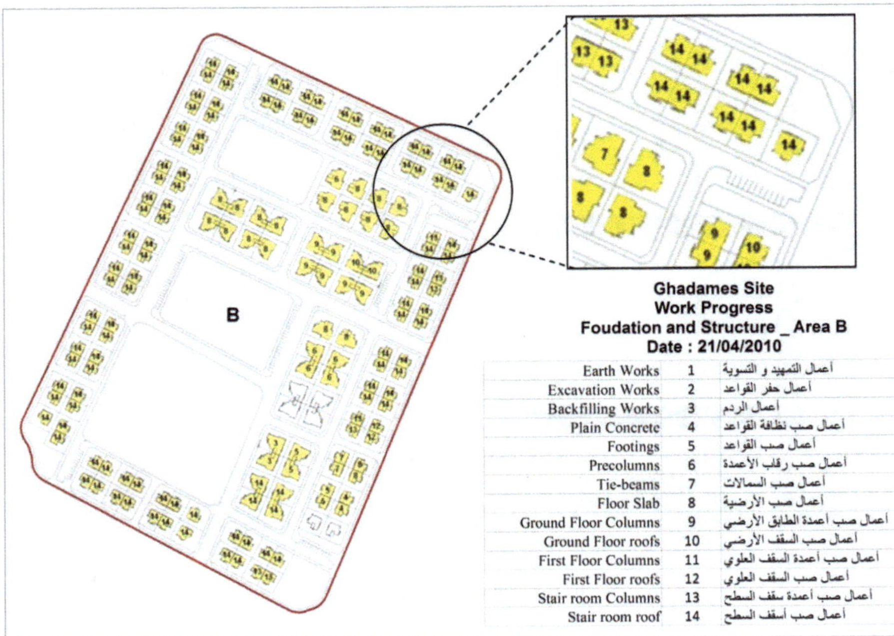

Fig. 9. Monitoring of works progress (Area B)

Fig. 10. Monitoring of works progress (Area A)

- ensuring a detailed works schedule with a daily updating;
- controlling the performance of the subcontractors and establishing there invoices according to their works progress;
- etc.

As shown in the Figs. 9 and 10 each work progress activity status at each house appears in different colors. These monthly geographic maps show the work progress situation in each house located at the specific zone. It permits to evaluate and to deduce so easily all the works quantities done for each activity and obviously the monthly invoice based on the Table 1.

4 Conclusions

This study concludes that GIS can be an efficient added supplement to project scheduling tools like Microsoft project and Primavera. It provides an inexperienced user a construction schedule visualization tool as well as easier understanding of the project through conveying what is being built when and where.

Acknowledgments. This work is undertaken as part of an applied research project funded by the Chaabane and Co Company. Helps, have been supplied by the executive team of the later, and are gratefully acknowledged.

The authors would also like to acknowledge the review of Mouna Zaabar Touil whose comments were great help and improved the quality of this paper.

Permission to publish has been granted by the Head Chief Manager of the Chaabane Company.

References

Naik, G.M., et al.: GIS based 4D model development for planning and scheduling of a construction project common structural and construction deficiencies of Nepalese. Int. J. Innov. Manag. Technol. **2**, 447 (2011)

Oloufa, A.R., et al.: Integrated GIS for construction site investigation. J. Constr. Eng. Manag. **120**(1), 211–222 (1994)

Rajesh, K., et al.: 4D model through GIS for planning and scheduling of residential construction projects. Res. J. Appl. Sci. **7**(4), 222–228 (2012)

Zhang, J.P., et al.: 4D visualization of construction site management. In: Proceedings of 5th International Conference Information Visualisation, pp. 382–387 (2001)

Stability Analysis of Souk-Tleta Earth Dam, North Algeria

Ryma Afiri[1(✉)], Saida Hadj Abderrahmane[1], Lynda Djerbal[2], and Smail Gabi[1]

[1] Geomaterials, Environment and Development Laboratory, Department of Civil Engineering, University Mouloud Mammeri of Tizi-Ouzou, Tizi Ouzou, Algeria
r.afiri@hotmail.fr

[2] Department of Hydraulic and Geotechnical Engineering, University of Science and Technology Houari Boumediene, Bab Ezzouar, Algeria

Abstract. Slope stability analysis of earth dam is very important to ascertain the stability of the structure. The stability of earth dam depends on its geometry, its components, materials, water pressure and the forces to which it is subjected.

This paper presents stability analysis carried out on 95 m high "Souk Tleta earth Dam" with a crest length of about 200 m situated on the river Bougdoura in north Algeria.

The slope stability analysis of Souk Tleta earth dam has been done by Plaxis software and is used under different conditions of ground water levels to evaluate slope stability. The factor of safety is calculated for each water level.

Keywords: Slope stability · Earth dam · Plaxis · Factor of safety

1 Introduction

The stability of earth dam depends on its geometry, its components, materials, properties of each component and the forces to which it is subjected. The design of earth dams involves many considerations that must be examined before initiating detailed stability analyses. Such as geological and subsurface explorations, the earth and/or rock-fill materials available for construction should be carefully studied.

The case study is Souk Tleta dam located on Bougdoura River immediately downstream of the confluence of the Tleta and TalaImedrane rivers, 20 km west of Tizi-Ouzou city in north Algeria (Fig. 1). This earth dam, currently under construction, hasa crest length of about 200 m, a maximum height above the river bed level of about 95 m and an estimated total storage capacity of 96 million m^3. The purpose of the dam is to store surface water for irrigation and domestic water supply.

This paper presents stability analysis carried out on Souk Tleta earth dam in Algeria. The values of safety factor, using finite element method analysis, were considered to this earth dam for three cases of reservoir filling: empty reservoir, steady-state water levels and low water level.

H. Shehata and Y. Rashed (eds.), *Numerical Analysis of Nonlinear Coupled Problems*,
Sustainable Civil Infrastructures, DOI 10.1007/978-3-319-61905-7_4

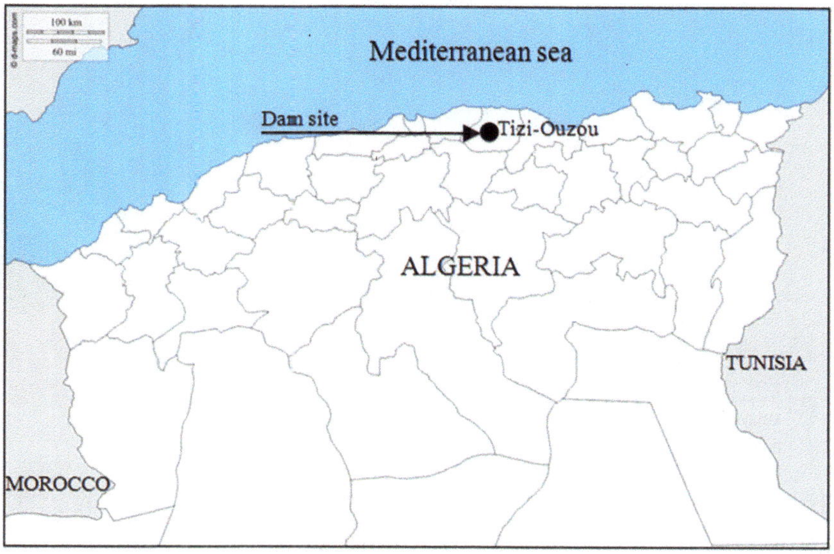

Fig. 1. Location of Souk Tleta dam

2 Geology

The regional geology of Souk Tleta dam and its reservoir are located on the meridional bank of sedimentary basin of Lower Miocene (Burdigalian) of Tizi-Ouzou city. In the area of the dam, Burdigalian sediments cover the site of under-Miocene with a transgressive facies which crops out at the upstream of the gorge of Souk Tleta valley, and covers a large surface right up to the foot of limestone chain of Djurdjura.

Geological overview of Fig. 2 shows the outcrops extension of lithofacies in the dam site.

3 Geometry of the Earth Dam

The general geometry and the foundation layers of the Souk Tleta Earth Embankment Dam profile are illustrated in Fig. 3 and Table 1. The study was performed on the profile of greater height.

4 Soil Properties

The geotechnical parameters of the slope layers are determined from laboratory and in-situ tests.

The necessary parameters for the slope stability analysis and design are presented in Table 2. Only the geotechnical parameters and properties which concern the slope stability analysis of the Souk Tleta dam are given.

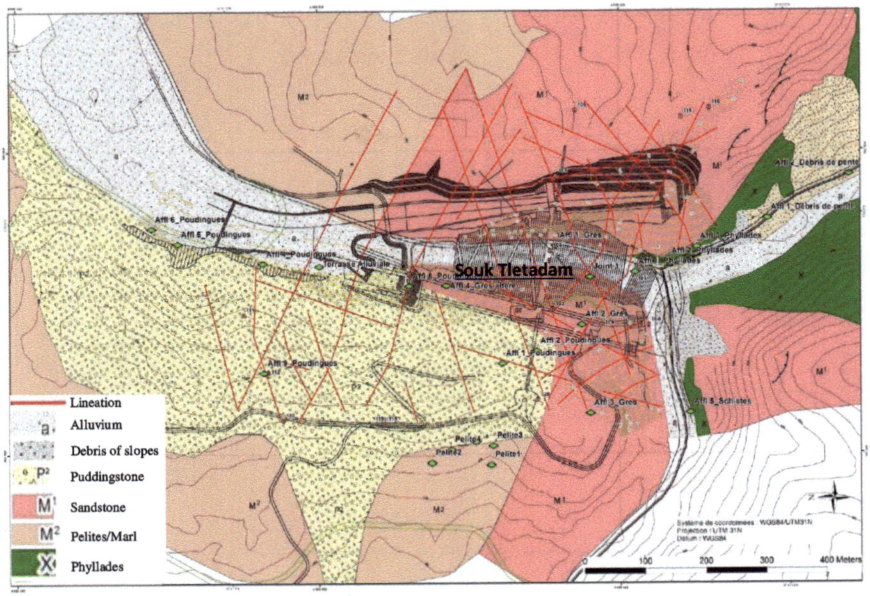

Fig. 2. Outcrops extension of lithofacies in the dam site

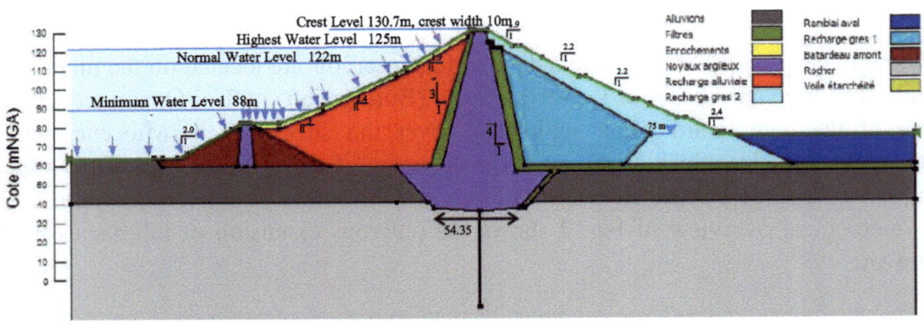

Fig. 3. Geometry of the cross section of the Souk Tleta earth dam

4.1 Stability Analysis and Loading Conditions

In this study, we present a slope stability analysis using a numerical modeling, based on the Mohr-coulomb model. To do so, we used the finite element software Plaxis 2D.

The code Plaxis developed by The Technical University of Delft (Holland) in 1987 covers all aspects and applications of geotechnical engineering simulation.

The Souk Tleta earth embankment dam and its foundation were designed and analyzed against the risk of slope stability loss.

In the finite element method a continuum is divided into a number of elements (volumes), each element consists of a number of nodes.

Table 1. The geometrical characteristics of Souk Tleta dam

Parameter	Value
Dam's height	95 m
Crest level	137.5 m NGA
Crest length	200 m
Crest width	10 m
Minimum water level	88 m NGA
Normal water level	122 m NGA
Highest water level	125 m NGA
Area of the watershed	465 km^2
Storage capacity	96 M m^3

Where NGA means general leveling of Algeria, with the 'zero level' situated at the port of Algiers.

Table 2. Parameters used for the Slope Stability Analysis of Souk Tleta dam

Material type	Unit weight γ_h (kN/m^3)	Saturated unit weight γ_{sat} (kN/m^3)	Internal angle of friction φ (°)	Cohesion of soil c (kPa)	Poisson's ratio ν	Young modulus E (MPa)	Permeability k (m/day)
Filter	19	21	34	0	0.25	100	8.64
Clay core	18	19	18	10	0.3	25	0.864×10^{-3}
Sandstone fill 1	19.5	20	30	0	0.23	90	0.0864
Sandstone fill 2	21	21.5	34	0	0.23	90	0.0864
Upstream backfill	16	19	10	0	0.3	90	0.0864
Rip rap	21	22	30	0	0.3	55	0.0864
Alluvium	20	21.5	32	0	0.23	35	0.864
Bedrock	21.3	21.8	28	20	0.3	150	8.6×10^{-3}

The finite element mesh used in the simulation of slope stability analysis is a very fine mesh, and consists of (12983) 15-nodes triangular elements. The 1587 elements have an average size of 8.915 m.

The 2D view for the finite element mesh of the dam and its surrounded soil mass is illustrated in Fig. 4.

Three conditions of reservoir filling were examined, as follows:

a. *Right after construction condition:*
   The calculation for the analysis of the dam after construction is divided into alternate construction phases until the end of construction. The critical condition to be analysed is at the completion of embankment dam construction but prior to filling with water.

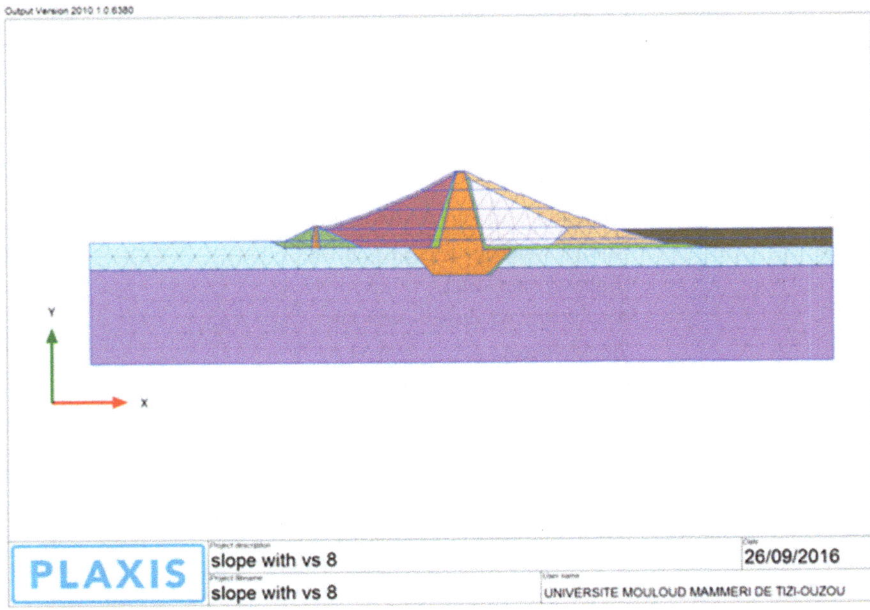

Fig. 4. Finite element mesh

b. *Two steady-state water levels of the dam:*
This stage involves the stability analysis of Souk Tleta dam when the water level is high and the pore pressure in the materials is at steady state condition.
Impoundment is carried out gradually over several phases until:

- Normal water level (122 m NGA),
- The highest water level (flood level: 125 m NGA)

c. *The minimum water level:*
Impoundment is carried out gradually until the minimum stored water level (88 m NGA).

5 Slope Stability Analysis Results

The factor of safety is commonly thought of as the ratio of the maximum load or stress that a soil can sustain to the actual load or stress that is applied. The factor of safety FS, with respect to strength, may be expressed as follows:

$$FS = \frac{\tau_{ff}}{\tau}$$

Where τ_{ff} is the maximum shear stress that the soil can sustain at the value of normal stress of $\sigma_{n,}$ and τ is the actual shear stress applied to the soil.

Typical minimum acceptable values of factor of safety are about 1.3 for end of construction and multistage loading, 1.5 for normal long-term loading conditions. The values of factor of safety listed provide guidance but are not prescribed for slopes other than the slopes of new embankment dams. (from US Army Corp EM 1110-2-1902)

5.1 Right After Construction Condition

In this first stage, the stability analyses have been performed for different upstream and downstream slopes for the purpose of optimizing the volume of the dam body and materials of construction. The analyses have been performed for the most critical cross-section taken along the highest part of the dam.

The stability after the construction of the dam must be justified, but does not exclude that deformation may not occur as shown in Fig. 5. Numerical analysis of dam in the end of construction shows a factor of safety equal to 1.62.

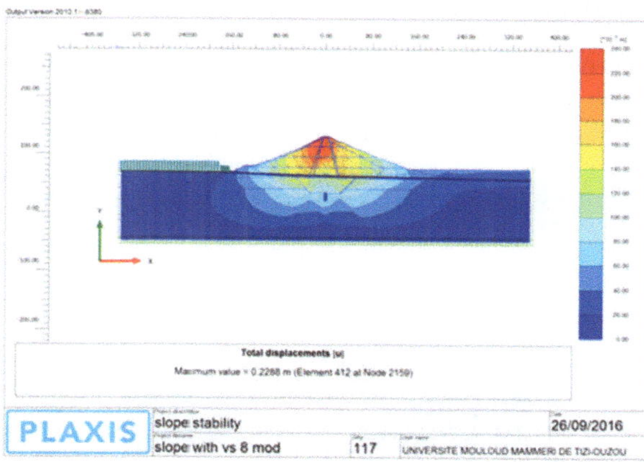

Fig. 5. Total displacements in the dam right after construction, FS = 1.62

5.2 Normal Water Level

For steady state condition with full water supply level at 122 m,

The case corresponds to the steady state seepage condition, the water table is at the top of 122 m and this is a critical condition. The total displacements obtained for both upstream and downstream are illustrated in Fig. 6, and the factor of safety obtained for this condition is 1.54.

The rise of water level in the reservoir causes water to penetrate into the dam body and increases both pore water pressures and weight of dam body.

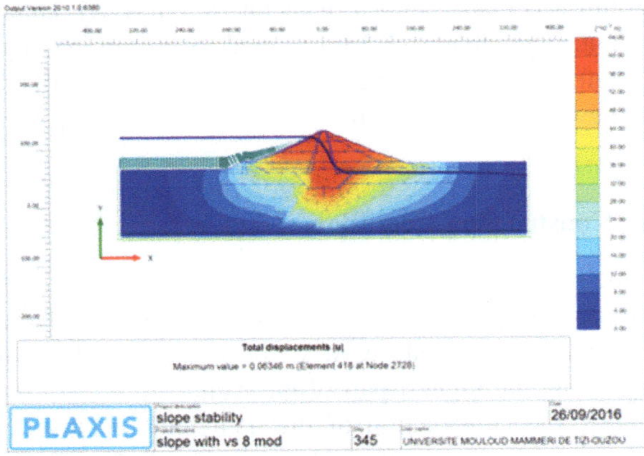

Fig. 6. Total displacements in the dam in normal steady-state water level of 122 m, FS = 1.54

5.3 The Highest Water Level (125 m)

The highest water level can be caused by rainfalls and wave's action. To evaluate how resistant materials are to water flow, total displacements obtained in case of steady state condition with the highest water level are illustrated in Fig. 7.

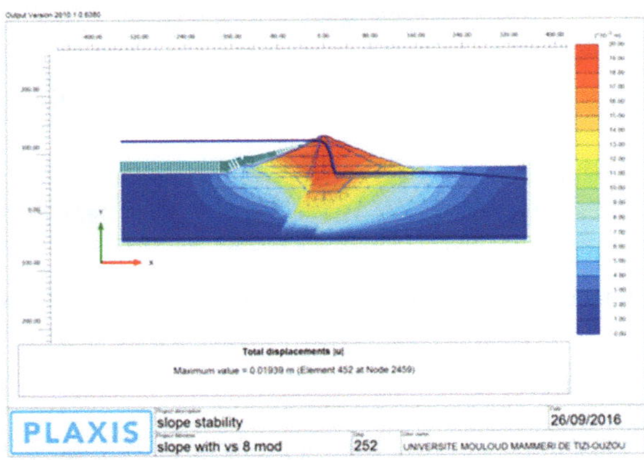

Fig. 7. Total displacements in the dam in steady-state high water level of 125 m, FS = 1.52

When the impounded water level reaches the highest level, the lowest factor of safety of 1.52 is observed.

5.4 The Minimum Water Level (88 m)

When the water level in the reservoir is at its lowest, total displacements in the dam in minimum steady-state water level are illustrated in Fig. 8. Numerical analysis of dam for the minimum water level shows a factor of safety equal to 1.62.

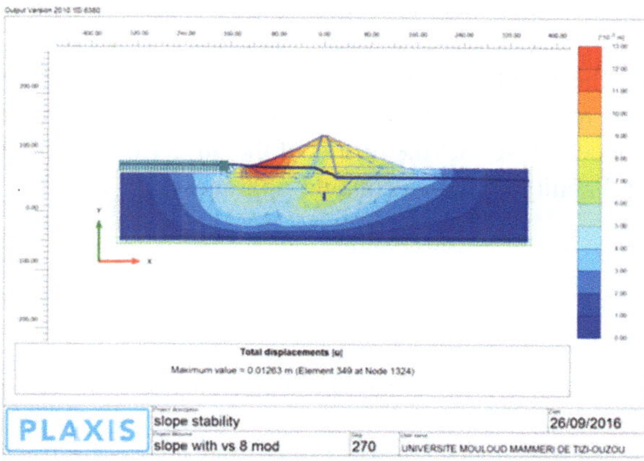

Fig. 8. Total displacements in the dam in minimum steady-state water level of 88 m, FS = 1.62

Factor of safety for all examined analysis conditions are summarized in Table 3.

Table 3. Factor of safety summary

Loading conditions	Factor of safety
Right after construction condition	1.62
Normal water level	1.54
High water level	1.52
Minimum water level	1.62

6 Conclusions and Discussion

The aim of the present study was to examine numerically the slope stability of Souk Tleta earth dam under static conditions.

The main conclusions derived from the presented study are:

The factor of safety evaluated using PLAXIS is found to be greater than 1.5 for right after construction, the full (high) reservoir condition and low reservoir condition.

The stability criteria have been well satisfied with the designed shape of the dam body during and right after the end of construction and also for all loading conditions that the dam may be subjected during its service life.

In the phase immediately after construction is expected that dam safety factor higher than other phases, because of no pore water pressure. In the steady-state leakage phase, due to pore water pressure in the dam is expected safety factor less than the phase immediately after construction.

In the three conditions, very small deformations are observed in the rock mass foundation due to weak quality of the rock mass.

The rock mass consists of Burdigalian sandstone, which forms the two supports of the dam and is classified as weathered and moderately weathered rock quality.

Reasonable crest movements will occur due to the reservoir's impounding of expected level.

When studying the problems related to slope stability in earth dams, in view of obtained results, it is recommended that observations should be carried out on the dam body during its service life, so that the results of the observation method serve to assess the accuracy and reliability of the calculation approaches.

References

Albtaineh, N.: Slope stability analysis using 2d and 3d methods. thesis of University of Akron (2006)

Aljairry, H.: 2D-flow analysis through zoned earth dam using finite element approach. Eng. Tech. J. **28**, 21 (2010)

Aminpoor, M.: Stability analysis of slope, seepage and dynamic behavior in Maroon Dam by Geo-Studio, secondary national lecture of dam (2008)

Griffoths, D.V., Lane, P.A.: Slope stability analysis by finite elements. Geotech. **49**(3), 387–403 (1999)

Gopal, P., Kiren Kumar, T.: Slope stability and seepage analysis of earthen dam of a summer storage tank: a case study by using different approaches. Int. J. Innovative Res. Adv. Eng. **1** (12), 131–134 (2014)

Huang, T.K.: Stability analysis of an earth dam under steady state seepage. Comput. Struct. **58** (6), 1075–1082 (1996)

Plaxis, 2D.: Tutorial Manual, Delft University of Technology & PLAXIS bv, The Netherlands (2010)

Sarma, S.K.: Stability analysis of embankments and slopes. J. Geotech. Eng. Div. ASCE **105** (12), 1511–1524 (1979)

Shivakumar, S.A., et al.: Seepage and stability analysis of earth dam using finite element method. Aquat. procedia **4**, 876–883 (2015). doi:10.1016/j.aqpro.2015.02.110. International conference on water resources, coastal and ocean encineering (ICWRCOE)

Tatewar S.P., Pawade, N.L.: Stability analysis of earth dam by geostudio software. Int. J. civ. Eng. Technol. **3** (2012)

Static Liquefaction Analysis of the Limonar Tailings Dam in Peru

Herbert M. Maturano Rafael[1](✉) and Celso Romanel[2]

[1] Geomat Ingenieria, Lima, Peru
mmaturano@geomatingenieria.com
[2] PUC-Rio, Rio de Janeiro, Brazil
romanel@puc-rio.br

Abstract. Tailings disposal has been a major concern for mining companies around the world with the overall goal to protect the environment and population from hazards associated with tailings storage. Large amounts of waste are produced daily in ore processing plants and depending on the industrial production waste the storage structure (tailings dam) needs to be redeveloped. A typical method for expanding the reservoir capacity is by raising the height of the dam body. The upstream method begins with the construction of a starting dike; after this step, tailings are deposited upstream, forming a beach that thickens over time, gradually increasing the waste strength and serving as a foundation for future dikes. This procedure goes on until the planned design size is reached. It is a simple and low cost method but its main drawback is that an excessive construction speed can induce static liquefaction, the main cause of the collapse of several tailings dams. This paper investigates the liquefaction potential of a copper tailings dam situated in Peru. Results from an empirical method, based on Standard Penetration Test data, are compared with those obtained with a more complex analysis carried out with a finite element program using an elastoplastic constitutive model. It has been concluded that both methodologies were suitable in order to predict the possible occurrence of static liquefaction.

1 Introduction

The phenomenon of static liquefaction in tailings dams has been widely investigated around the world because of its potential destructive consequences, often involving loss of life as well as economic, social and environmental damages. The technical literature presents several historical cases of collapse of tailings dams due to liquefaction as the Merriespruit dam (South Africa), Sullivan mine dam (Canada), Los Frailes dam (Spain), among others.

The concept of liquefaction can be summarized as the loss of shear strength of the material induced by pore pressure increments. Soils susceptible to the occurrence of this phenomenon are those with tendency of volume contraction under shear, as loose sands. Mine tailings are generally granular or non-plastic fine materials that form layers of relatively low density material in hydraulically filled deposits, with high saturation susceptible to trigger liquefaction by the application of non-drained loads.

H. Shehata and Y. Rashed (eds.), *Numerical Analysis of Nonlinear Coupled Problems*,
Sustainable Civil Infrastructures, DOI 10.1007/978-3-319-61905-7_5

Assessment of static liquefaction potential can be made through empirical methods, based on back analysis of historical cases, or numerically by the finite element method incorporating a specific constitutive model for simulation of liquefaction of the material. In this work, the liquefaction potential of the copper tailings dam of Limonar, situated in Peru, is investigated using the empirical method proposed by Olson (2001), through correlations with field test data (SPT), and the finite element method incorporating the UBCS and constitutive model (Byrne et al. 2004) implemented in the computer program Plaxis 2D.

2 Liquefaction Suscetibility

There are many criteria published in the literature to assess the liquefaction susceptibility of soils, among which:

(a) Material composition criterion - for many years it was believed that liquefaction was restricted to sand deposits only. Fine grained soils were admitted unable to generate high pore pressure values, commonly associated with liquefaction, while high permeable coarse materials were, in turn, considered unable to keep pore pressure increments for a sufficient time for the liquefaction process to develop. However, the limits of particle size criteria were expanded since liquefaction was observed in no plastic silts (Ishihara 1984, 1985), under laboratory and field conditions, indicating that the plasticity characteristics are more influential than particle size distribution in fine soils. According to Wang (1979), fine soils that satisfy each of the following conditions of the Chinese criterion may be considered susceptible to liquefaction: (i) fraction of soil finer than $5\mu \leq 15\%$; (ii) Liquidity Limit $LL \leq 35\%$; (iii) water content w $0.9LL$; (iv) Liquidity Index ≤ 0.75. To consider differences in American practice, the U.S. Army Corps of Engineers recommended the Chinese criterion with the following modifications: (i) decrease of the fine fraction by 5%; (ii) Liquidity Limit increased by 1%; (iii) natural water content increased by 2% (Finn 1994).

(b) Status Criteria - even if a soil satisfies the material composition criterion the liquefaction may or may not occur. The susceptibility relies heavily on a state criterion, depending on the relative density and the initial stresses within the soil mass. Historically, the following state criteria were presented:

(b.1) Critical void ratio criterion - Casagrande (1936) experimentally verified that in drained triaxial tests (controlled deformation) loose and dense sand samples under the same effective stress reached a constant value of relative density as the samples were sheared under large deformation. The void ratio corresponding to this final state of constant volume was called critical void ratio. By performing additional tests under different confining pressures, Casagrande found that the critical void ratio could be uniquely related to the confining pressure through a critical void ratio line, interpreting it as a boundary between regions of positive porepressure (volume contraction, loose soils) and negative porepressure (volume expansion, dense soils).

(b.2) Deformation state criterion - Castro (1969) performed a series of stress controlled undrained triaxial tests on isotropically and anisotropically consolidated sand samples. This test program indicated the existence of a unique relationship between void ratio and confining pressure under large deformations that graphically is plotted parallel to but below the critical void ratio line obtained by Casagrande (1936). The state in which the soil flows continuously under constant shear stress, constant volume and at constant rate was then defined as the steady state line SSL (Castro and Poulos 1977, Poulos 1981). The SSL line is useful for identifying conditions under which the material may be susceptible to liquefaction. A soil whose state is plotted below the SSL line is not considered susceptible, while for another depicted above it liquefaction may occurs if the shear stresses necessary to ensure static equilibrium are greater than the available residual shear strength.

(b.3) Status parameter criterion - relative density or void ratio have only limited applicability for estimation of the susceptibility of soil to liquefaction. A soil element with a particular void ratio (or determined relative density) may be susceptible to liquefaction under high confining stress but not susceptible if they are low. Been and Jefferies (1985) introduced the state parameter concept, defined by

$$\psi = e_0 - e_{ss} \tag{1}$$

where e_0 is the initial void ratio and e_{ss} the void ratio on SSL line under the effective confining pressure of interest. When ψ is positive, the soil exhibits contractive behavior and may be susceptible to liquefaction, while for negative values of ψ volume variation is negative (expansion) and the material is not considered susceptible. The state parameter was related to the soil friction angle, dilatancy angle, field test results (CPT, SPT, DMT) by Been et al (1986, 1987), Sladen (1985), Ishihara (1993), among others.

3 Liquefaction Potential

The fact that a soil deposit is susceptible to liquefaction does not necessarily mean that it will occur, since its triggering depends on the characteristics of the applied loading. In mining, the analysis of the liquefaction potential is important due to the geotechnical characteristics of granular waste, since the materials is often hydraulically deposited in tailings dams, under saturated and low density conditions. The triggering mechanism to static liquefaction may be caused by high construction rate, pore pressure changes caused by intense rains as well as overtopping of waste over the dam crest. The construction rate in the upstream method must be carefully controlled to prevent any significant increase in porepressure. Vick (1983) and Mittal & Morgenstern (1976) suggested a raising rate of 4.6 m/year to 9.1 m/year in order to dissipate the excess porepressure due to loading while Smith (1972) presented two practical recommendations based on static liquefaction analyses of several tailings dams: (i) it must be ensured that the relative density of the tailings dikes be greater than the critical relative density; (ii) the facility must have an efficient drainage system to prevent that the waste inside the containment structure remain in the saturated condition.

4 Olson's Method (2001)

In the literature there are several empirical methods for analysis of soil liquefaction, such as Campanella & Robertson (1985), Seed & Harder (1990) and Olson (2001). Olson's method is briefly described in this paper, a proposition based on 33 back analyses of liquefaction case studies in clean sands, silty sands, sandy silts and sandy tailings. It involves three basic steps: (i) assessment of liquefaction susceptibility; (ii) analysis of liquefaction triggering mechanism; (iii) post-liquefaction slope stability analysis.

4.1 Liquefaction Susceptibility

In this first step it is determined whether the soil has a contractive or dilatant behavior under the in situ stress state. The liquefaction susceptibility is assessed based on the initial effective vertical stress and the number of blow counts in SPT tests carried out at given depth. Figure 1 presents curves of susceptibility recommended by Olson (2001) and Olson and Stark (2003). Values plotted to the left of them indicate that the soil has a tendency of volume contraction under shear, so they are susceptible to liquefaction; values shown to the right of the displayed curves correspond to dilatant soils, not susceptible to the phenomenon.

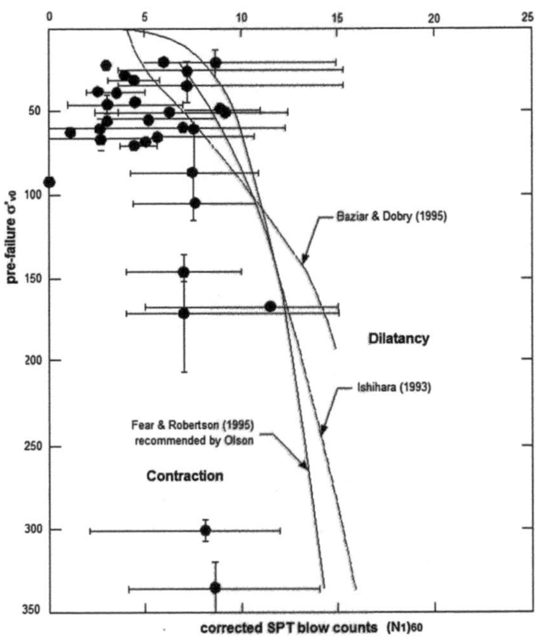

Fig. 1. Comparison of penetration test results with contractive/dilative boundaries (Olson 2001)

The curve suggested by Fear and Robertson (1995) in Fig. 1, is mathematically expressed as

$$\sigma'_{v0} = 9.5812 \times 10^{-4}\left[(N_1)_{60}\right]^{4.7863} \tag{2}$$

Where $(N_1)_{60}$ is the corrected SPT blow count.

4.2 Liquefaction Potential

If soil regions are identified as susceptible to liquefaction, then analysis of the liquefaction potential should be performed in order to determine whether the load rate is sufficient to trigger the phenomenon. Hanzawa et al. (1979) measured undrained shear strength in triaxial compression tests performed on contractive sands under different confining pressures and equal void ratios. They observed that the undrained shear strength ratio S_u^{peak}/σ'_{v0} approximately define a linear shear strength envelope mathematically expressed as

$$S_u^{peak}/\sigma'_{v0} = \tan(\phi_m) \tag{3}$$

where σ'_{v0} is the initial effective vertical stress and ϕ_m the mobilized friction angle.

Hence, failure caused by static loading could be used to estimate the undrained shear strength ratio according to Eq. (3).

The steps for analysis of the liquefaction potential are the following:

- Carry out a slope stability analysis for estimation of static shear stress in soils susceptible to liquefaction. Initial shear strength values are assumed for these soils, modifying them gradually until the safety factor FS = 1 is reached. For soils not susceptible to liquefaction the actual values of drained or undrained shear strength should be assigned.
- Divide the soil region into slices bounded by the critical failure surface previously determined. According to Olson and Stark (2003) a number of 10 to 15 slices is generally sufficient.
- Determine the corrected SPT blow count $(N_1)_{60}$.
- Estimate the shear strength ratio S_u^{peak}/σ'_{v0} from the average curve shown in Fig. 2, whose equation is given by:

$$\frac{S_u^{peak}}{\sigma'_{v0}} = 0.205 + 0.0075\left[(N_1)_{60}\right] \pm 0.04 \text{ for } (N_1)_{60} \leq 12 \tag{4}$$

- Multiply the shear strength ratio by the initial effective vertical stress to calculate the corresponding value of S_u^{peak}.

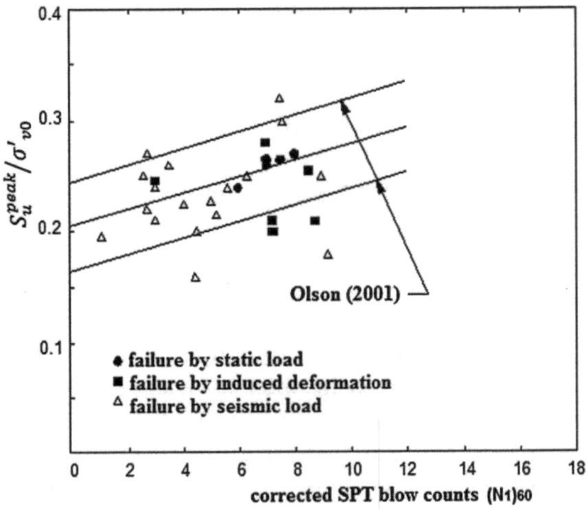

Fig. 2. Correlation between undrained shear strength ratio and SPT blow counts (adapted from Olson 2001)

- Determine the safety factor against liquefaction triggering for each slice of the susceptible soils considering

$$FS^{liq} = \frac{S_u^{peak}}{\tau^{static}} \tag{5}$$

If all the safety factor values are greater than 1 then a post-liquefaction stability analysis is not required. If for some slices $FS^{liq} < 1$ then assign them the value of the shear strength of the liquefied soil and proceed to a post-liquefaction stability analysis.

4.3 Post-liquefaction Stability Analysis

If liquefaction occurs, a post-liquefaction stability analysis should be performed to verify whether the static shear stresses exceed the shear strength available to the liquefied soils. The shear strength ratio in this step takes the value corresponding to the liquefied condition, which can be obtained from graphs recommended by Olson (2001), shown in Fig. 3, whose mathematical expression for the average line is given by

$$\frac{S_u^{liq}}{\sigma'_{v0}} = 0.03 + 0.0075\left[(N_1)_{60}\right] \pm 0.03 \text{ for } (N_1)_{60} \leq 12 \tag{6}$$

Values of the liquefied shear strength are assigned to the slices of susceptible soils whenever $\mathrm{FS}^{\mathrm{liq}} < 1$. If in this stability analysis the safety factor FS^{flow} results less than or equal to 1 then liquefaction flow should occur. If $1 < FS^{flow} < 1.1$ some deformation may occur and a new stability analysis may be performed assigning values of liquefied

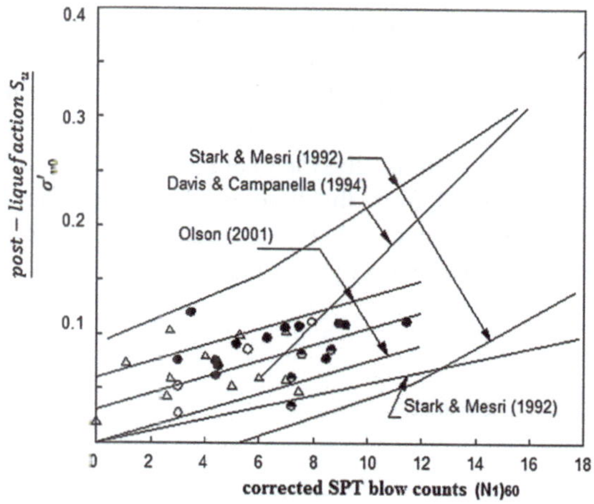

Fig. 3. Correlation between post-liquefaction shear strength and SPT blow counts (adapted from Olson 2001)

shear strength to all slices where $FS^{liq} < 1.1$, which takes into account the possibility of a progressive slope failure. The minimum value of FS^{flow} will be determined when in all contractive soil slices the liquefaction occurs and in the post-liquefaction stability analysis the liquefied shear strength is assigned to all of them.

5 The Limonar Tailings Dam

The Limonar tailings dam was developed for Cobriza copper mine (Fig. 4), with an average daily processing capacity of 2,000 cubic meters of ore. The mine is located 2300 m above sea level, in the department of Huancavelica, Peru, at an approximate distance of 480 km from the capital Lima.

Fig. 4. Limonar tailings dam in Peru

The expansion of the waste storage capacity was done by the upstream method, discarding other construction techniques since industrial facilities are located near the toe of the initial dike. Three additional levels for increasing tailings storage were designed (Fig. 5), supported by dikes built with the compacted waste material and slopes with inclination 1.5:1. A drainage system at the base of the dam and on the inner slopes should capture the fluid from the tailings, decreasing the water content and improving densification of the material. Pipes, strategically distributed, are covered with a geotextile to avoid obstruction by fine material.

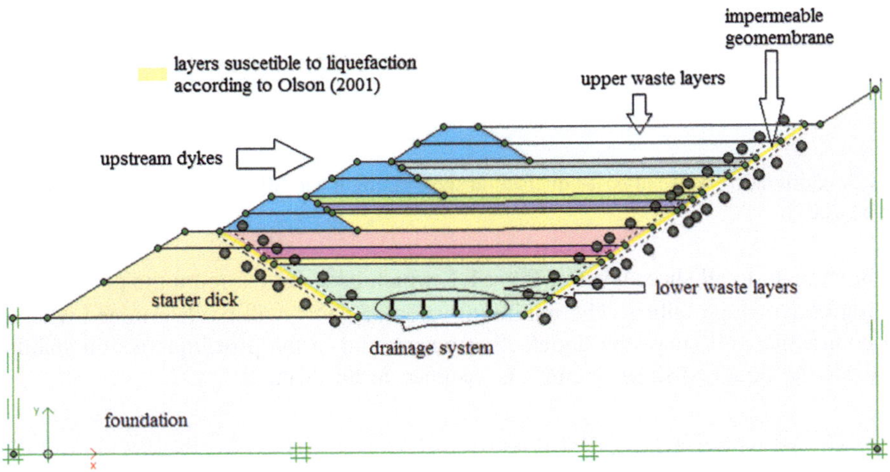

Fig. 5. Scheme of tailing dam construction

5.1 Material Properties

The soil constitutive model UBCSand was used to represent the material behavior of tailings dam and the raising dikes (built with compacted waste) while the Mohr-Coulomb model was used for the starter dyke and the foundation soil. The parameters of the UBCSand model were evaluated from SPT tests. The values of the material properties are listed in Tables 1, 2 and 3.

Table 1. Geotechnical material properties

Material	Unit weight γ (kN/m^3)	Cohesion c (kPa)	Friction angle (ϕ^0)	Coefficient of permeability (m/day)
Foundation	18.5	20	33	0.78
Starter dyke	21	20	36	0.50
Waste	24	0	32	2.59
Compacted waste	24	0	36	0.03

Table 2. Parameters for the UBCSand model

Parameter	Symbol	Unit	Waste	Compacted waste
Elastic shear modulus number	K_G^e	–	Table 3	800
Elastic bulk modulus modulus	K_B^e	–	Table 3	2000
Plastic shear modulus number	K_G^p	–	Table 3	1200
Elastic shear modulus index	ne	–	0.5	0.5
Elastic bulk modulus index	me	–	0.5	0.5
Plastic shear modulus index	mp	–	0.5	0.5
Constant volume friction angle	ϕ_{cv}	[°]	30	32
Peak friction angle	ϕ_p	[°]	32	36
Cohesion	c	[kPa]	0	0
Failure Ratio	Rf	–	0.95	0.95
Tension cut-off	σ_t	[kPa]	0	0

Table 3. Parameters for the UBCSand model (waste material)

Dyke	Depth (m) from top of the dyke	$(N_1)_{60}$	K_G^e	K_B^e	K_G^p
First	10–13	39.0	1472	942	4956
	13–18	22.5	1225	784	1446
	18–20 (*)	6.0	789	505	162
Second	10–12	9.0	903	578	259
	12–14	24.0	1252	801	1664
	14–20 (*)	5.9	784	502	161
Third	10–12	33.0	1392	891	3389
	12–15	14.0	1046	669	545
	15–20 (*)	6.0	789	505	162
Upper layers			200	400	100
Lower layers			500	1300	600

(*) Layers suscetible to liquefaction according to Olson (2001)

6 Liquefaction Potential by the Finite Element Method

The UBCSand model, developed by Byrne et al. (2004), is able to simulate the stress-strain behavior of sands under static or cyclic loading under drained or undrained conditions. It is a very appropriate model to simulate the static liquefaction of soils, whose parameters may be determined from results of field tests (SPT, CPT). The UBCSand model modifies the classic criterion of Mohr Coulomb in order to capture plastic deformations that occur in all stages of loading. Finite element analyses were carried out using Plaxis 2D software.

The beginning of the liquefaction is controlled by the maximum values of the r_u porepressure parameter defined as

$$r_u = \frac{\Delta u}{\sigma'_v} \tag{7}$$

where, at a given point of the dam, r_u represents the excess porepressure caused by the a waste layer and σ'_v the effective vertical stress.

Figures 6 and 7 show, immediately after the construction of the second dyke, the distribution of porepressures and the effective stresses, whereas Fig. 8 shows the points that during the dam raising reached the highest values of ru = 0.85, with a safety factor against hydraulic failure FS = 1/0.85 = 1.18. Therefore, according to the numerical results the possibility of static liquefaction in Limonar dam is somewhat low.

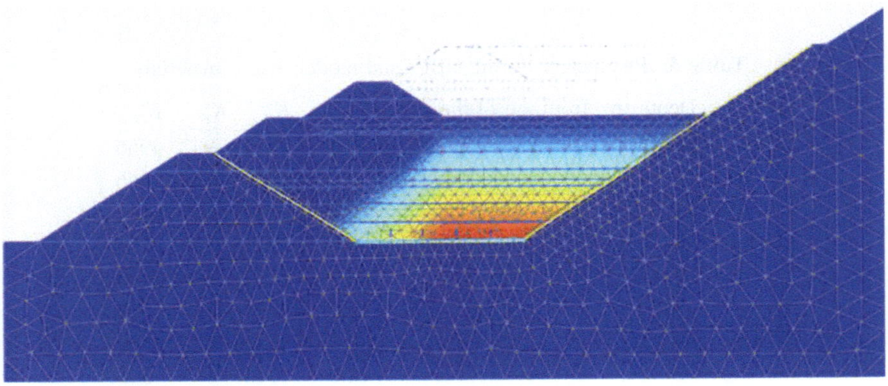

Fig. 6. Porepressure distribution after construction of the second dyke

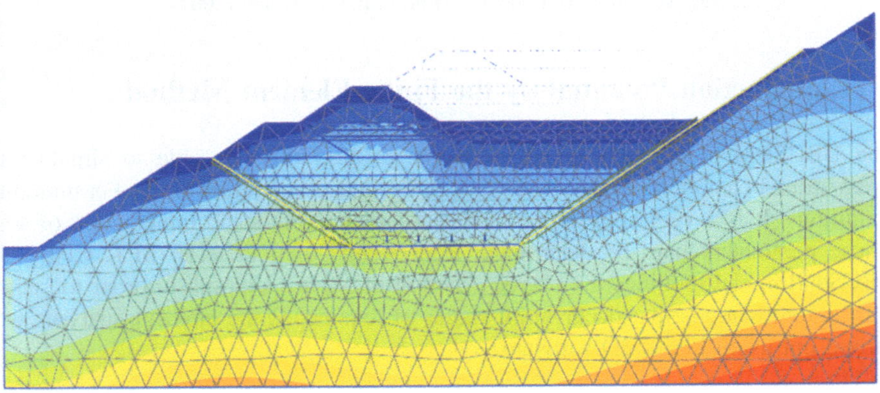

Fig. 7. Effective vertical stress distribution after construction of the second dyke

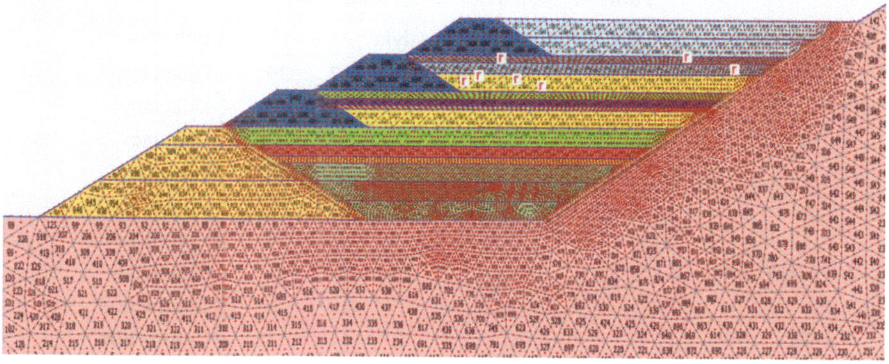

Fig. 8. Nodal points, marked with r, where porepressure parameter reached the highest value $r_u = 0.85$.

7 Liquefaction Potential by Olson's Method

Four SPT tests were carried out (SPT-05R, SPT-06R1, SPT-06R2 and SPT-06R3) on the crests of the dykes, as shown in Fig. 9, with the N-SPT blow counts corrected to account for the overburden stress and the fall energy of the hammer. The susceptibility to liquefaction was then verified on basis of Fig. 1, concluding that some layers are susceptible, as shown in Table 4. Verified the existence of three layers susceptible to liquefaction, marked in yellow color in Fig. 9, the analysis continued with the determination of the undrained shear strength ratio based in Fig. 2 or Eq. 3 to obtain the results listed in Table 5.

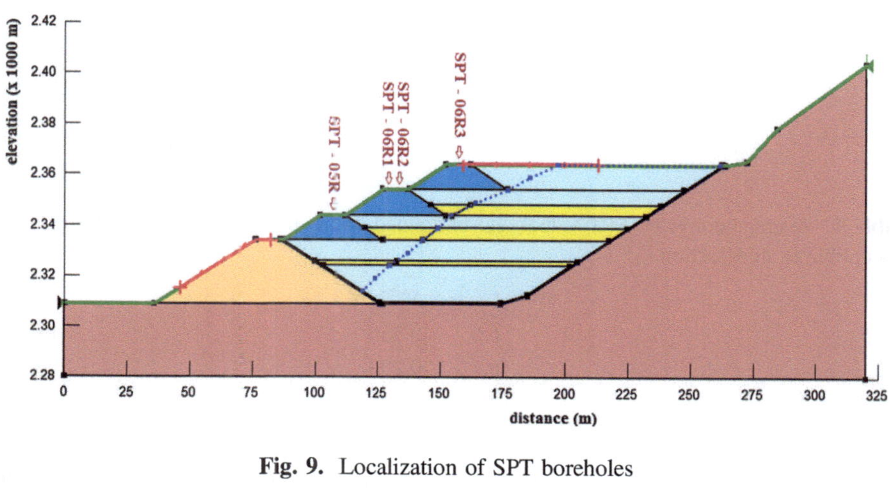

Fig. 9. Localization of SPT boreholes

Table 4. Results of susceptibility analysis

Borehole	Depth (m) from top of the dyke	N_{SPT}	$(N_1)_{60}$	Behavior
SPT-05R	11	47	39	Dilatancy
	13	>50	–	Dilatancy
	14	36	27	Dilatancy
	15	24	18	Dilatancy
	18	8	6	Contraction
	20	9	6	Contraction
SPT-06R1	13	24	19	Dilatancy
	15	10	8	Contraction
	17	11	8	Contraction
	19	4	3	Contraction
	21	7	5	Contraction
	23	15	9	Contraction
	24	>50	–	Dilatancy
	26.5	13	7	Contraction
SPT-06-R2	12	27	22	Dilatancy
	14	12	9	Contraction
	16	7	5	Contraction
	18	6	4	Contraction
	20	4	3	Contraction
	24	29	17	Dilatancy
	26	41	23	Dilatancy
	30	35	18	Dilatancy
SPT-06-R3	12	41	33	Dilatancy
	14	19	14	Dilatancy
	16	4	3	Contraction
	18	6	4	Contraction
	20	8	5	Contraction
	23	49	29	Dilatancy
	26	29	16	Dilatancy

Table 5. Average corrected SPT values and undrained shear strength ratio for the three layers susceptible to liquefaction

Layer	$(N_1)_{60}$	S_u^{peak}/σ'_{v0}
1	6.0	0.250
2	6.1	0.250
3	4.0	0.235

Table 6 shows the values of geotechnical parameters for the foundation soil, the starter dyke, the compacted material for the following dykes, the material susceptible to liquefaction and the stable waste.

Table 6. Geotechnical parameters for liquefaction potential analysis

Material	γ (kN/m³)	C (kPa)	ϕ (°)	S_u^{peak}/σ'_{v0}		
				Layer 1	Layer 2	Layer 3
Foundation	18.5	20	33	-	-	-
Starter dyke	21	20	36	-	-	-
Waste	24	0	32	-	-	-
Suscetible waste to liquefaction	24	0	32	0.250	0.250	0.235
Compacted waste	24	0	36			

According to Olson's empirical method for cases of static load involving soils exhibiting contractive behavior, the shear stress mobilized immediately before failure is approximately equal to the values of undrained shear strength. A slope stability analysis can provide a reasonable estimate of these mobilized shear stresses and, in this work, the Simplified Bishop Method was used with the computer program SLOPE / W (GeoStudio). For the susceptible layers 1 (the upper layer marked in yellow color in Fig. 9), 2 and 3 (the lower layers) the initial values of shear strength were gradually decreased until the safety factor against failure reached the value FS = 1. In this process, the shear strength of not susceptible materials was kept constant.

Figures 10, 11 and 12 show the position of the failures surfaces after raising the tailings dam to the first, second and third designed levels, respectively, and Tables 7, 8 and 9 list the values of shear stresses and the initial effective vertical stress at the bottom of the slices.

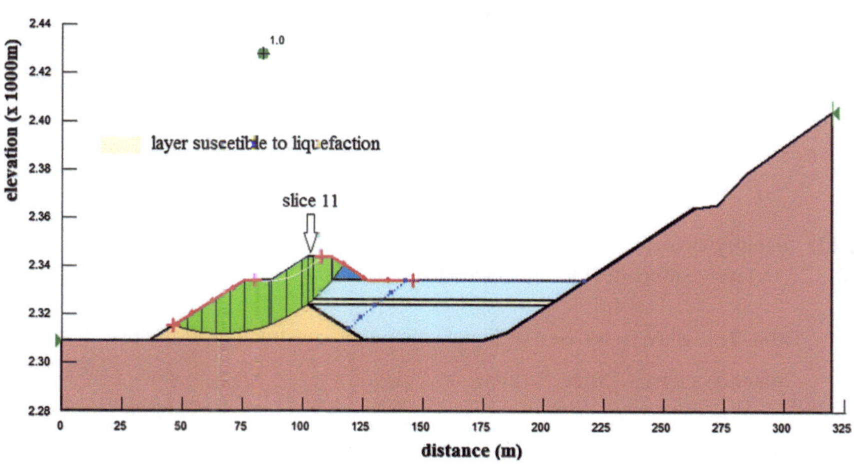

Fig. 10. Stability analysis after construction of the first dyke (FS = 1) with base of slice 11 in layer susceptible to liquefaction

The values of the safety factors against liquefaction Eq. (5) were all determined greater than 1.1, which minimizes the possibility of static liquefaction occurrence and

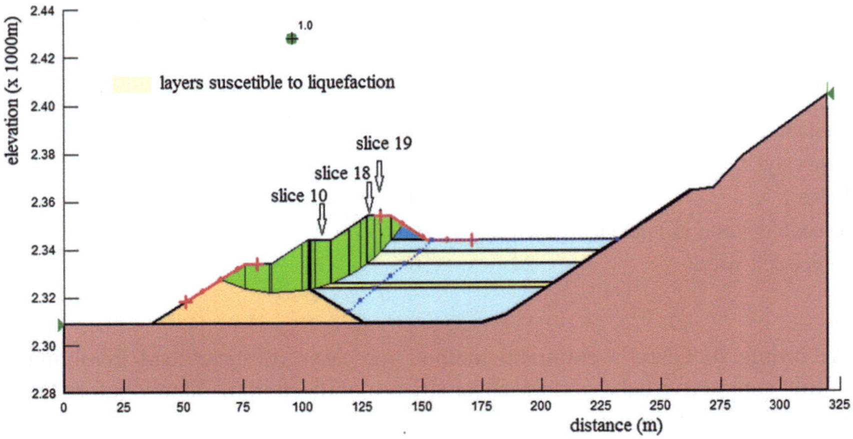

Fig. 11. Stability analysis after construction of the second dyke (FS = 1) with base of slices 10, 18, 19 in layers susceptible to liquefaction

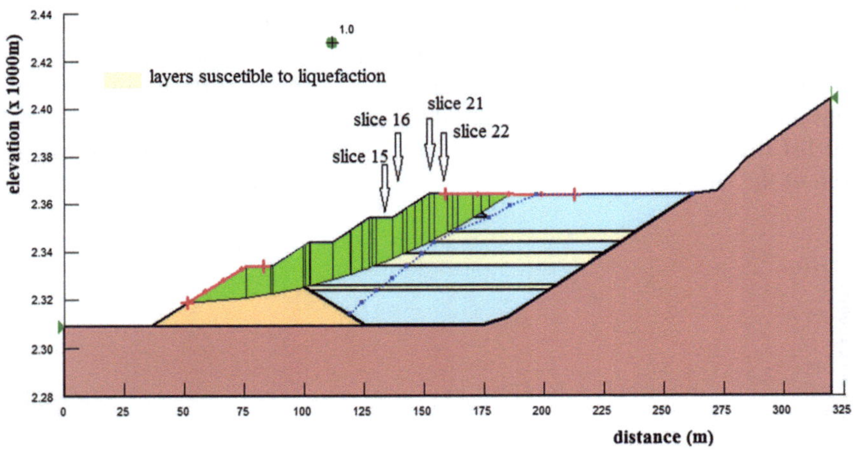

Fig. 12. Stability analysis after construction of the third dyke (FS = 1) with base of slices 15, 16, 21, 22 in layers susceptible to liquefaction

Table 7. Safety factor against liquefaction after construction of the first dyke

Stability analysis	Slice	τ (kPa)	σ'$_{v0}$ (kPa)	S_u^{peak}/σ'_{v0} (kPa)	S_u^{peak} (kPa)	FS^{liq}
Dick 1	11	57.31	456	0.250	114	2.0

Table 8. Safety factor against liquefaction after construction of the second dyke

Stability analysis	Slice	τ (kPa)	σ'$_{v0}$ (kPa)	S_u^{peak}/σ'_{v0} (kPa)	S_u^{peak} (kPa)	FS^{liq}
Dick 1	10	57.67	456.21	0.250	114.1	1.9
Dick 2	18	43.11	442.73	0.250	110.7	2.6
	19	40.44	387.04	0.250	96.8	2.4

Table 9. Safety factor against liquefaction after construction of the third dyke

Stability analysis	Slice	τ (kPa)	σ'$_{v0}$ (kPa)	S_u^{peak}/σ'_{v0} (kPa)	$S_u^{peak}n$ (kPa)	FS^{liq}
Dick 2	15	58.3	434.6	0.250	108.6	1.9
	16	53.7	413.92	0.250	103.5	1.9
Dick 3	21	43.4	465.88	0.235	109.5	2.5
	22	38.6	417.49	0.235	98.1	2.5

the need for further post-liquefaction investigation. The results of the Olson's empirical model confirm the results previously obtained with the finite element method and the UBCSand constitutive strain-stress law.

8 Conclusion

Two forecasting methods for static liquefaction in tailings dams were used in this study: an empirical approach proposed by Olson (2001), based on back analysis of 33 historical failure cases by liquefaction, and a numerical method incorporating a specific constitutive model (UBCSand model) for this particular kind of problem. Both methods used the N-SPT blow counts for definition of geotechnical parameters. In the case of Limonar tailings dam the static liquefaction hazard was proven low by both approaches.

References

Been, K., Jefferies, M.G.: A state parameter for sands. Géotechnique **35**(2), 99–112 (1985)

Been, K., Crooks, J.H.A., Becker, D.E., Jefferies, M.G.: The Cone penetration test in sands. Part I: States parameter interpretation. Géotechnique **36**(2), 239–249 (1986)

Been, K., Jefferies, M.G., Crooks, J.H.A., Rothenburg, L.: The Cone penetration test in sands. part II: general inference of states. Géotechnique **37**(3), 285–300 (1987)

Byrne, P.M., Park, S.S., Beaty, M., Sharp, M.K., Gonzalez, L., Abdoun, T.: Numerical modeling of liquefaction and comparison with centrifuge tests. Canadian Geotech. J. **41**(2), 193–211 (2004)

Casagrande, A.: Characteristic of cohesionless soils affecting the stability of slope and earth fill. J. Boston Soc. Civil Eng. **23**, 13–32 (1936)

Castro, G.: Liquefaction of Sand. Harvard Soil Mechanic Series, vol. 87. Harvard University, Cambridge (1969)

Castro, G., Poulos, S.J.: Factors affecting liquefaction and cyclic mobility. J. Geotech. Eng. Div. ASCE **103**, 501–516 (1977)

Fears, C.E., Robertson, P.K.: Estimating the undrained strength of sand: a theoretical framework. Canadian Geotech. J. **32**(4), 859–870 (1995)

Finn, W.D.L., Ledbetter, R.H., Wu, G.: Liquefaction in silty soils: design and analysis. In: Prakash, S., Dakoulas, P. (eds.) Ground Failures under seismic conditions, pp. 51–76. American Society of Civil Engineers, Reston (1994)

Hanzawa, H., Itoh, Y., Suzuki, K.: Shear characteristics of a quick sand in the Arabian Gulf. Soils Found. **19**(4), 1–15 (1979)
Ishihara, K.: Liquefaction and flow failure during earthquakes. Géotechnique **43**(3), 351–415 (1993)
Ishihara, K.: Post-earthquake failure of a tailing dam due to liquefaction of de pound deposit. In: International Conference on the Case Histories in Geotechnical Engineering, vol. 1, pp. 1129–1143. University of Missouri, St. Louis (1984)
Ishihara, K.: Stability of natural deposits during earthquakes. In: 11th International Conference on Soil Mechanics and Foundation Engineering, San Francisco, vol. 1, pp. 321–376 (1985)
Mittal, H.K., Morgenstern, N.R.: Seepage control in tailings dams. Canadian Geotech. J. **13**(3), 277–293 (1976)
Olson, S.M.: Liquefaction analysis of level and sloping ground using field case histories and penetration resistance. Ph.D. thesis, University of Illinois at Urbana-Champaign (2001)
Olson, S., Stark, T.: Yield strength ratio and liquefaction analysis of slopes and embankments. J. Geotech. Geoenviron. Eng. ASCE **129**(8), 727–737 (2003)
Poulos, S.J.: The steady state of deformation. J. Geotech. Eng. Div. ASCE **107**(GT5), 553–562 (1981)
Robertson, P.K., Campanella, R.G.: Liquefaction potential of sands using the CPT. J. Geotech. Eng. Div. ASCE **111**(3), 384–403 (1985)
Seed, R.B., Harder, L.F.: SPT-based analysis of cyclic pore pressure generation and undrained residual strength. In: Proceedings of H.B. Seed Memorial Symposium, vol. 2, pp. 351–376. Hi-Tech Publishing Ltd. (1990)
Sladen, J.A., d'Hollander, R.D., Krahn, J.: The liquefaction of sands, a collapse surface approach. Canadian Geotechn. J. **22**(1), 11–27 (1985)
Smith, E.S.: Tailings disposal – failures and lessons. In: Proceedings of 1st International Symposium Tailings Disposal Today, Tucson, AZ (1972)
Vick, R.: Planning, Design and Analysis of Tailings Dam, Department of Civil Engineering. MIT, Cambridge (1983)
Wang, W.: Some finding in soil liquefaction. Water Conservancy and Hydroelectric Power Scientific Research Institute, Beijing (1979)

Probabilistic Seismic Hazard and Dynamic Stability Assessment of a Tailings Dam Located in Jamaica

Frank Perez[1(✉)] and Celso Romanel[2]

[1] Geotechnical Engineering, USIL, Lima, Peru
frankpcz@hotmail.com
[2] Pontifical Catholic University of Rio de Janeiro, Rio de Janeiro, Brazil
romanel@puc-rio.br

Abstract. This research presents a probabilistic seismic hazard analysis (PSHA) and evaluation of the dynamic response of a tailings storage facility (TSF) located in Jamaica, country of high seismic activity situated along the North American and Caribbean tectonic plates. PSHA was performed in order to study the regional seismicity and to obtain the appropriate ground motion parameters. The stability of the dam and the permanent displacements caused by the design earthquake were estimated by pseudo-static method and a more complex solution using 2D finite element method (FEM). The signal treatment for the design earthquake and the definition of the Rayleigh damping based on the 1D equivalent linear method (ELM) are also discussed.

1 Introduction

The Jamaican territory is an elevated portion above the sea level of the underwater volcanic platform of Nicaragua (Nicaragua Bank). Seismicity conditions of the country are associated with the interaction between the North American plate, the Gonave microplate, and the Caribbean plate (on which lies the island). Major regional earthquake generation sources are associated with the movement of Gonave microplate, which is bounded on the north by the *Oriente* fault zone (OFZ), west by the Cayman Spreading Center (CSC), and south by Walton (WFZ) and Enriquillo (EFZ) fault zones (Fig. 1).

Based on ERN (2009), were identified 13 seismogenic sources in the Jamaica Region; in this study, only 7 were considered taking into account its proximity to the dam site (Fig. 2), which naturally present more influence in the seismic hazard estimation. Table 1 lists the seismic parameters λ and β, following Gutenberg-Richter law (1944), with correlation coefficients of linear regression over 0.95.

H. Shehata and Y. Rashed (eds.), *Numerical Analysis of Nonlinear Coupled Problems*, Sustainable Civil Infrastructures, DOI 10.1007/978-3-319-61905-7_6

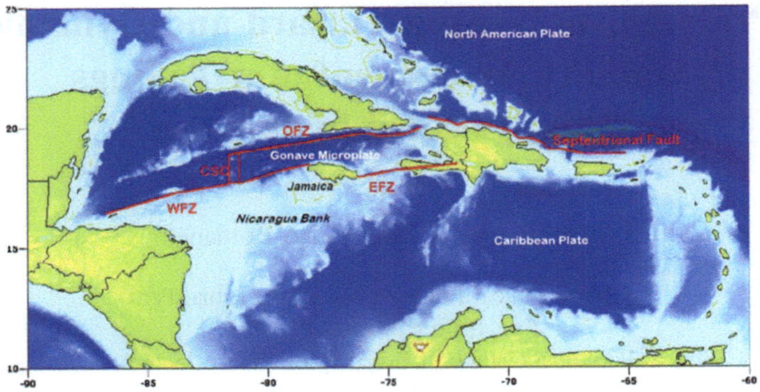

Fig. 1. Tectonics and seismicity of Jamaica (ERN 2009)

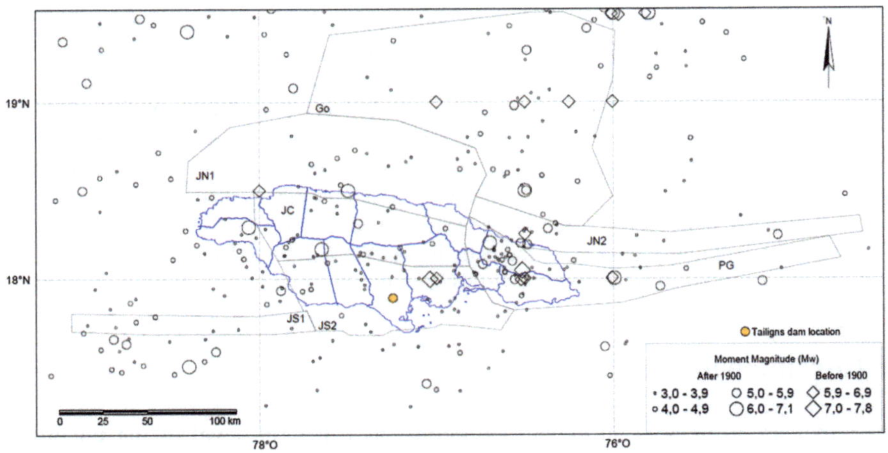

Fig. 2. Jamaican seismogenic sources (ERN 2009)

Table 1. Seismic sources parameters

j	Seismic source	Symbol	Depth (km)	M_u	λ_j	β	M_0
1	Jamaica 1 North Faults	JN1	11	6	0.50	1.3	3
2	Jamaica 2 North Faults	JN2	11	6	0.50	1.3	3
3	Gonâve Fault System	GO	13	6.5	0.77	1.0	3
4	Jamaica Center Faults System	JC	13	6	1.04	1.2	3
5	Jamaica 1 South Faults System	JS1	12	6.3	0.49	1.2	3
6	Jamaica 2 South Faults System	JS2	12	7.8	0.74	1.2	3
7	Plantain Garden Faults	PG	14	7	1.22	1.1	3

2 Ground Motion Model

Jamaica doesn't have yet an own attenuation law and most of the former PSHA (Shepherd et al. 1997, USAI-OAS 1997) relationship is applicable. This attenuation law was developed based on analysis of acceleration-time histories recorded from earthquakes in the East coast of the United Estates, as presented in Eq. 1.

$$\ln(Sa(T)) = B_{1ALL} + B_2(M-6) + B_3(M-6)^2 + B_5 \cdot \ln(r) + B_V \cdot \ln\left(\frac{V_s}{V_A}\right) \tag{1}$$

$$r = \sqrt{r_{jb}^2 + h^2} \tag{2}$$

where $Sa(T)$ is the ground acceleration (in terms of gravity g), M is moment magnitude scale, V_s is average velocity of shear wave propagation S (m/s), V_A is a reference velocity for S (m/s) measured from downhole tests in A class soil and r_{jb} is the nearest horizontal distance measured form the station to the projection of the rupture surface (km), as expressed in Eq. 2.

The threshold magnitude considered in the seismic catalogue was $M_0 = 3$ and the upper limit M_u varies according to the seismic source considered, as shown in Table 1. Once defined the seismic sources, their parameters and the attenuation law, then the PSHA was performed using the software CRISIS 2007, developed by Ordaz *et al.* (2007). The results in terms of spectral accelerations were calculated in the upper 30 m of the site geology with a shear wave velocity V_{s30} of 650 m/s. In order to obtain the uniform hazard response spectra (UHRS) it is necessary the determination of seismic hazard curves for different spectral periods. Figure 3 presents the seismic hazard curve in function of the exceedance rate of acceleration for 50 years of dam exposition and period T = 0. Figure 4 shows the UHRS for a 475 year return period and 10% exceedance probability in 50 years.

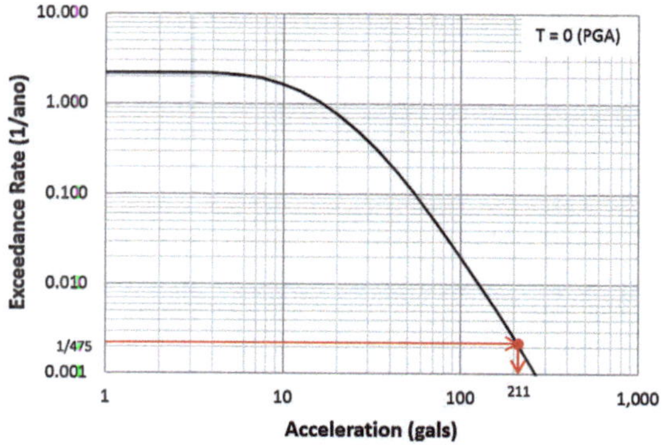

Fig. 3. Probabilistic seismic hazard curve

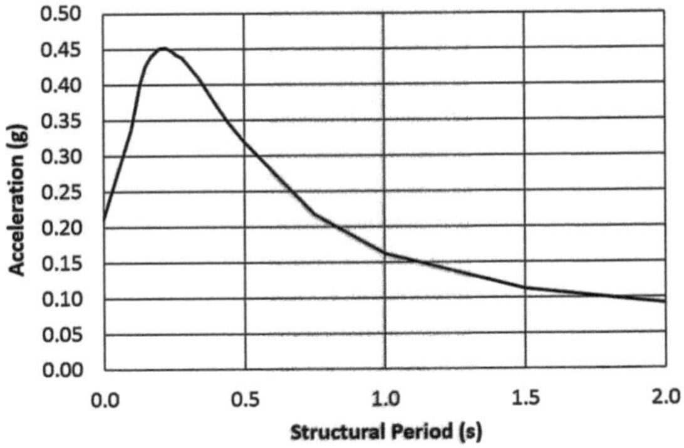

Fig. 4. Uniform seismic hazard spectra for a 475 year return period earthquake

3 Geotechnical Caracterization

The TSF design is composed by two dikes, the first one already existent and the second dike of 3 m height projected over the existent tailings, as presented in Figs. 5 and 6. The new tailings will be disposed with a slope of 3% and the existent tailings were divided in 11 regions, designed with letters A to K, according to the values of undrained shear strength S_u measured from field and laboratory tests. The S_u variation with depth z from the surface of the existent tailings (Table 2) is illustrated in Fig. 7. The existent tailings have a depth of 23 m, and they consist of clayey material of plastic index (PI) of about 55%, in the upper 10 m, and PI of 65%, in the lower 13 m. Below the existent tailings is the foundation soil, conformed by alluvial soil (12.4 m thick) and a rigid clayey soil (2.8 m thick) followed by a bedrock. Phreatic surface is located at 30 m depth.

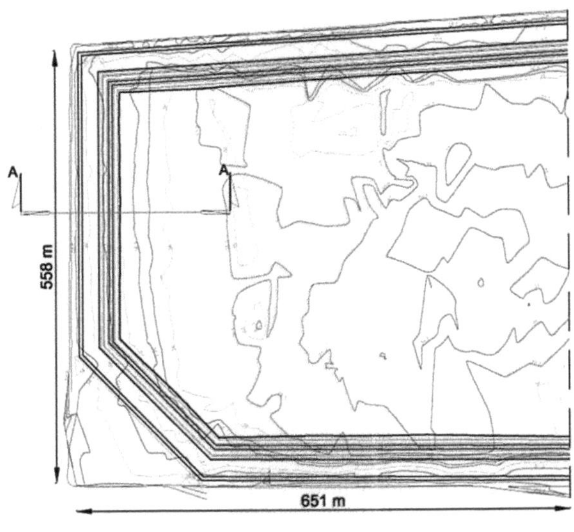

Fig. 5. Plan view of tailings storage facility

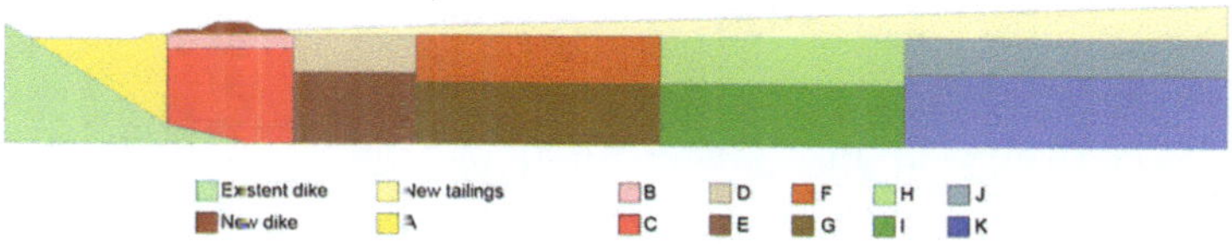

Fig. 6. A-A cross section of tailing storage facility

Table 2. Undrained shear strength variation in existent tailings as function of depth z

Existent tailings	S_u (kPa)	Existent tailings	S_u (kPa)	Existent tailings	S_u (kPa)
A	2.5 + 1.43z	E	8 + 0.902z	I	18.36 + 2.588z
B	6.99 + 0.897z	F	8.28 + 0.955z	J	13.36 + 2.588z
C	2.5 + 1.43z	G	17.8 + 2.188z	K	19.01 + 2.693z
D	9.45 + 1.696z	H	11.38 + 0.76z	–	–

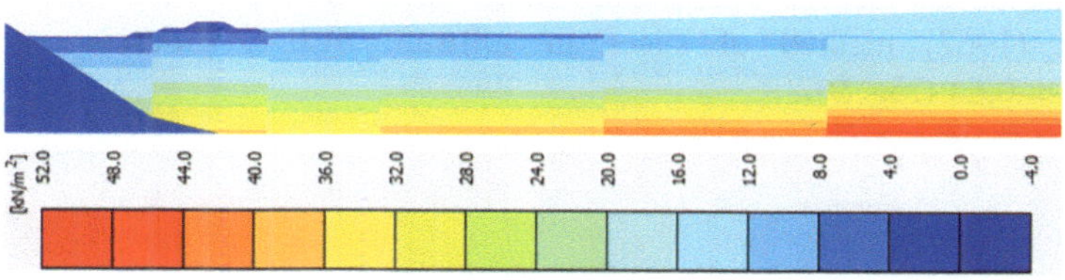

Fig. 7. Undrained strength variation with depth

4 Factors of Safety

4.1 Static Factor of Safety

Initial stresses have been determined using the finite element program PLAXIS v.2011 considering the geometry shown in Fig. 6 and three stages of construction: (a) existent dam; (b) new dike; (c) tailings deposition. The finite element mesh was composed by quadratic triangular elements (6 nodes) and the strength reduction method was used to obtain the static factor of safety (*FoS*) using Eqs. (3) and (4).

$$c^* = \frac{c}{M} \tag{3}$$

$$\tan^* \phi = \frac{\tan \phi}{M} \tag{4}$$

where M is a parameter that reduces the Mohr Coulomb parameters c and $tan\phi$ in consecutives non-linear analysis until reaching the slope collapse, when finally $M = FoS$. The static factor of safety determined in this analysis was $FoS = 1.083$ and the potential failure surface involves the new dike and existent tailings, as presented in Fig. 8.

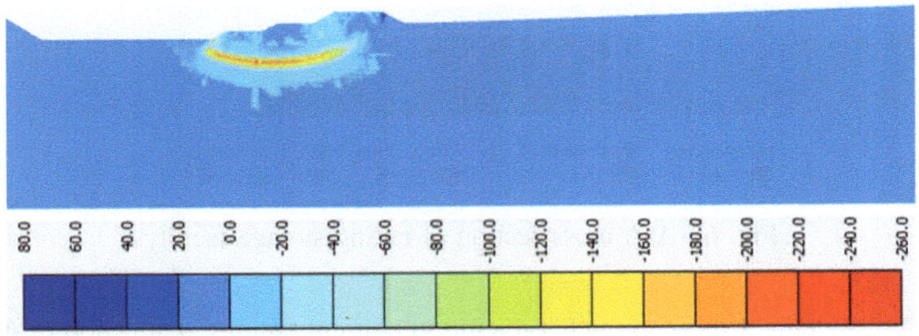

Fig. 8. Critical slip surface in static analysis based on incremental shear strain $(\Delta\gamma_{xy})$

4.2 Pseudo-Static Factor of Safety

The horizontal seismic coefficient k was determined in accordance to Hynes-Griffin and Franklin (1984) recommendation, that considers k as half of the peak ground acceleration (PGA) $k = 0.5PGA^{rocha}/g$. From the PSHA it was determined $PGA^{bedrock} = 0.21\ g$, hence $k = 0.105$. The pseudo-static factor of safety was obtained using Spencer's (1967) method of slices in software SLIDE (*Rocscience*) resulting $FoS^{pseudo} = 0.64$ (Fig. 9).

Fig. 9. Critical slip surface in pseudo-static analysis

5 Non-linear Dynamic Analysis

5.1 Selection of a Real Earthquake

For the non-linear dynamic analysis it was selected an earthquake with epicenter 18 km away from TSF, occurred in March 10th, 2012, with duration of 60.96 s and moment magnitude $M_w = 5.1$. The event was registered by the onshore accelerograph station SMAD, from the Jamaican seismographic network, located 156 km from the seismic event.

5.2 Spectral Matching

In order to obtain a design earthquake compatible with the local seismicity, an earthquake signal treatment was performed, consisting of base line correction, filtering and spectral matching in time domain, using the computational program SEISMOMATCH

(Seismosoft Ltd 2012). The objective of the spectral matching is to correct the original signal iteratively in the time domain, until reaching a signal compatible with the target spectrum. This process has been done through the addition of wavelets, preserving the non-stationary character of the reference acceleration-time history. This procedure was first proposed by Kaul (1978) and was extended to simultaneously match spectra with multiple damping values by Lilhanand and Tseng (1987). Figure 10 presents the target spectrum, and the original and matched acceleration spectra.

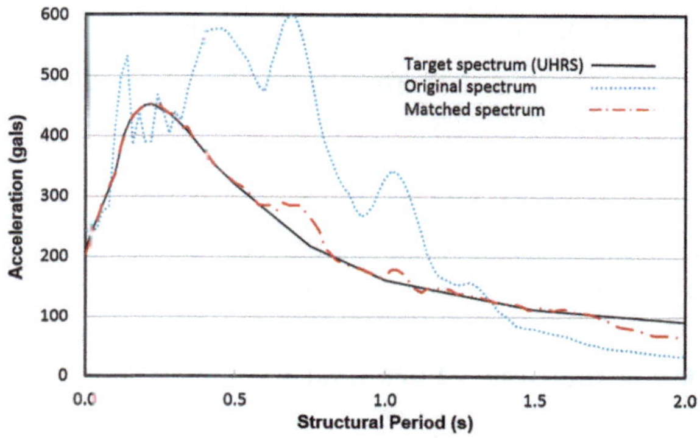

Fig. 10. Spectral matching in time domain

5.3 Cutoff Frequency

The cutoff frequency was estimated from the power spectrum function (PSFD); in this case a cutoff frequency $f_c = 5$ Hz was obtained (Fig. 11) which corresponds to 98% of the original earthquake power. The calculation of this frequency is important in order to determine the maximum allowable size of finite elements.

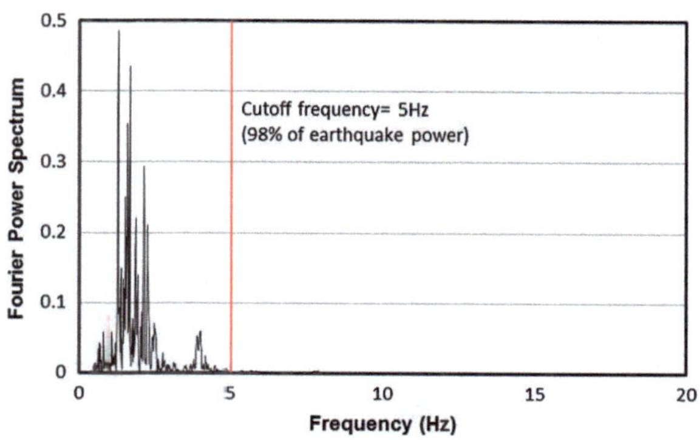

Fig. 11. Cutoff frequency

5.4 Finite Element Size

In order to ensure correct shear wave propagation through the elements of the model, the maximum element size in each material region has to be obtained. The maximum element size is defined as 1/8 of the shear wave length for each material considering f_c = 5 Hz and the shear wave velocities listed in Table 3. The maximum and actual element sizes for each region are also presented in Table 3. The finite element mesh for the TSF model was composed by 40.109 quadratic triangular elements (6 nodes).

Table 3. Maximum and actual finite element sizes

Material	New dike	Existent dam	New tailings	Existent tilings
Shear Wave Velocity v_s (m/s)	73.3	71.4	16.5	15.6
Cutoff Frequency fc (Hz)	5			
Wave Length λ (m)	14.7	14.3	3.3	3.1
Maximum Element Size $\lambda/8$ (m)	1.83	1.79	0.41	0.39
Actual Element Size (m)	0.50	0.90	0.41	0.39

5.5 Amplification of the Design Earthquake

The bedrock level, where the design earthquake was determined, is located at depth 38.2 m and the base of the numerical model encompasses only the existent tailings at depth 23 m. A one-dimension equivalent linear method (ELM) was performed, using program SHAKE2000 (Ordonez 2011), to obtain the acceleration history from the

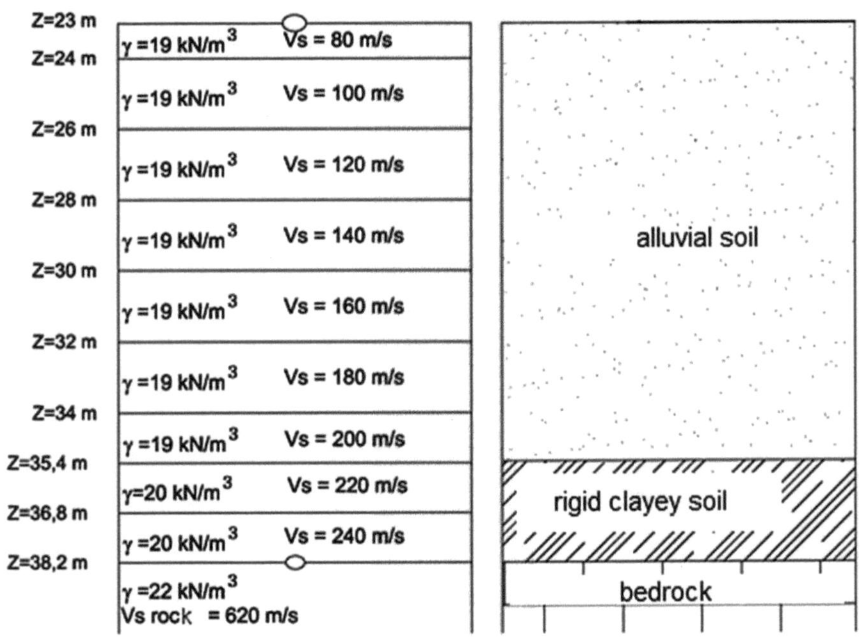

Fig. 12. One-dimensional shear wave propagation analysis in ELM

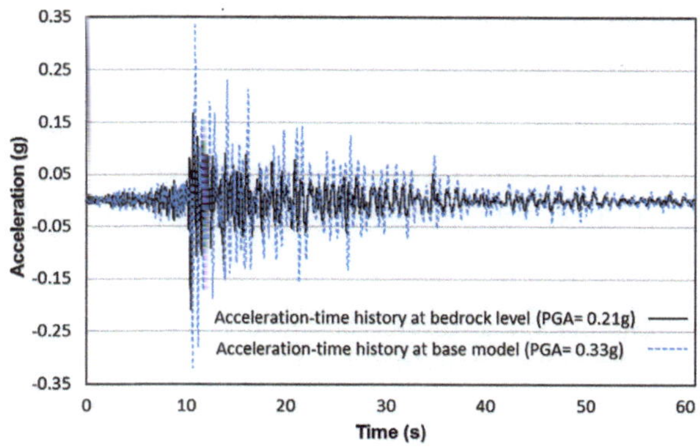

Fig. 13. Results of 1D propagation of the matched seismic input

bedrock level to the base of the model, as shown in Fig. 12. The corresponding acceleration – time history are indicated in Fig. 13, with PGA = 0.21 g at the bedrock level and PGA = 0.33 g at the base of the finite element mesh. Since the numerical model admits a compliant base (soil material), only the incident shear waves were taken into account, which corresponds to half of the seismic record determined in the previous section (Mejia and Dawson 2006).

5.6 Material Damping Calibration

As the PLAXIS 2D program does not include in the formulation the hysteretic damping, usually employed in the study of dynamic behavior of geological materials, it was necessary to establish an equivalence between the frequency-dependent Rayleigh damping, available in PLAXIS 2D, and the hysteretic model, available in the SHAKE2000 program. The technique to correlate Rayleigh and hysteretic damping is also suggested by FLAC 2D program (Itasca 2015) for dynamic analysis.

The comparison was made considering the responses of one-dimensional models according to Fig. 14, depicting a soil column for the SHAKE2000 computer program and a very large soil region for the Plaxis program, to assure a 1D deformation state, with the lateral boundaries of the finite element mesh specified as free fields and the lower boundary admitted as a flexible base.

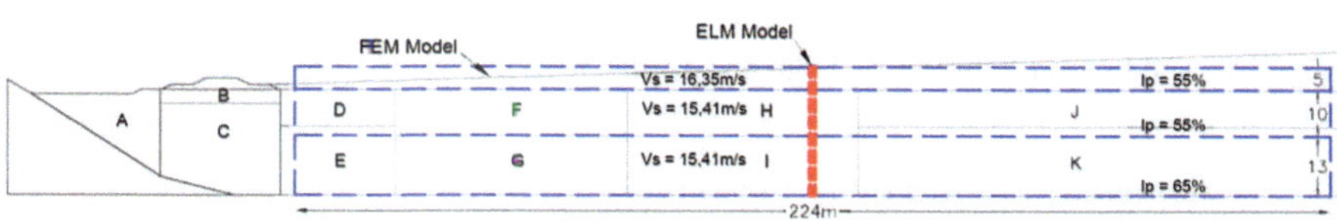

Fig. 14. Models for calibration between rayleigh (FEM) and hysteretic (ELM) damping

In the SHAKE2000 analysis the hysteretic damping curves for the tailings materials were those proposed by Vucetic and Dobry (1991), which take into account the effects of soil plasticity during a cyclic load. In the FEM model the Rayleigh damping rates were admitted $\xi = 5\%, 6\%$ and 7% (typical values for geological materials) considering a numerical damping of the system $\gamma = 0.1$.

Results from such comparisons are shown in Figs. 15 e 16 with a good correlation in terms of maximum acceleration and Fourier spectra for $\xi = 6\%$, although the maximum shear stresses were (Fig. 17) were higher than those obtained with the ELM model. Based on these numerical results the value $\xi = 6\%$ was assigned for tailings and $\xi = 4\%$ for the new dike, a typical value for compacted materials.

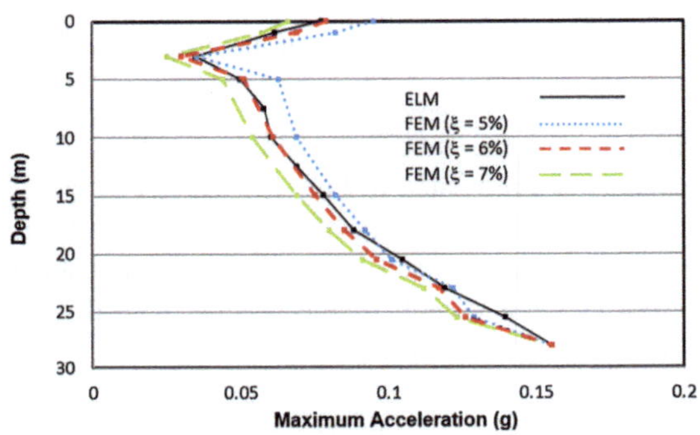

Fig. 15. Calibration between rayleigh and hysteretic damping in terms of maximum acceleration

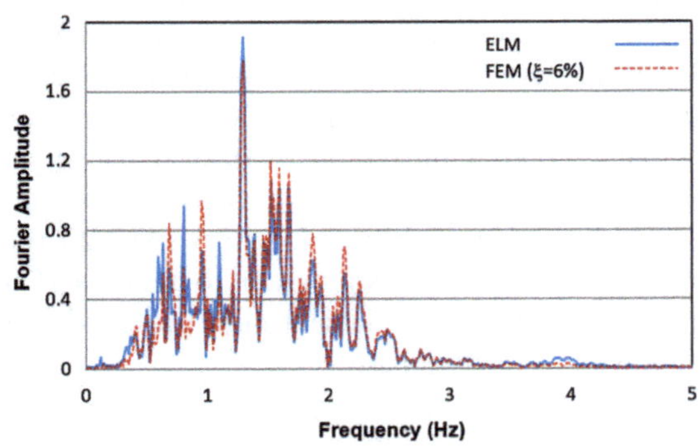

Fig. 16. Calibration between rayleigh and hysteretic damping in terms of fourier spectra

5.7 Fundamental Frequency of the Dam

The first and second predominant frequencies of the tailings dam were calculated from an undamped elastic analysis. The power spectra from horizontal accelerations were calculated in some control points (Fig. 18) establishing a predominant frequency of 0.705 Hz (point H) for the new dike and 2.116 Hz for other materials (Fig. 19).

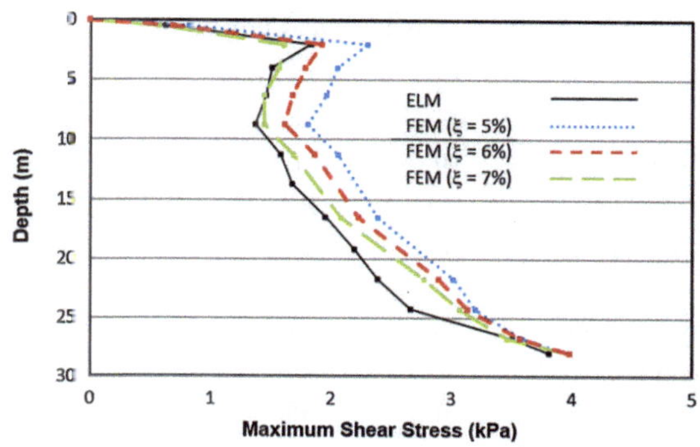

Fig. 17. Calibration between rayleigh and hysteretic damping in terms of maximum shear stress

Fig. 18. Control nodes in dynamic analysis

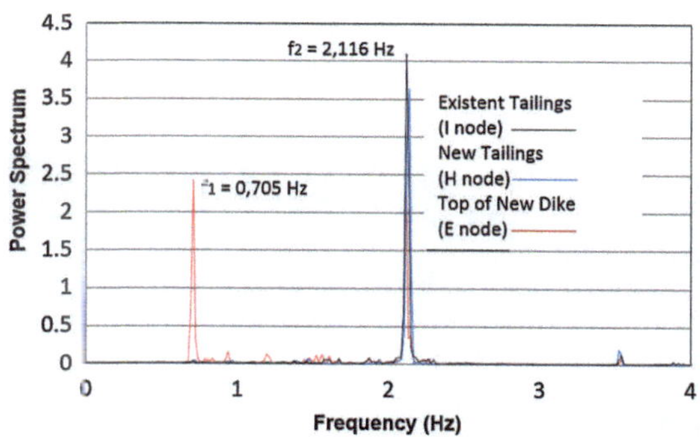

Fig. 19. Power spectra from horizontal accelerations in undamped elastic analysis

5.8 Permanent Displacements

Once obtained the geotechnical properties, the model characteristics and the design earthquake, the dynamic analysis was performed based on the finite element method. The mechanical behavior of the soils was simulated considering the Mohr Coulomb constitutive model with parameters listed in Table 4. Control nodes (Fig. 18) were assigned in order to record displacements and accelerations due to earthquake. Horizontal and vertical displacements are presented in Table 5 as well as in Figs. 20 and 21. Maximum vertical displacements occurred at the slope toe and crest of the new dike. Figures 22 and 23 show the evolution in time of the horizontal and vertical displacements for nodes B, C, D and E. From Table 5 it can be observed that permanent displacements obtained in nodes G and H belonging to the new tailings are the lowest values.

Table 4. Geotechnical Properties

Properties	New dike	Existent dam	New tailings	Existent tailings
Unit Weight γ (kN/m^3)	19	20	16.5	16.5
Poisson Ratio ν	0.4	0.4	0.49	0.49
Small Strain Shear Modulus G_{max} (MPa)	10.2	10.2	0.45	0.4
Shear Wave Velocity v_s (m/s)	73.3	71.4	16.5	15.6
Undrained Shear Strength S_u (kPa)	–	–	16.0	Table 2
Friction Angle ϕ (°)	34	35	–	–

Table 5. Post-earthquake permanent displacements

Point	Horizontal displacement (m)	Vertical displacement (m)	Point	Horizontal displacement (m)	Vertical displacement (m)
A	−0.069	0.062	F	−0.143	−0.113
B	−0.619	0.030	G	−0.029	−0.005
C	−0.611	−0.098	H	−0.008	−0.003
D	−0.363	−0.380	I	−0.015	−0.003
E	−0.391	−0.393	J	−0.020	0.014

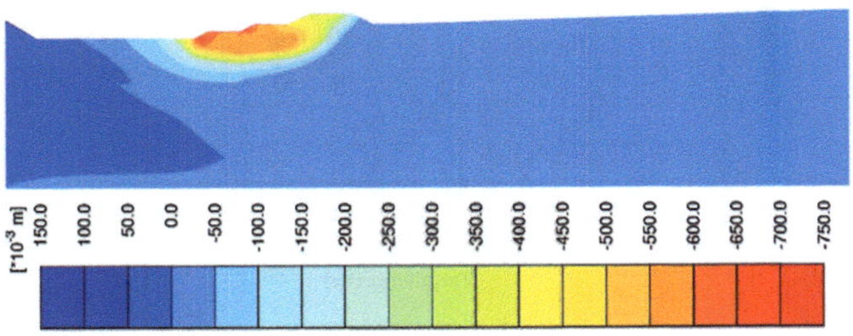

Fig. 20. Post-earthquake permanent horizontal displacements

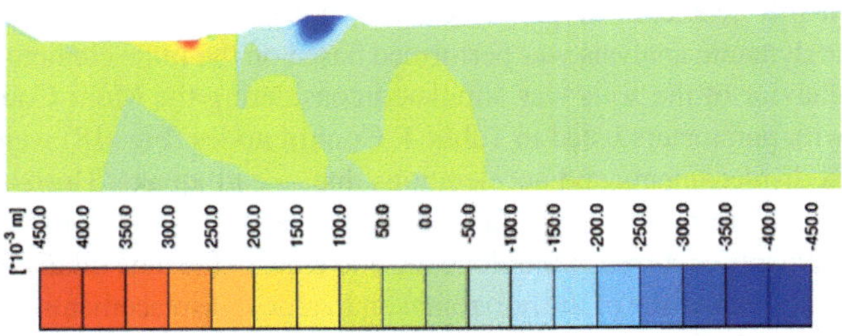

Fig. 21. Post-earthquake permanent vertical displacements

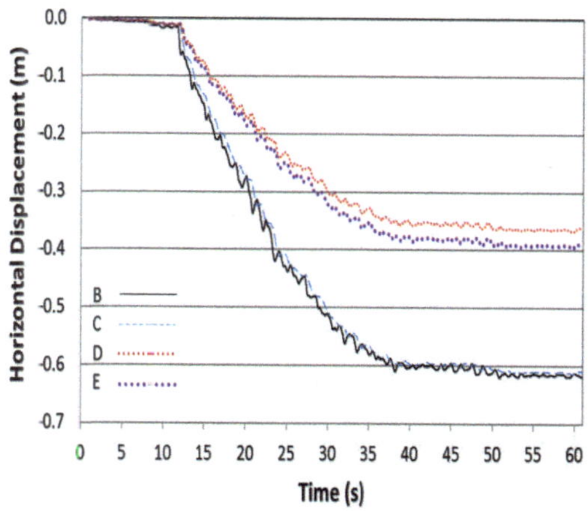

Fig. 22. Permanent horizontal displacement time history.

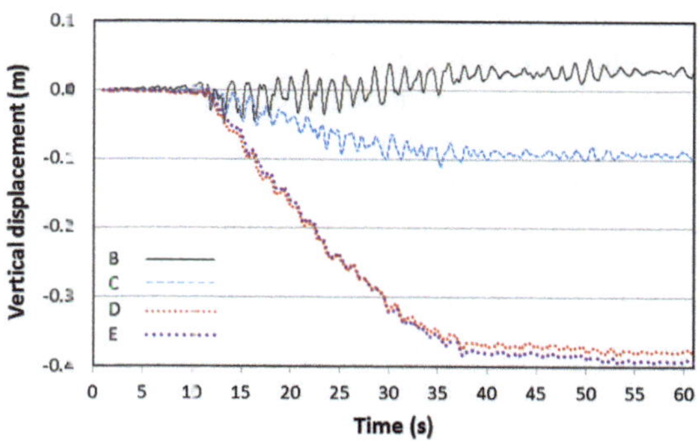

Fig. 23. Permanent vertical displacement time history.

5.9 Comparison of Acceleration Response Spectra

An additional comparison between the acceleration response spectra for the control point H was made considering the time domain (FEM) and the frequency domain (ELM) analyses. Point H was selected since due to its localization the expected response is mainly 1D. Figure 24 shows the comparison of both spectra, very similar to each other, with slightly higher amplitudes computed in the FEM analysis, especially within the range 2 Hz $< f <$ 4 Hz. Moreover, the FEM results represent better the spectrum peaks.

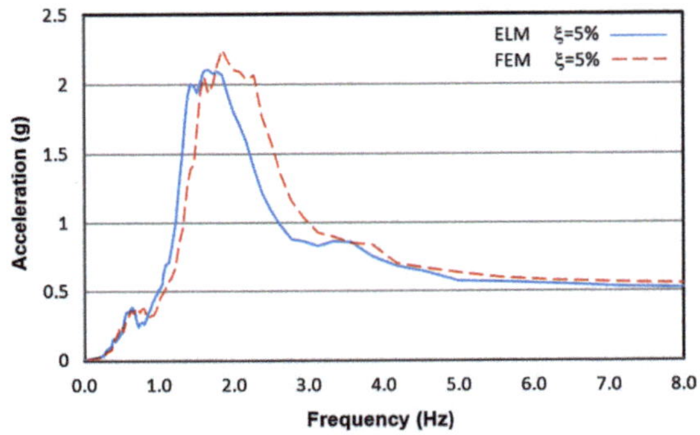

Fig. 24. Comparison between response spectra in new tailings

6 Conclusion

Pseudo-static slope stability analysis carried out with a limit equilibrium method determined a factor of safety $FS^{pseudo} = 0.64$, which indicates an instability condition for the new dike configuration. A more complex dynamic investigation, based on the FEM method considering a design earthquake determined from PSHA, suggests that despite the new dike may suffer significant permanent displacements the collapse anticipated by the pseudo-static analysis would not occur.

Post-earthquake permanent displacements reach their maximum horizontal magnitude of 0.619 m at control point B (slope toe) and maximum settlement of 0.393 m at control point E (crest). For the new tailings region the maximum magnitudes at control point G are 0.029 m and −0.005 m, respectively.

References

ERN: Country-specific risk evaluation for Bolivia, Guatemala, Jamaica and Peru. ATN/JF-9349-RS, Catastrophe Risk Profile Jamaica, Inter-American Development Bank, D. C. (2009)

Gutenberg, R., Richter, C.F.: Frequency of earthquakes in california. Bull. Seismol. Soc. Am. **34**, 185–188 (1944)

Hynes-Griffin, M., Franklin, A.: Rationalizing the seismic coefficient method. U.S. Army Engineer Waterways Experiment Station, Miscellaneous Paper GL-84–13, Vicksburg, MS (1984)

Itasca Consulting Group Inc.: FLAC 2D: fast lagrangian analysis of continua, vol. 8 (2015)

Kaul, M.K.: Spectrum consistent time-history generation. ASCE J. Eng. Mech. **EM4**, 781–788 (1978)

Lilhanand, K., Tseng, W.S.: Generation of synthetic time histories compatible with multiple-damping response spectra, SMIRT-9, Lausanne (1987)

Mejia, L.H., Dawson, E.M.: Earthquake deconvolution for FLAC. In: Hart & Verona (eds.) 4th International FLAC Symposium on Numerical Modeling in Geomechanics – 2006, pp. 04–10. Itasca Consulting Group Inc., Minneapolis (2006). ISBN 0-9767577-0-2

Ordaz, M., Aguilar, A., Arboleda, J.: CRISIS2007: program for computing seismic hazard. Instituto de Ingeniería, UNAM (2007)

Ordóñez, A.G.: SHAKE2000 User's manual, p. 252 (2011)

Seismomatch: User's manual. Seismosoft Ltd. (2012). http://www.seismosoft.com/

Shepherd, J.B., Tanner, J.G., McQueen, C.M., Lynch, L.L.: Final report – Seismic hazard in Latin America and the Caribbean Seismic Hazard maps for the Caribbean (1997)

USAID – OAS: Kingston metropolitan area seismic hazard assessment. Caribbean disaster mitigation project, Final Report (2001). http://www.oas.org/cdmp/document/kma/seismic/kma1.htm

Vucetic, M., Dobry, R.: Effects of the soil plasticity on cyclic response. ASCE. J. Geotech. Eng. Div. **117**(1), 89–107 (1991)

Analysis and Recovery Proposal for Erosion Process Located in the City of Planaltina-GO

Rideci Farias[1,2,3,4(✉)], Rhael Maycon Noronha Ribeiro[1], Haroldo Paranhos[1,2,3,4], Itamar de Souza Bezerra[5], and Roberto Pimentel[6]

[1] Universidade Católica de Brasília (UCB), Brasília, Brazil
rideci.reforsolo@gmail.com, rhaelmaycon@gmail.com, reforsolo@gmail.com
[2] Reforsolo Engenharia, Brasília, Brazil
[3] UniCEUB, Brasília, Brazil
[4] IesPlan, Brasília, Brazil
[5] Maccaferri, Goiânia, Brazil
itamar@maccaferri.com.br
[6] UnB, Brasília, Brazil
eng.robertopimentel@gmail.com

Abstract. The rapid process of growth of urban areas in the country, with disorderly occupations with little care to the physical environment, has caused serious erosion and numerous problems. In the central region of Brazil, in particular, in the Planaltina city in the state of Goiás, the situation is not very different, requiring urgent interventions aimed at the recovery of degraded areas. As striking consequence of the disorderly occupation is the erosion, which occurs without distinction on the various geomorphological domains present in the Midwest. Thus, erosions generate various economic social consequences such as loss of inhabitable or cultivable areas, pathways interruption, silting of the riverbed of watercourses, exposure of the public and private equity and risks to nearby communities, common facts also to other cities and constantly aired in the media. used urban occupations practices have been one of the main causes for the intensification of erosion processes around these areas. The removal of native vegetation, soil sealing and destination rainwater without necessary care can modify the flow regime active with the resulting upwelling of erosive processes. In this context, an erosive process large located in the city of Planaltina / GO evolves over two decades mainly due to the advancement of human occupation combined with little drainage systems consistent with the local situation, but also the lack of appropriate interventions that have contributed for converting slopes with high risk of instability and to surrounding areas. Thus, it is of paramount importance to know, among other factors, the origin and evolution of the erosive process in question, the type of local soil, environmental changes caused over time in the study area with a view to proposing interventions can be efficient recovery of the area degraded by erosion. Therefore, this article seeks to present an analysis and proposed recovery for the erosive process in question in order to use the area to be recovered for leisure, sport, education, among others.

H. Shehata and Y. Rashed (eds.), *Numerical Analysis of Nonlinear Coupled Problems*,
Sustainable Civil Infrastructures, DOI 10.1007/978-3-319-61905-7_7

1 Introduction

The constant problems related to erosion processes cause various concerns so that the authors of this article have been following an erosion in the municipality of Planaltina, Goiás State, since 2007. In this context, this paper aims to present an analysis of the region, such as vegetation, climate, river system and the model evolutionary erosion, togheter with the summary there is a proposal recovery for erosive process explaining the engineering works and others actions to solve the problem, taking advantage of the area to be recovered with destination to leisure, sports, education, among others.

2 Characteristics of Municipality of Planaltina / GO

The city of Planaltina / GO (15 Coordinates ° 27′10″S and 47 ° 36′ 50″) located in Mesoregion East Goiano in the micro-region surrounding the Federal District, with limits to the municipalities of Formosa, Goias Agua Fria, Mimoso de Goiás and Father Bernardo. The distance to the Federal Capital is close to 63 km. The city has, according to IBGE, the area of 2,543 square kilometers, estimated population in 2015 of 87,474 inhabitants and density of 32,10 hab./km^2. The approximate average altitude is 1035 m.

The municipality is in the Integrated Development Region of the Federal District and Surrounding Areas (RIDE), integrated area of economic development, created by Complementary Law No. 94 of 19/02/1998. This region is formed by the Federal District and the municipalities of Abadiania, Goiás Fria, Aguas Lindas de Goias, Alexânia, headboards, West Town, Cocalzinho of Goias Corumba de Goias, Cristalina, Taiwan, Luziânia, Mimoso de Goiás, Novo Gama, Father Bernardo, Pirenópolis, Planaltina, St. Anthony's discovered, Valparaiso and Vila Boa, State of Goiás, and Unai and Buritis, in the state of Minas Gerais. It occupies an area of 55,000 square kilometers and a population approaching 3.7 million inhabitants.

3 General

3.1 Regional Climate Summary

The climate, according to Köppen climate classification comprises: tropical savannas, tropical altitudes between 1,000 and 1,200 m with the coldest month temperature below 18 °C and months warmer with average above 22 °C, summer rain and dry in winter, and the tropical climate. The average annual rainfall is around 1,600 mm. In October begins the rainy season, in which the mechanical action of raindrops on the soil surface dried out by prolonged dry season can cause erosion with greater intensity on the steepest areas. In the months that follow, the storm water runoff starts to act more intensively until the month of May, when the rains are already scarce. predominant information on the site of Wikipedia, in March 2016.

3.2 Region Vegetation Summary

The cerrado biome covers the entire area of the municipality, the Federal District, as well as the states of Goiás, Tocantins, Mato Grosso do Sul, part of Mato Grosso, West of Bahia and West of Minas Gerais. In this biome the occurrence of various types of vegetation are recorded as: cerradão; typical cerrado; cerrado savanna or drain field; dirty field - have similar floristic composition of typical cerrado and cerrado drain; clean camp; riparian vegetation; paths; and rock fields.

3.3 Summary of River System

The water system of the city is characterized by waterways that have characteristics typical plateau area of drainage which are frequent gaps and enclosed valleys.

3.4 Model of Evolutionary Erosions to the Region

There are several models that attempt to translate the evolutionary processes of erosion for many different locations. In the Midwest, several studies have been developed for the Region of the Federal District. Thus the proximity of the region, the following are some models that attempt to translate the type of occurrence.

Costa (1981) studied about erosions in the city of Gama (DF) and classified two main types of occurrence, laminar erosion and gullies. The analysis of evolution of erosions, classified the development of erosion in four phases:

I - in the first stage is the formation of surface erosion and grooves;
II - in the second phase there is deepening in "V" section until the decomposed rock;
III - the third stage is the development in decomposed rock with excavations in the horizontal direction providing the formation of a section in the form of "U"; and
IV - the fourth stage to the base level of bedrock, based on enlargement and the emergence of new erosions on the flanks.

Mortari (1994) proposed a "Model Embedded" for development of erosions in the Federal District as a result of geological and geotechnical and structural conditions of the region, especially the orientation of the layers of saprolite diving and metasediments of the local geological domain. By "Embedded Model" at the beginning of the erosion gullies usually have the form "V" and evolve in depth, width and length depending on the water conditions and geotechnical characteristics of the soil.

The process evolves to reach the bedrock, which in the region of the Federal District consists mostly of slate and *metarritimitos* that due to active tectonics, present their strata rather inclined, with the layers dip in the order of 40 to 60.

The water flow to achieve this contact, tends to "fit" and following flow about its orientation and tending to deepen, following the dive itself the less resistant layers.

As the groove is deepened, it becomes more evident and the material becomes more resilient, tending to stabilize the bottom of the erosion wear with base then be considered a normal geological erosion process. This type of behavior makes it difficult to meander background channel, preventing erosion with the side extension of the base (trapezoidal shape) to an equilibrium profile with the subsequent development of vegetation. In the Federal District and Surrounding Region, as the city of Planaltina, typically erosions occur in the form of "V" and the depth is limited to the existence of saprolite.

Another erosion process is the preferential flow of the rocky and especially clayey massifs, but this process is not applied to this case, because of the kind of start of erosion, caused by superficial rainfall flow.

4 Characteristics of Erosion Studied

Erosive process consisting at first of two branches with an approximate length of 180 m each. Then the two branches, erosion consists of a main body with the next extension of 2,160 m and slopes ranging up to over 30 m. It is inserted in the region Mansions Sector West Sector, Courts 01, 02, 05, 6:11.

Erosion is characterized by being a large gully, caused mainly by the outflow of two existing drainages in its headwaters, and this forms a water erosion. This drainpipe is constituted of a 1,200 mm pipe, a branch, and two 1500 mm pipes, in another branch. The pellet then generated erosion has caused the constant silting of a stream called Lambari Stream, existing at the end of the main erosion process. Approximately at point 1080 m there is a lateral erosion on the left, with an approximate length of 54 m and an average width of 14 m. Close to 1680 m there is a lateral erosion on the right side with an approximate length of 320 m and an average width of 5 m. Figure 1 shows the overview of the erosion process.

Fig. 1. Location of the erosion process in the municipality.

For better understanding, divide into four (4) main parts, as shown in Fig. 2.

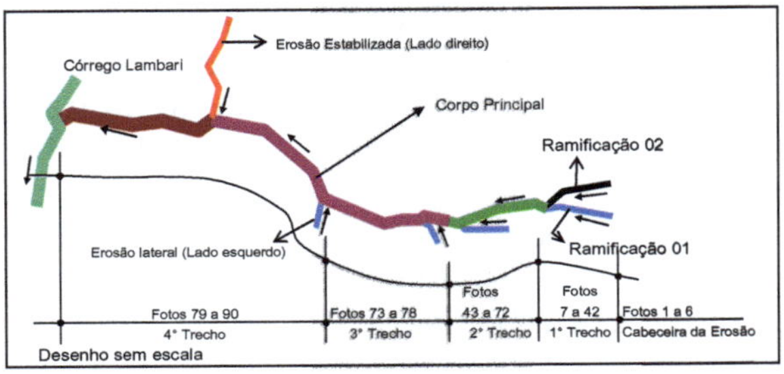

Fig. 2. Sketching erosion.

4.1 Description Each Excerpts Erosion

4.1.1 First Excerpt

It extends from the start of erosion by the approximate point 200 m and consists of two branches with approximately 180 m each. Presence of vegetation formed by species of medium to small (mamonas, hoses, etc.) and creeping species such as Brachiaria. Additionally, we have:

(a) branch 01 - Presence drainage tube with 1200 mm; soil transported with small stone blocks. The bases of slopes consist of hard rocks and medium-sized rocks fractured.
(b) 02 branch - Presence drain with two tubes each 1500 mm; stratified rock hard and small stone blocks.

Figures 3, 4, 5, 6 and 7 show the area of the beginning of the erosion process, drains and views of the branches.

Fig. 3. Area of the beginning of the erosion process.

4.1.2 Second Excerpt

It extends 20 m after the junction of the two branches, around 200 m from the start of erosion, and goes up to about 680 m. soft friable rock presence and predominance of

Fig. 4. Desagüe the branch 1.

Fig. 5. View of the branch 1.

Fig. 6. Desagüe the branch 2.

Fig. 7. View of the branch 2.

oxisol, yellow soil and in cambissolo, blocks of rock - based hard rock at the base of the slopes. Such rock layer to a height of 2 m. Presence of embankments in fractured rocks. Figures 8 and 9 show the erosion process in the second section views.

Fig. 8. Erosion in the second sentence.

4.1.3 Third Excerpt

This section extends approximately from the distance of 680 m from the beginning erosion and continues until the approximate distance of 980 m.

There is a predominance of embankments in oxisol, yellow soil and in cambisols, and water upwelling in the valley of erosion with the consequent saturation of the bases of slopes. Figure 10 shows a view of the erosion process in the third section.

4.1.4 Fourth Excerpt

This section extends approximately 980 m away from the beginning of the erosion, and goes to the distance of 2.339,11 m. There is a predominance change soil with latosols switching, and water upwelling in the valley of erosion with the consequent saturation

Fig. 9. Erosion in the second sentence.

Fig. 10. Erosion in the third sentence.

Fig. 11. Erosion process in the fourth section (Photo 2007. The residence view has been consumed by erosion).

Fig. 12. Stream silted at the end of the erosive process.

of the bases of slopes. Figures 11 and 12 show views of the erosive process in the fourth section.

5 Risks in the Surrounding of Erosion

In addition to erosion as a whole constitute as a risk area, there are specific points that deserve special attention, such as shown in Figs. 13 and 14.

Fig. 13. Residences near to erosion.

6 Design / Project Description

The project design basically part with the recovery of degraded the correct discipline area of rainwater and the preparation of infrastructure for future installation of a linear park consisting of several elements structured as reforestation, lighting, amphitheater, sports courts, trails for skate among others, to benefit the greatest possible number of

Fig. 14. Living close to erosion.

people. In general, linear parks are urban interventions to recover for citizens awareness of the natural site in which they live, gradually expanding the green areas, leisure, recreation, etc. In order to establish a set of actions, under the coordination of the Executive, and the participation of owners, residents, users and investors in general, to promote structural urban transformations and progressive enhancement and improvement of environmental quality in the municipality incorporated into Areas System Green city. The proposed infrastructure for the area to be recovered is described below.

7 Population to Be Granted

The population to be benefited comprises the whole of the city of Planaltina, specifically the Mansions of the West Sector Sector district and its adjoining areas estimated at more than 10,000 people, according to estimates of the City of Planaltina.

8 Proposed Project

The following is the summary of the intended project.

8.1 Stormwater

In general the project consists of the capture of surface and driving the same waters, in a disciplined manner, to the stream Lambari, located at the end of erosion. Along the stretch of erosion, other releases can be observed: one BSTC network 1200-570 m; one BSTC network 800-740 m and other network BSTC 800-1090 m from the source.

A contribution in the form of erosion, stabilized, comprising 15% of the total flow is observed at 1680 m from the source.

In order to regulate the flow, it was proposed the extension of releases BDTC and BSTC until convergence thereof. Starting this convergence, (200 m from the beginning), in order to make aproveitavel area for collective purposes is suggested driving water through gallery.

In order to reduce the speed of the system and optimize the topography, they are suggested some energy dissipation.

From 680 m, the drainage structure passes gallery for gabion, this going until the desague in stream Lambari.

The project foresees the receipt of future releases of Building neighboring areas to erosion.

8.2 Landfills and Contentions

After careful review of the types present along the eroded soils were set appropriate geotechnical solutions each profile.

containment solutions take into account the reduced use of imported soil. For this practice should make a tradeoff between the volumes of cut and fill.

The unstable embankments must be retaludados with gradient compatible with the type of soil and revegetated, thus preventing the initiation of new surface erosion.

As the erosion reached the horizon "C" soil, silted surface vegetation layer, there is the need to import most fertile soil ("topsoil") for the surface treatment of slopes.

For the embankment without the possibility of retaludamento to a secure tilt suggest the inclusion of geosynthetic reinforcement, so that the reinforced structures are incorporated into the area, without visual intrusion.

In areas for collective use and sports practices teraplenagem should be in platores. Draining of the crests and toes of embankments should be forwarded through channels to sluice gates.

8.3 Sanitation

All waste from internal intalações park will eventually receive proper treatment, before launching the network.

8.4 Floors and Flooring

The areas of parking lots should be interlocked block. The paths and walks are coated in concrete. The water from unpaved areas should be targeted for retention basins to be postriormente loops in the network.

8.5 Reforestation and Treatment Landscape

The species to be planted are native primarily medium-sized trees of the savannah and municipality characteristics, as well as palm trees and shrubs that may be alive or type climbing fences.

It is recommended, in environmental education activities to be undertaken by the Executive, according to cincunvizinhos residents so that they plant trees on their properties, allowing the common use. In this case, it is recommended to give preference to fruit species.

As for landscaping, this is the structuring element of designed spaces, defining their limits, their routes and improving environmental quality.

8.6 Linear Park Proposed

The future installation of the Proposed Park, in addition to being mounted infrastructure, also aims to motivate educational programs targeting the good care of household waste, cleaning of public spaces, the permanent reorganization of watercourses and the supervision of these spaces.

9 Implementation of Works

The implementation of the proposed works should be given as the project and following the technical specifications for infrastructure works (collection of surface and driving the same waters, in a disciplined manner, to the stream Lambari, located at the end of erosion, is including also platores and retaining embankments) to be described in the Executive Project, with a planned period of eight (8) months for its completion.

One should take advantage of the dry season to better progress of ground handling services and drainage. As for the works of Revegetation, surface treatment of embankments and landscaping should take the start of the rainy season.

The regularization of the main channel will be performed using conventional method. The existence of side roads facilitate the logistics of the work, since they may be used as a service road.

All excavated material will be used in landfills (volumes compensation). It should be preferred initially for the use of stone materials as foundation of landfill structures. For revegetation of the area should be appropriate imported topsoil.

The section to be protected with gabion blanket should be performed as designed, respecting the technical specifications.

The passages in low bearing capacity soil should be removed to areas where these soils do not engage the bearing function.

10 Conclusions

Summarizing these erosion recovery works, they will transform a devalued region into a beautiful and convivial area for the region's residents, showing a possibility of allied engineering work with improved people's lives.

Acknowledgments. The Reforsolo Engenharia Ltda., Planaltina City of Goiás, Catholic University of Brasilia (UCB), IesPlan and UniCEUB, with important contributions that made possible the realization of this work, but also to the Graduate Program in Geotechnical Engineering from the University of Brasilia with the availability of PRONEX Project.

References

Costa, W.D.: Taludes Naturais: "Caso Histórico de Erosão na Cidade do Gama, DF". Curso de Extensão Universitária - Obras de Terra e Fundações Especiais. ABMS e UnB, Brasília, pp. CI/01-CI/46 (1981)

https://pt.wikipedia.org/wiki/Planaltina_(Goi%C3%A1s)

http://www.cidades.ibge.gov.br/xtras/perfil.php?lang=&codmun=521760&search=||infogr%E1ficos:-informa%E7%F5es-completas

Mortari, D.: Caracterização Geotécnica e Análise do Processo Evolutivo das Erosões no Distrito Federal. Dissertação de Mestrado, Publicação G.DM-010A/94, Departamento de Engenharia Civil, Universidade de Brasília, Brasília, DF, p. 200 (1994)

Soil Structure Interaction Studies with Use of Geosynthetics in Soils Beneath Footings

R. Shivashankar[1(✉)], Nalini E. Rebello[2], V.R. Sastry[3], and B.R. Jayalekshmi[1]

[1] Department of Civil Engineering, National Institute of Technology Karnataka, Mangalore 575025, India
shivashankar.surathkal@gmail.com, br.jaya@gmail.com
[2] St. Joseph Engineering College, Mangalore 575028, India
nalinirebello@rediffmail.com
[3] Department of Mining Engineering, National Institute of Technology Karnataka, Mangalore 575025, India
vedala_sastry@yahoo.co.in

Abstract. The present study consists of two parts. In the first part, the effect of using a geomembrane (slip) layer in soil beneath footings under seismic excitations from shake table tests is being investigated. A triaxial shaker system is used to carry out the tests on a one-third scaled model of single storey, single bay RC space frame. The structure with different base conditions are subjected to sine sweep tests and simulated seismic excitation corresponding to the design spectrum for Zone III as per the Indian standard code (IS 1893 (Part 1): 2002). It is observed that the natural frequency of the structure decreases with increase in the flexibility of supporting soil In the second part, effects of using geosynthetic reinforced soil beneath footings of multistoried structures with tunneling operations beneath are being looked into. Numerical investigations on shallow depth tunnels like metro tunnels in granular soils, response due to tunneling itself (single and twin tunnels) and also their impact on the buildings above are carried out, using 3DEC software. Tunneling for metro causes innumerable changes in the form of distortion taking place in strata surrounding the tunnel, and also affect the member forces of framed structures on the surface. Height of superstructure and building eccentricities from the tunnel centre line are also varied. Results reveal that the presence of geosynthetic reinforcement in soil considerably reduces the displacements under footings.

1 Introduction

1.1 Dynamic Soil-Structure Interaction and Geomembrane Slip Layer

The interaction among structures, their foundations and the soil medium below the foundations alter the actual behavior of the structure considerably than what is obtained from the consideration of the structure alone. Thus the flexibility of the support reduces the stiffness of the structure and increases the period of the system. In this study, which is in two parts, in the first part, the effect of a geomembrane slip layer in soil beneath footing subject to seismic excitations, from shake table tests, is being looked into, i.e. the

H. Shehata and Y. Rashed (eds.), *Numerical Analysis of Nonlinear Coupled Problems*, Sustainable Civil Infrastructures, DOI 10.1007/978-3-319-61905-7_8

effect of geomembrane under a building which is known to act as a base isolator under dynamic loads is being investigated in this part of the study through triaxial shake table tests. The use of geomembranes for base isolation has been advocated by many researchers (Yegian and Lahlaf 1992a, b; Thurston 2007; Jayalekshmi et al. 2011 etc.). The liner (slip layer) will allow dissipation of earthquake energy in sliding friction.

1.2 Geosynthetics in Foundations and Soil-Structure-Tunnel Interaction

Geosynthetics are very widely used nowadays in ground modification and in several other civil engineering applications. Geosynthetics include geotextiles, geogrids, geomembranes and others. Geomembranes are impermeable and smooth polymeric sheets. In the second part of this study, the effect of tunneling beneath a structure (with footings resting on geosynthetic reinforced soil) is being studied. Underground structure like tunnels for roads, metros and sewage transportation are constructed in most parts of the world. When tunneling operations are carried out at shallow depths, overburden displacements generated at the crown are reflected as displacements at the surface and the surface displacement profile generated is in the form of a Gaussian curve. The nature of displacements predominantly depends on the type of strata, method of excavation. However, one of the most important factors governing the nature of displacements is the existence of buildings on surface at top. Previous researches, with the presence of buildings at top, have been conducted using numerical modeling investigations by several researchers such as Potts and Addenbrooke (1997) and Franzius (2003). Their analyses indicate that the existence of building at top might alter the displacement profile and hence reduce displacements. Mroueh and Shahrour (2003) conducted studies on a symmetrically placed, two-storied framed structure, to predict changes due to excavation-induced ground movements. Overall reduction in displacements, across the transverse settlement profile and longitudinal profile were observed. Furthermore, studies on the effect of multi faced tunneling in water bearing soft ground by Yoo and Kim (2008), revealed that the presence of a structure at top tends to reduce surface settlement by 15–25% compared to green field conditions. This is because a stiff structure makes surrounding soil stiff and thus reduces settlements.

Previous research on tunnelling has been limited to modelling of a tunnel in the presence or absence of the building, under mixed soil conditions or under clayey soils. The effect of geosynthetic in soil beneath footing acting as stiffener under static loads is being investigated in this part of the study to predict the displacements on tunnelling. Studies on the effect of varying building loads, under single and twin tunnelling and in the presence of a geosynthetic layer are being investigated and analysed in order to make predictions of the transverse displacement profile in general and displacements under footings in particular.

2 Effect of Geomembrane Slip Layer in Soil Beneath Footing Subject to Seismic Excitations

2.1 Shake Table Tests

In the first part of this study, the performance of a geomembrane slip layer in soil beneath an isolated footing as a base isolator is being studied. The geomembrane used in the slip layer is an ultra-high molecular weight polyethylene, with a static friction coefficient of about 0.1 and a dynamic friction coefficient of about 0.07 (Patil and Reddy). Base isolation shifts the fundamental frequency of a structure away from the damaging frequency range of the most probable earthquake. The reduction in the natural frequency of structures will cause the structural period to be shifted to the right of the falling curve of response spectra as given in IS 1893: (2002), resulting in a reduced seismic structural response compared to a fixed base structure.

Shake table studies of single bay, single storey, three dimensional model with isolated footing, with (a) fixed base, and resting on various soil conditions such as (b) dry sand, (c) dry sand reinforced with geomembrane, (d) saturated sand and (e) saturated sand reinforced with geomembrane are conducted to determine the variation in the natural frequencies and also to study the variation of maximum acceleration at the roof level for base conditions such as (a), (b) and (c) above. Shake table tests are carried out in a triaxial shaker system on a one-third scaled model of a single storey, single bay RC space frame resting on soil or soil with geomembrane (Fig. 1). These structures with different base conditions are subjected to sine sweep tests to obtain the natural frequencies and are also subjected to the motion corresponding to the response spectrum of Zone III as per IS 1893 (Part I): (2002). Sufficient mass was added to the structure to maintain the fundamental frequency in the range of 3 Hz to 10 Hz, which corresponds to structures with maximum spectral acceleration according to the response spectra for rock and soil site for 5% damping as given in IS 1893 (Part I): (2002).

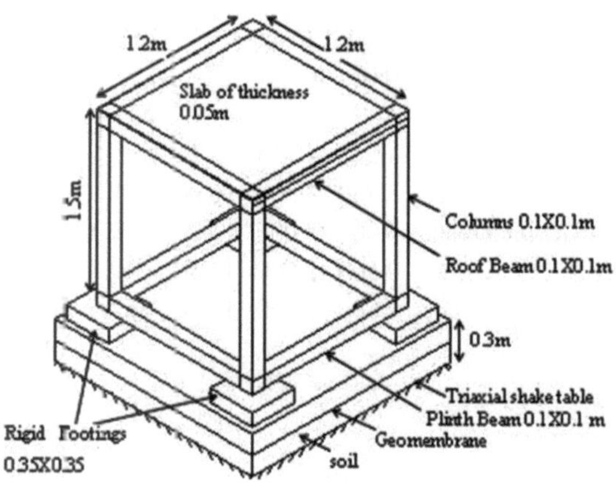

Fig. 1. RC model space frame resting on soil with geomembrane

A steel tank of size 2 m X 2 m fixed to the shake table (3 m X 3 m) was used as a container for the supporting soil and soil with geomembrane (Fig. 2). The soil in both cases was filled to a depth of 300 mm in the steel tank. Polyethylene geomembrane with very low friction coefficient was embedded horizontally in the soil at one third height from the top of soil layer. The shake table had six degrees of freedom, three translational and three rotational, and a maximum payload of 100 kN. Experimental setup on the triaxial shaker is shown in Fig. 2. The variation in the natural frequency of the structure, from experimental results, is shown in Table 1. It is observed that the natural frequency decreases with increase in the flexibility of supporting soil. Comparison of time history responses, of base conditions (a), (b) and (c) above, showed a reduction in the maximum acceleration value from 7.41 m/sec^2 for a fixed base to 5.45 m/sec^2 for base with dry sand, and to 4.66 m/sec^2 for base with dry sand and geomembrane. More details of the shake table testing and results are reported in Jayalekshmi et al. (2011).

Fig. 2. Experimental set-up on the triaxial shaker

Table 1. Variation in natural frequency

Base conditions	Fundamental natural frequency (Hz)			Percentage variation of natural frequency		
	X	Y	Z	X	Y	Z
Fixed base	4.218	4.125	23.25	–	–	–
Dry sand	3.750	3.500	16.00	−11.1	−15.2	−31.2
Dry sand + Geomembrane	3.000	3.000	13.00	−28.9	−27.3	−44.1
Wet sand	2.500	2.500	12.50	−40.7	−39.4	−46.2
Wet sand + geomembrane	2.000	2.000	12.00	−52.6	−51.5	−48.4

3 Effect of Tunneling Beneath a Structure (with Footings Resting on Geosynthetic Reinforced Soil)

3.1 Details of Simulation/ Parametric Study

A parametric study, using 3DEC software, was carried out involving structural parameters of building, various geometrical variables in single layer of coarse grained soil of moderate density to study the following:

a. Effect of varying building storey and varying eccentricities i.e. distances from the tunnel centre line on displacements, in the case of a single tunnel.
b. Effect of varying building storey and varying eccentricities in the case of twin tunnelling.

3.2 Material Properties of Strata

A single layer of strata is being considered. The material was granular soil of uniform density 20 kN/m^3 and zero cohesion. *Ko* value of 0.5 is taken throughout the analysis. The changes in displacements due to excavation, followed by installation of tunnel lining, are noted. The state of stress/ displacement prior to excavation is known and comparison is made with the changes on creation of opening with subsequent installation of lining. The properties assigned to the strata are given in Table 2.

Table 2. Properties assigned to strata

Density (kN/m^3)	Bulk modulus (MPa)	Rigidity modulus (MPa)	Angle of internal friction
20.0	330.0	110.0	32°

3.3 Details About Single and Twin Tunneling in Strata

In the first stage a single tunnel of Diameter 6.1 m and lining thickness of 0.25 m was considered. A constant depth of overburden of 8 m from the surface was taken into consideration. The centre line of the building was varied with respect to the centre line of the tunnel in the transverse direction. The centre line distance of the building were varied as e = 0 m, 5 m, 10 m, 15 m and 20 m. Number of storeys of the building was varied from 2 storey to 4 storey and 8 storey (Fig. 3)

In the second stage, twin tunnels located at a depth 8 m from the surface were considered for analysis. The centre to centre distance between the tunnels was 12 m. External diameter of each tunnel was 6.1 m, with a lining thickness of 0.25 m. Buildings with varying storey of 2, 4 and 8 storey were introduced on the centre line of the two tunnels. Changes in the surface settlement were noted at different building eccentricities of 0 m, 5 m and 10 m from the centre line of the two tunnels. Typical models generated with 2 storey building placed at 0 m, 5 m, and 10 m eccentricities are illustrated in Fig. 4.

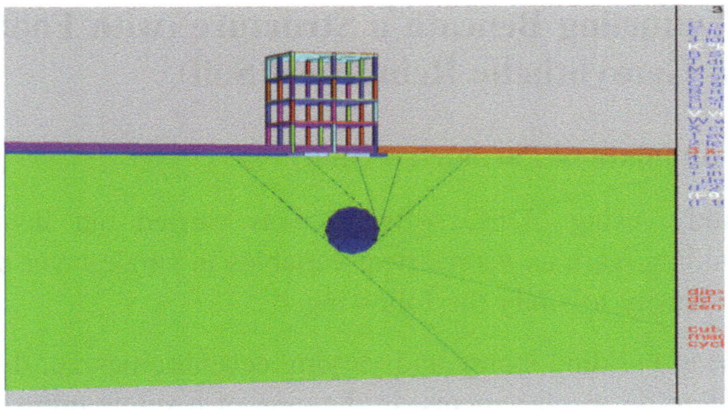

Fig. 3. Elevation of building with single tunnel

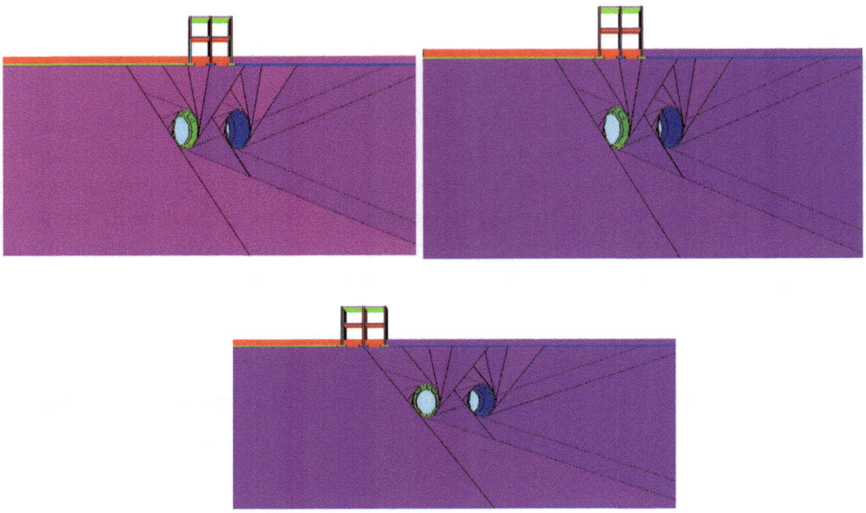

Fig. 4. Models with twin tunnels with 2 storey building placed with (a) eccentricity of 0 m (b) eccentricity of 5 m (c) eccentricity of 10 m from centre line

3.4 Details of the Building

A framed building without brick-infill walls was considered for the analysis (Fig. 3). Columns are of size 0.35 m X 0.45 m with an axial stiffness of 128 MN. Slab is assigned a thickness of 0.15 m. Beams have cross-sectional dimension of 0.3 m X 0.35 m. The footings are of dimensions 2 m X 2 m with a thickness of 0.5 m. A centre to centre distance of 4 m is being considered between the footings, both in the transverse as well as in the longitudinal directions. Bearing pressures exerted by each footing are about 24.6 kN/m^2, 49.0 kN/m^2 and 98.5 kN/m^2 for 2, 4 and 8 storey building respectively. To further stiffen the strata under footings, a sand layer of 10 cm thickness sandwiched between two geosynthetic layers is provided under each footing and an angle of friction of 20° was assigned between the soil and geosynthetic (Nalini 2015; Nalini et al. 2016).

3.5 Analysis of Displacements on Excavation of Single Tunnel

Effect of varying building storey on vertical displacements: The buildings considered were of 2, 4 and 8 storeys. The change in displacements upon excavation, without the presence of building (greenfield condition) was −10.37 mm at the surface above the tunnel (Fig. 5). Inclusion of a building reduced the displacements at the surface as compared to the case without building loads i.e. greenfield condition. The displacements reduced by 6.26%, 10.6% and 16.97% respectively in 2 storey, 4 storey and 8 storeyed buildings, respectively. The main reason for reduction in displacements is that the soil surrounding the footing stiffens upon inclusion of building weight and as a result, overall displacements reduce. Presence of the geosynthetic layer further compounded the effect. With an increase in building load and with a *Ko* value of 0.5, greater magnitude of displacements is transferred to the sides of the tunnel and this therefore, reduces displacements at the crown. The greater the number of storeys the greater will be the transfer of stress to either sides of the tunnel. Thus, lesser displacements will be noticed at the crown and surface in dense granular soils. The reduction in vertical movement is in agreement with the results obtained by Potts and Addenbrooke (1997), Mroueh and Shahrour (2003), Franzius (2003) and Yoo and Kim (2008).

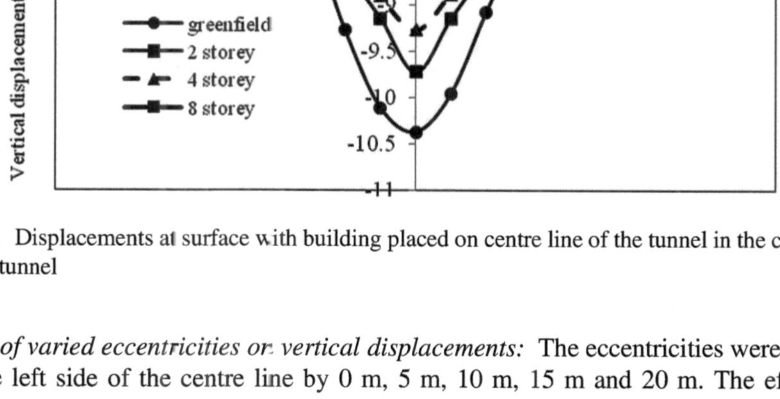

Fig. 5. Displacements at surface with building placed on centre line of the tunnel in the case of a single tunnel

Effect of varied eccentricities on vertical displacements: The eccentricities were varied on the left side of the centre line by 0 m, 5 m, 10 m, 15 m and 20 m. The effect of varied eccentricities on displacements is studied for all the three building storeys. Asymmetric application of load led to asymmetric vertical strata movements at surface. From Figs. 6, 7 and 8, it can be observed that the magnitude of displacements was lesser at the location of the footing. The displacements at 3 m (0.49D) on the left side of the centerline, when building was placed at 5 m (0.819D) from centre line due to 2 storey, 4 storey and 8 storey was −9.02 mm, −8.82 mm and −8.2 mm respectively. Similarly at −15 m (2.45D) from the tunnel centerline, displacements were −7.64 mm, −7.57 mm and −7.31 mm for 2, 4 and 8 storeys respectively. As the building eccentricity increased, the displacements at the centre line, on the surface, increased and thus displacements matched the transverse displacement profile created in the case without a building. Even though there is an overall reduction in displacements due to inclusion of

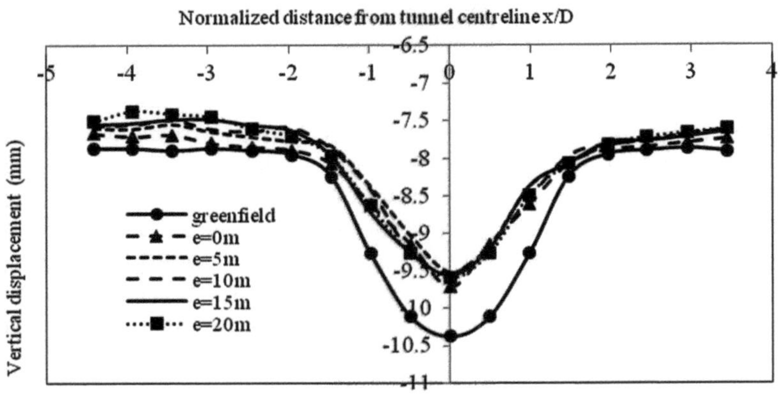

Fig. 6. Displacements at surface with 2-storey building placed at varying eccentricities from centre line in the case of a single tunnel

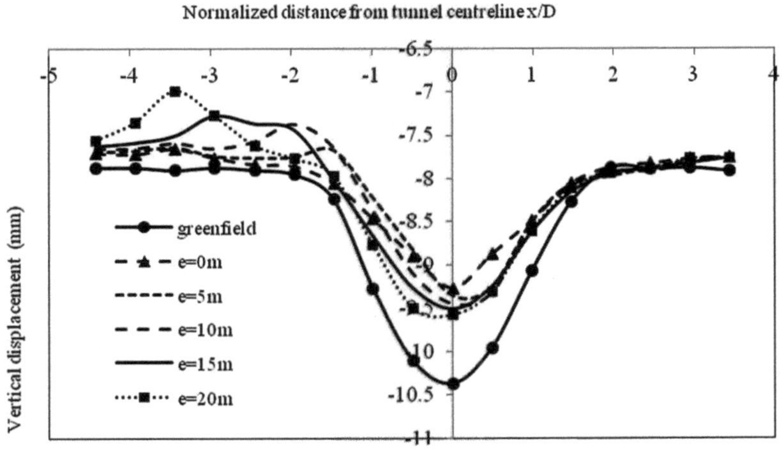

Fig. 7. Displacements at surface with 4-storey building placed at varying eccentricities from centre line in the case of a single tunnel

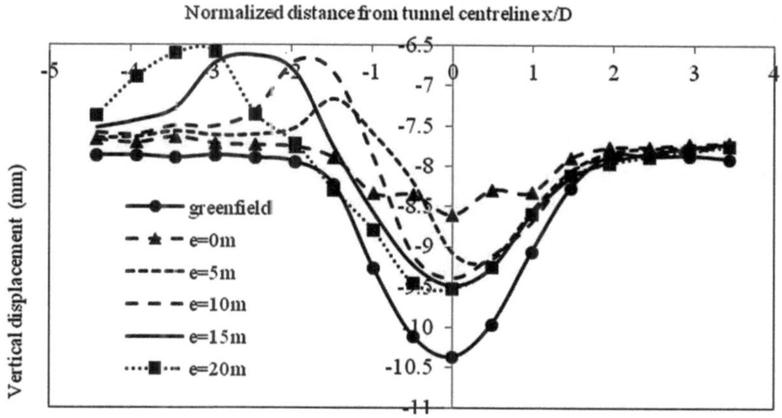

Fig. 8. Displacements at surface with 8-storey building placed at varying eccentricities from centre line in the case of a single tunnel

Table 3. Vertical displacements (mm) on the centre line of single tunnel due to varying building eccentricities in single layered granular soil

	gf *	e = 0 m	e = 5 m	e = 10 m	e = 15 m	e = 20 m
2 storey	−10.37	−9.72	−9.52	−9.53	−9.57	−9.6
4 storey	−10.37	−9.27	−9.35	−9.45	−9.50	−9.58
8 storey	−10.37	−8.61	−9.07	−9.39	−9.48	−9.52

* gf: under Greenfield condition; e: eccentricity of building

building weight, displacements at the footing were of lesser magnitude compared to other points in the transverse direction (Figs. 6, 7, 8, and Table 3).

Displacement at footing, reduced by 1.14% and 1.65% when the 4 storey and 8 storey building was placed at 5 m from centre line as compared to displacements generated by 2 storey structure. Similarly, displacement reduced by 0.39%, 0.91% when the 4 and 8 storey structure was placed at 15 m from centre line. Thus it is evident that the presence of the building led to reduction in displacements near the tunnel. Greater reduction was observed in taller structures than smaller storeys. Further, the geo-synthetic layer under the footings led to significant reduction in displacements at locations of the building.

3.6 Analysis of Displacements on Twin Tunneling

Analysis of displacements of twin tunnels in single layered granular soil indicated that excavation of tunnels led to a displacement of −14.21 mm, at the surface on the centre line of both the tunnels. Maximum displacement above the tunnels was −14.52 mm. Vertical displacement generated on single tunnelling, in single layer of granular strata, was of magnitude −10.37 mm. This indicates that simultaneous excavation of tunnels led to greater magnitude of displacements compared excavation of a single tunnel.

Change in vertical displacement at −6 m (−0.98D) and 6 m (0.98D) from the centre line of both tunnels was −14.29 mm, −14.45 mm and −14.37 mm for 2, 4 and 8 storied buildings. It can be concluded that displacement decreased upon including building storey/load which is similar to the observations of displacements generated due to single tunnelling. With a 12 m separation distance between the tunnels, the settlements were also not independent of each other. Presence of the building led to overall reduction in displacements along the transverse direction. Further increase in building storey would have increased the displacements. Displacements of −14.2 mm, −14.14 mm and −14.12 mm were noticed at −0.983D, when the 2, 4 and 8 storey

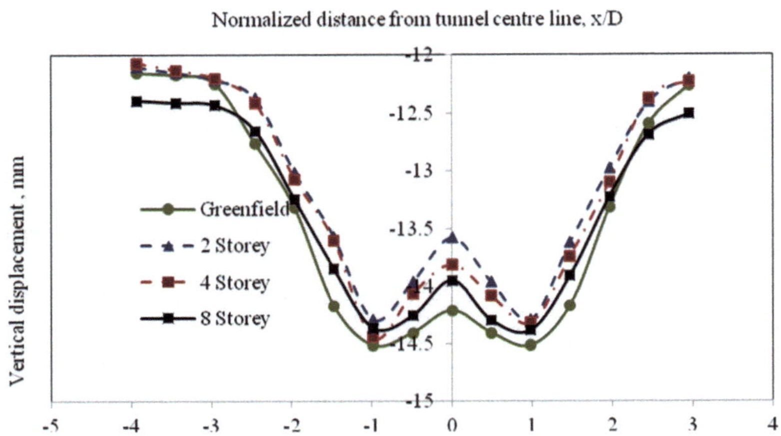

Fig. 9. Displacement on surface with 2 storey, 4 storey and 8 storey building placed at 0 m eccentricity from centre line of both tunnels in single layered strata

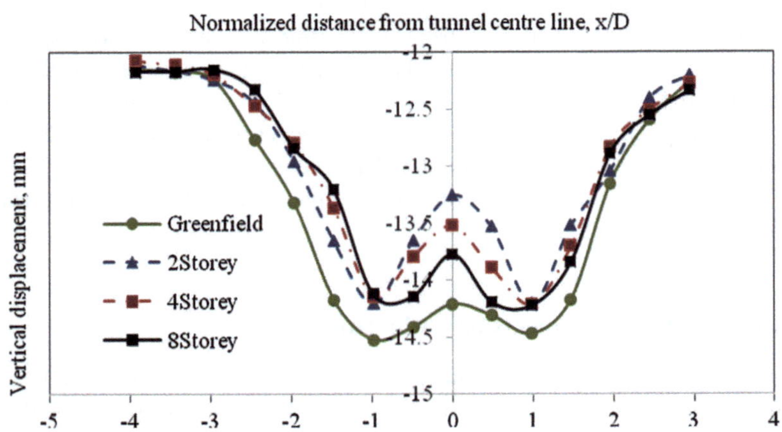

Fig. 10. Displacement on surface with 2 storey, 4 storey and 8 storey building placed at 5 m eccentricity from centre line of both tunnels in single layered strata

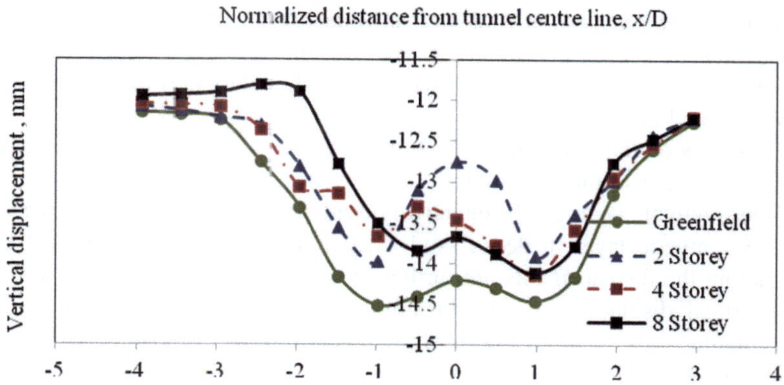

Fig. 11. Displacement on surface with 2 storey, 4 storey and 8 storey building placed at 10 m eccentricity from centre line of both tunnels in single layered strata

Table 4. Vertical displacements (mm) in case of twin tunneling with varying building eccentricities in single layered granular soil

	gf	e = 0 m	e = 5 m	e = 10 m
2 Storey	−14.21	−13.58	−13.25	−12.76
4 Storey	−14.21	−13.81	−13.52	−13.47
8 Storey	−14.21	−13.95	−13.77	−13.68

gf under Greenfield condition; e: Eccentricity of building

buildings were shifted to −5 m (0.819D) from the centre line and displacements of −13.99 mm, −13.68 mm and −13.51 mm were noticed when the building was shifted to −10 m (−1.64D) from the centre line of tunnel (Figs. 9, 10, 11 and Table 4).

As can be seen in Table 4, displacements of magnitude −13.58 mm, −13.81 mm and −13.95 mm were noticed at the centre line of both tunnels, when the 2, 4 and 8 storey building was placed at '0' m from the centre line of both tunnels. When the tunnel was shifted to 5 m from the centre line, vertical displacements at centre line got reduced to −13.25 mm, −13.52 mm and −13.77 mm respectively.

4 Conclusions

(A) Results of shake table tests indicate that geosynthetic layer can be effectively used under footings since

1. Natural frequency of the structure decreases with increase in flexibility of the supporting soil. In the case of wet sand with geomembrane the natural frequency recorded was 2 Hz as compared to 4.218 Hz in case of fixed base.

2. Maximum accelerations at roof level also decrease with increase in flexibility of the supporting soil. Maximum acceleration value reduced from 7.41 m/sec^2 for a fixed base to 5.45 m/sec^2 for base with dry sand, and to 4.66 m/sec^2 for base with dry sand and geomembrane.

(B) From the modeling of a tunnel/twin tunnels in the presence of a building at the top at the surface also indicates that

1. Geosynthetic layer can be effectively used under footings since the presence of geosynthetic layer in dense sand strata reduces the settlements and on tunneling the displacements are further reduced especially under footings.
2. Vertical displacements due to simultaneous excavation of twin tunnels in granular soils led to greater displacement at surface, compared to the condition with a single tunnel.
3. In the case of twin tunnels, inclusion of building loads, with 2, 4 and 8 stories, led to a considerable decrease of vertical displacements at surface.

Acknowledgments. The shake table tests were done under a BARC funded research project at the National Institute of Technology Karnataka, India. Shake table tests were performed at the Central Power Research Institute at Bangalore, India. The support of everyone is acknowledged.

References

Franzius, J.N.: Behaviour of Buildings due to tunnel induced subsidence. Ph.D. thesis. Imperial college of Science, Technology and Medicine, London, 358 pages (2003)

IS 1893 (Part I).:Criteria for Earthquake Resistant Design of Structures - General provisions and Buildings. Bureau of Indian Standards, New Delhi (2002)

Jayalekshmi, B. R., Shivashankar, R., Venkataramana, K., Ramesh Babu, R., Reddy, G.R., Parulekar, Y.M., Patil, S.J., Gundlapalli, P.: Shake table tests to investigate the efficacy of geomembranes for soil isolation in a space frame with isolated footing. In: 13th International Conference of International Association for Computer Methods and Advances in Geomechanics, vol. 2, pp. 834-838, 9–11 May, Melbourne, Australia (2011). ISBN 978-098082442-1

Mroueh, H., Shahrour, I.: A full 3-D finite element analysis of tunneling-adjacent structures interaction. Comput. Geotech. **30**, 245–255 (2003)

Rebello, NE.: Studies on tunnels subjected to static loads and blalst induced vibrations. PhD thesis. Department of Civil Engineering, National Institute of Technology Karnataka, India. 228 pages (2015)

Rebello, N.E., Shivashankar, R., Sastry, V.R.: Response of strata and buildings to blast induced vibrations in the presence and absence of a tunnel. Geotech. Geol. Eng. **34**, 1013–1028 (2016). doi:10.1007/s10706-016-0021-y. Springer publishers

Potts, D.M., Addenbrooke, T.I.: A structure's influence on tunnelling-induced ground movements. Proc. Instn. Civ. Engrs. Geotech. Eng. **125**, 109–125 (1997)

Thurston, S.J.: Base isolation of low-rise buildings using synthetic liner. BRANZ study report SR17, BRANZ Ltd., Judgeford, New Zealand, 18 pages (2007)

Yegian, M.K., Lahlaf, A.M.: Geomembranes as base isolation. Geotechnical Fabrics Report, September 1992, St. Paul, MN 55101–1088, USA (1992a)

Yegian, M.K., Lahlaf, A.M.: Dynamic interface shear strength properties of geomembranes and geotextiles. J. Geotech. Eng. **118**(5), 760–779 (1992b)

Yoo, C., Kim, S.B.: 3-D numerical investigations of multifaced tunneling in water bearing soft ground. Can. Geotech. J. **45**, 1467–1486 (2008)

A Posteriori Error Estimation for the Non-associated Plasticity Drucker-Prager Model with Hardening

Dao Duy Lam(✉)

University of Transport and Communications, Hanoi, Vietnam
daoduylam@utc.edu.vn

Abstract. The numerical solution of non-associated elastoplasticity is still a key aspect of research and development in computational plasticity. Approximate solution procedures are based, in the context of a displacement method, on a weak form of the equilibrium and reply upon two main ingredients: the numerical integration of the rate constitutive relations over a generic time step (local stage) and the iterative algorithm exploited to solve the nonlinear equilibrium equations (global stage). The fully discrete problem is the obtained by performing a spatial discretization of the field equations and a time-integration of the evolution rule. The interest is here given to the discretization errors, which are caused by the numerical discretization of the continuous mathematical model in order to define an adaptive strategy.

The aim of this paper is to extend the concept of error in the constitutive equations to non-associated plasticity Drucker-Prager model to handle non-associative rate-independent plasticity problems solved by employing the incremental displacement conforming finite element method.

Numerical examples by PLSAER2D (a Matlab program) for both the associated and the non-associated cases for Drucker-Prager model with hardening are also presented.

Keywords: Non-associated plasticity · Drucker-Prager model · Hardening · Error estimation

1 Introduction

The numerical solution of nonlinear boundary value problems arising in rate-independent elastoplasticity with prescribed accuracy is still a key aspect of research and development in computational plasticity. Approximate solution procedures are based, in the context of a displacement method, on a weak form of the equilibrium and rely upon two main ingredients: the numerical integration of the rate constitutive relations over a generic time step (local stage) and the iterative algorithm exploited to solve the nonlinear equilibrium equations (global stage). The fully discrete problem is the obtained by performing a spatial discretization of the field equations and a time-integration of the evolution rules. The interest is here given to the discretization errors, which are caused by the numerical discretization of the continuous mathematical model. Both time and space discretization produce errors which need to be estimated in

H. Shehata and Y. Rashed (eds.), *Numerical Analysis of Nonlinear Coupled Problems*,
Sustainable Civil Infrastructures, DOI 10.1007/978-3-319-61905-7_9

order to devise an adaptive strategy. Since a priori error estimates are not available for nonlinear time-dependent problems, useful error estimation must employ *a posteriori* technique to predict the overall discretization error. *A posteriori* error estimates are computed locally and contain significant information about the distribution of error among individual elements referred to as error indicators, which can form the basis of adaptive procedures. Today, the procedures available for error estimation can be essentially reduced to two kinds. In the first class, error measures are constructed based on local residuals of the numerical solution as introduced originally by Babuska and Rheinbolt [3]. In the second approach, a more accurate representation of the unknowns is sought prior estimating the error. This stage is known as the recovery process where emphasis has been put on self-equilibrating patches [19].

For nonlinear time-dependent problems, techniques devised for linear problems or time-independent nonlinear problems have been used at each time instant [1, 8] then suggested a posteriori error estimates for the primal variational formulation of elasto-plasticity with linear hardening [5, 9, 13, 18].

A family of error measures with a clear physical meaning and capable to account for effects of time and space discretization is given by the error in the constitutive equations. The concept, introduced by Ladevèze in 1975 for linear problems, was then extended to history-dependent materials modeled with functional constitutive relations [13] and to constitutive models with internal variables and described by a standard evolution rule. For the latter formulation, the potential structure and the convexity property of the state laws and evolution laws is exploited by adopting equivalent scalar formulations of the tensorial constitutive equations. More precisely both state and evolution equations are formulated using a scalar equation which involves two convex scalar “pseudo” potentials. The key property is the positiveness of the residual. The concept of dissipation error, introduced by Ladevèze and Moës [12], exploits this scalar nature of the constitutive relations. The error is quantified by the residual in the evolution law produced by time continuous admissible solutions that satisfy the compatibility relations, the equilibrium equations, the state law and the initial conditions. It was shown that the proposed error measure is able to account for effects of time and space discretization.

The dissipation error was then extended in [11] and [5, 8] by removing the state laws from the admissibility conditions. Applications of this error measure were given for a coupled elastoplastic- damage model [11] where the non-incremental LATIN method was used to solve the discrete problem [14]. The extended dissipation error was successfully applied by Orlando [17] to the finite element solution of elastoplastic problems discretized in time with the classical incremental finite element method with change of finite element mesh from one-time step to the other, which introduces a discontinuity jump in the time linear interpolation of the discrete values of the solution.

In this paper, we propose a natural extension of the concept of error in the constitutive equations to non-standard plasticity Drucker-Prager models [4–6, 8, 9] whose evolution is also described by a scalar equation which involve a unique scalar-valued function of both the velocity vector and the generalized stresses. This function is called bi-potential developed by de Saxcé et al. [7, 9], standard models are particular cases where the bi-potential can be represented as the sum of two scalar functions. Both the generalized evolution rule as well as its inverse are derived from this function using normality. Within this framework, the positiveness of the residual is kept which permit the extension of the generalized dissipation error to non-standard models described by a bi-potential.

2 Pseudo-Potentials and Bi-Potentials

The notation used here will be one in which symmetric second-order tensors, denoted by double underlined bold letters $(\boldsymbol{\pi}, \boldsymbol{\kappa})$, are represented as six-dimensional vectors. The vectors are bolded and more complex operators are double capitalized (e.g. $\mathbb{C}$ for Hooke's tensor).

In research investigating the structure of mechanical law, Moreau proposed to relax the differentiability condition by allowing potential to be non-differentiable as an appropriate way to take into account some multivalued laws within the same formalism then the concept of pseudo-potential, which is merely a scalar-valued non-differentiable function and made use of non-smooth analysis tools to set up his formulation, since classical tools of differential calculus are no longer applicable [16]. However, the pseudo-potential can be derived only for a standard plastic model.

The function is said to be a *bi-potential* if the following inequality holds:

$$b_p(\dot{\boldsymbol{\kappa}}', \boldsymbol{\pi}') \geq \boldsymbol{\pi}' \cdot \dot{\boldsymbol{\kappa}}', \quad \forall(\dot{\boldsymbol{\kappa}}', \boldsymbol{\pi}') \in \mathbf{F} \times \mathbf{V} \tag{1}$$

A pair $(\dot{\boldsymbol{\kappa}}, \boldsymbol{\pi})$ is said to be extremal if the equality is achieved for this pair:

$$b_p(\dot{\boldsymbol{\kappa}}, \boldsymbol{\pi}) = \boldsymbol{\pi} \cdot \dot{\boldsymbol{\kappa}} \tag{2}$$

Then, any extremal pair is characterized by the following relations:

$$\forall \dot{\boldsymbol{\kappa}}' \in \mathbf{V} \quad : \quad b_p(\dot{\boldsymbol{\kappa}}', \boldsymbol{\pi}) - b_p(\dot{\boldsymbol{\kappa}}, \pi) \geq \boldsymbol{\pi} \cdot (\dot{\boldsymbol{\kappa}}' - \dot{\boldsymbol{\kappa}}) \tag{3}$$

$$\forall \boldsymbol{\pi}' \in \mathbf{F} \quad : \quad b_p(\dot{\boldsymbol{\kappa}}, \pi') - b_p(\dot{\boldsymbol{\kappa}}, \pi) \geq \dot{\boldsymbol{\kappa}} \cdot (\boldsymbol{\pi}' - \boldsymbol{\pi}) \tag{4}$$

Let us remark that the extremal pairs satisfy

$$b_p(\dot{\boldsymbol{\kappa}}, \pi) - \boldsymbol{\pi} \cdot \dot{\boldsymbol{\kappa}} = \inf_{\dot{\boldsymbol{\kappa}}'} \left[b_p(\dot{\boldsymbol{\kappa}}', \boldsymbol{\pi}) - \dot{\boldsymbol{\kappa}}' \cdot \boldsymbol{\pi} \right] = \inf_{\pi'} \left[b_p(\dot{\boldsymbol{\kappa}}, \pi') - \dot{\boldsymbol{\kappa}} \cdot \boldsymbol{\pi}' \right] \tag{5}$$

These relations represent a multivalued constitutive relationship which is now implicit in the sense of the implicit function theorem. This explains the name of *Implicit Standard Materials* proposed by de Saxcé [7] for this class of materials. Of course, explicit standard materials are particular cases of implicit standard ones with separable bi-potentials [9]:

$$b_p(\dot{\boldsymbol{\kappa}}, \boldsymbol{\pi}) = \varphi(\dot{\boldsymbol{\kappa}}) + \varphi^*(\boldsymbol{\pi}) \tag{6}$$

3 Non-associated Drucker-Prager Model with Hardening

It is well known that the soil materials have a very complicated behavior. Idealizations are, therefore, often necessary in order to develop simple mathematical constitutive laws for practical applications. Several models can be found in the literature, most of them are complex and require many parameters, where some can be physically meaningless. The relative simplicity of the Drucker-Prager model [4–6, 8, 9], which can reflect some characteristics of soils behavior with only three parameters, explains why this model is widely used. Hence, the model is described by the dual variables (ε^p, p), (σ, R) where

$$\text{stress } \boldsymbol{\sigma} = (\mathbf{s}, s_m) \text{ with } s_m = \frac{1}{3}\mathrm{Tr}(\boldsymbol{\sigma}) \text{ deviator } \mathbf{s} = \boldsymbol{\sigma} - s_m \mathbf{1}$$

$$\text{strain } \dot{\varepsilon}^p = (\dot{\mathbf{e}}^p, \dot{e}^p_m) \text{ with } \dot{e}^p_m = \mathrm{Tr}(\dot{\varepsilon}^p) \text{ deviator } \dot{\mathbf{e}}^p = \dot{\varepsilon}^p - \frac{\dot{e}^p_m}{3}\mathbf{1}$$

$$\mathbf{1} = \{1 \quad 1 \quad 1 \quad 0 \quad 0 \quad 0\}^{\mathrm{T}}$$

At each time the stress must belong to the set convex K_ϕ of the plastically admissible stresses defined by (Fig. 1)

$$\mathrm{K}_\phi = \left\{\pi \in \mathbb{R}^m \middle| \frac{1}{\mathrm{k}_d}\|\mathbf{s}\| + \mathrm{s}_m \tan\phi - (\mathrm{c} + R) \leq 0\right\} \tag{7}$$

where c is the cohesion, ϕ the friction angle and k_d a constant. Its boundary defines the yield function. Assuming an isotropic hardening law, the evolution of the elastic domain is governed by the relation R = h.p with p is the cumulated plastic deformation and h is the hardening modulus.

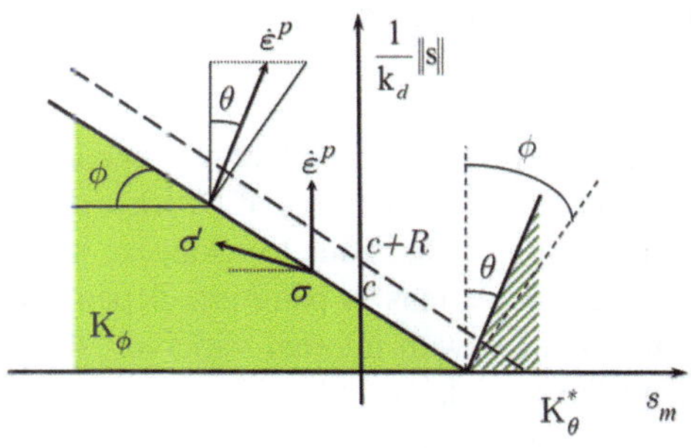

Fig. 1. Drucker-Prager with hardening criterion

The flow rule restricts the plastic strain rate to belong to the cone defined by

$$\mathrm{K}^*_\theta = \left\{(\dot{\mathbf{e}}^p, \dot{e}^p_m) \,\middle|\, \dot{e}^p_m \geq \mathrm{k}_d \tan\theta \, \|\dot{\mathbf{e}}^p\|\right\} \tag{8}$$

in which θ is the dilatancy angle; K_ϕ and K^*_θ are dual cones.

For any stress state situated on the regular part of the yield surface, the plastic strain rate vector inclination with respect to the vertical axis, is constant and less than the frictional angle ϕ. At the apex, this vector belongs to a cone of vectors defined by inclinations equal or greater than $\tan\theta$. Simple geometrical considerations allow us to write the flow rule in the following compact form:

$$\left(\dot{e}^p_m + \mathrm{k}_d(\tan\phi - \tan\theta)\|\dot{\mathbf{e}}^p\|, \dot{\mathbf{e}}^p\right) \in \partial_\sigma \mathbf{I}_{K_\phi}(\boldsymbol{\pi}) \tag{9}$$

where $\mathrm{I}_K(\pi)$ is the indicator function of the closed convex set K.

Although the preceding considerations seem quite obvious, it has the advantage to rule out definitively the question of existence of a convex pseudo-potential in the present situation. When the pseudo-potential exists, the cyclic monotony condition gives a means to construct its expression. Then, this law does not admit a convex pseudo-potential. Nevertheless, it admits a bi-potential defined by

$$b_p(\dot{\boldsymbol{\kappa}}, \pi) = \frac{c}{tan\phi}\dot{e}^p_m + k_d(\tan\phi - \tan\theta)\left[\frac{c+R}{\tan\phi} - \mathrm{s}_m\right]\|\dot{\mathbf{e}}^p\| + \mathrm{I}_{\mathrm{K}_\phi}(\boldsymbol{\pi}) + \mathrm{I}_{\mathrm{K}^*_\theta}(\dot{\boldsymbol{\kappa}}) + \mathrm{I}_{\mathrm{C}}(\dot{\boldsymbol{\kappa}}) \tag{10}$$

When $\theta = \phi$, the mixed term disappears and the bi-potential reduces to the sum of two dual pseudo-potentials:

$$b_p(\dot{\boldsymbol{\kappa}}, \pi) = \underbrace{\frac{c}{\tan\phi}\dot{e}^p_m + \mathrm{I}_{\mathrm{K}^*_\phi}(\dot{\boldsymbol{\kappa}}) + \mathrm{I}_{\mathrm{C}}(\dot{\boldsymbol{\kappa}})}_{\varphi(\dot{\boldsymbol{\kappa}})} + \underbrace{\mathrm{I}_{\mathrm{K}_\phi}(\boldsymbol{\pi})}_{\varphi^*(\boldsymbol{\pi})} \tag{11}$$

4 A Posteriori Error Estimation

The first error measure for *nonlinear evolution problems* is based on the Drucker inequality. It applies to dissipative problems where the functional formalism is adopted for the constitutive modeling and the conditions of Drucker stability [4] are fulfilled by the material. The definition of this error is due to Ladevèze in 1985 and has be used to control the accuracy of F.E. calculations in path/time dependent problems [13, 14].

In the constitutive formulation with internal variables, the potential structure and the convexity property of the state laws and evolution laws is exploited by adopting equivalent scalar formulations of the tensorial constitutive equations. The first use of the error in the constitutive equations to standard constitutive models described in terms of internal variables is due to Ladevèze [13], the state equations have been added to the first group of equations. The admissibility conditions then combine the kinematic compatibility relations, the equilibrium equations, the state equation and the initial conditions. Accordingly, the only equation which is left out is the evolution law that governs the dissipative process. By exploiting the convexity structure of this equation, Ladevèze introduced the concept of dissipation error given by the residual in the evolution law, which is appropriately reformulated in terms of the dissipation pseudo-potential and its Legendre-Fenchel transform. The dissipation error has been applied by Ladevèze and Moës [11, 12] to assess the accuracy of incremental finite element solutions of evolution problems of standard materials. Furthermore, error indicators which separate the sources of the different discretization errors have been defined and used to drive the adaptive process in time and space.

By removing the state laws from the admissibility conditions, new measure of the error in the constitutive equations, called extended dissipation error, has been introduced in Ladevèze [14] and Hjiaj [8]. This new error measure is a natural extension of the dissipation error. Applications of this error measure were given for an elastic-damage coupled model [14] solved with the non-incremental LATIN method and to the Prandtl-Reuss plasticity model in Orlando and Peric [17] solved with the classical incremental finite element method. The extension of this error measure to non-standard behaviour has been given by Hjiaj and Dao [6, 8]. This extension is based on a generalization of the Fenchel inequality, proposed by de Saxcé, to cover a broader range of behaviour including non-standard models.

For this new measure of error, the admissibility conditions include the kinematic compatibility relations, the equilibrium equations and the initial conditions. As a result, the only equations left apart are the complete constitutive equations, *i.e.* the state laws and the evolution law. The residual produced in the state equations and the evolution law is used as a direct measure of the discretization error.

Definition of Error The equations that are not satisfied by an admissible solution are the state law and the evolution rule. The residual in the evolution law and the residual in the state equation are considered to assess the quality of the approximation associated with the given admissible solution. A natural measure of this residual is provided by the equivalent formulation of the state equations which exploits the convexity properties of the law. Furthermore, given the nature of the state laws that relate the current value of the kinematic variables to the corresponding static ones, a global measure of the error is obtained by integrating the pointwise residual over the domain and over the time interval. Therefore, the relative error associated with an element is given by:

$$\mathbf{e}^2_{ext}(T) = \sup_{t\leq T}\left\{\int_\Omega {}^s\zeta^2_{x,t}(\boldsymbol{\eta}_{ad}-\boldsymbol{\kappa}_{ad},\boldsymbol{\pi}_{ad})\mathrm{d}\Omega + \int_\Omega\int_0^t {}^d\zeta^2_{x,t}(\dot{\boldsymbol{\kappa}}_{ad},\boldsymbol{\pi}_{ad})\,\mathrm{d}\tau\,\mathrm{d}\Omega\right\} \tag{12}$$

The extended dissipation error $\mathbf{e}^2_{ext}(T)$ has a finite value if and only if

$$\begin{array}{ll} (\dot{\boldsymbol{\kappa}}_{ad},\boldsymbol{\pi}_{ad}) \in dom\, b_p & \text{non - standard model} \\ (\boldsymbol{\sigma}_{ad},R_{ad}) \in dom\,\boldsymbol{\varphi}^* \text{ et } (\dot{\boldsymbol{\varepsilon}}^p_{ad},\dot{p}_{ad}) \in dom\,\boldsymbol{\varphi} & \text{standard model} \end{array} \tag{13}$$

Given an admissible solution $s_{ad} = s_{ad}(\boldsymbol{x},t)$ we have the following fundamental property of the extended dissipation error:

$$\begin{array}{l} \mathbf{e}^2_{ext}(T) \geq 0 \\ \text{and} \\ \mathbf{e}^2_{ext}(T) = 0 \Leftrightarrow s_{ad}(\boldsymbol{x},t) = s_{ex}(\boldsymbol{x},t) \quad \forall \boldsymbol{x} \in \Omega \text{ et } \forall t \in [0,\mathrm{T}] \end{array} \tag{14}$$

Since most a posteriori error estimates are computed locally, they contain significant information about the distribution of error among individual elements providing useful error indicators than can used to devise adaptive strategies. The extended dissipation error measure defines a global error measure that account of both time and space discretization. The relative extended dissipation error is defined as follow:

$$\varepsilon^2 = \frac{\mathrm{e}^2_{ext}(T)}{\sup\limits_{t\in[0,T]} \mathrm{D}^2(t)} \tag{15}$$

In which

$$\mathrm{D}^2(t) = \int_\Omega {}^s\mathrm{D}^2_{x,t}(\boldsymbol{\eta}_{ad} - \boldsymbol{\kappa}_{ad}, \boldsymbol{\pi}_{ad})\mathrm{d}\Omega + \int_\Omega \int_0^t {}^d\mathrm{D}^2_{x,t}(\dot{\boldsymbol{\kappa}}_{ad}, \boldsymbol{\pi}_{ad})\,\mathrm{d}\tau\,\mathrm{d}\Omega \tag{16}$$

With

$${}^s\mathrm{D}^2_{x,t}(\boldsymbol{\eta}_{ad} - \boldsymbol{\kappa}_{ad}, \boldsymbol{\pi}_{ad}) = \psi\, \boldsymbol{\eta}_{ad}(\boldsymbol{x}, t) - \boldsymbol{\kappa}_{ad}(\boldsymbol{x}, t) + \psi^* \boldsymbol{\pi}_{ad}(\boldsymbol{x}, t) \tag{17}$$

And

$${}^d\mathrm{D}^2_{x,t}(\dot{\boldsymbol{\kappa}}_{ad}, \boldsymbol{\pi}_{ad}) = b_p(\dot{\boldsymbol{\kappa}}_{ad}(\boldsymbol{x}, t), \boldsymbol{\pi}_{ad}(\boldsymbol{x}, t)) \tag{18}$$

The contribution of an element E to the extended dissipation error is obtained by integrating the pointwise residual over the considered element. The relative error associated with an element is given by:

$$\varepsilon^2_E = \frac{\mathrm{e}^2_{ext,E}(T)}{\sup\limits_{t\in[0,T]} \mathrm{D}^2_E(t)} \tag{19}$$

With

$$\mathrm{e}^2_{ext,E}(T) = \underbrace{\sup}_{0\le t\le T} \left\{ {}^s\zeta^2_{\Omega_E}(t) + {}^d\zeta^2_{\Omega_E}(t) \right\} \tag{20}$$

In the definition of an admissible solution for the computation of the extended dissipation error, the statically admissible variables are not constrained to their conjugate variables by means of the state laws as it happens in the dissipation error. This allows more information from the finite element solution to be included in building the corresponding admissible solution and strengthen the link between the two solutions. Likewise, conforming finite element displacements can be used as part of the admissible solution and do not need to be modified, unlike for the definition of the admissible solution to compute the dissipation error [5, 8, 10, 15].

5 Numerical Application

The numerical application investigates the load-displacement of two rigid strip footings resting on a homogeneous cohesive-frictional soil of great depth using PLSAER2D - a Matlab [20] program proposed by Dao [5]. Each footing is having a width B and the space between the footings is denoted S. Each footing is subjected to a vertical load at its centerline. It is assumed that the footings are smooth and the soil is weightless. The length of each footing, denoted B, is supposed large enough such that a condition of plane strain will exist in the soil mass supporting the foundations. The cohesive frictional soil is assumed to be a linear elastic-plastic material obeying the Drucker-Prager yield criterion with a non-associated flow rule and linear hardening. This model requires the specification of the Young Modulus E, the Poisson ratio ν, a cohesion c, a friction angle φ, a dilatancy angle θ and a hardening modulus H. All calculations are based on values of B = 2 m, S = 6 m, E = 2.5×10^5 kN/m^2, ν = 1/3, c = 2kN/m^2 and $\varphi = 35^\circ$. The constant k_d is obtained by imposing that the Drucker-Prager yield surface is inscribed in the Coulomb one. The load-displacement curve is computed for the following value of the angle of dilatation θ: 35°, 30°, 25° and 20°. Any restriction on the soil to flow freely in the vicinity of failure may have a significant influence on the computed load- displacement curve. Hence, the boundaries of the domain should be far away from the footings in order to be able to accommodate the plastic zone without any restriction on the soil displacement (Fig. 2).

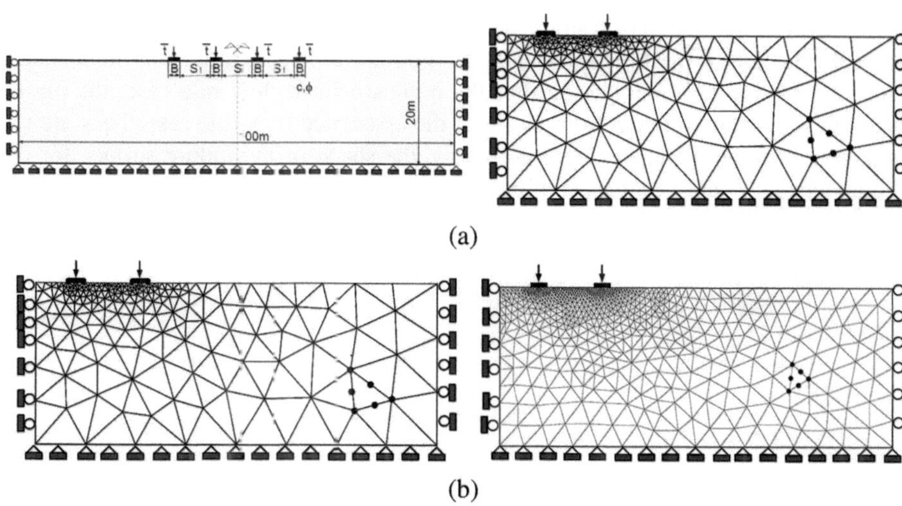

Fig. 2. a. Finite element model for four strips footings (Mesh N1: 339 elements), b. Mesh N2 (1332 elements) and Mesh N3 (1332 elements)

Load-displacement curves. For each dilation angle, three hardening modulus are considered: h = 100, h = 1 and h = 0.01. The last value of h has been considered in order to simulate the perfectly plastic case and therefore estimate the bearing capacity of the footing. In Fig. 3, the load-displacement curves predicted by the Drucker-Prager model with isotropic hardening for the considered values of the dilatancy angle θ are shown for the coarse mesh. The load represents the force applied to a single footing and is calculated as the sum of nodal forces at the footing nodes. As can be seen from Fig. 3, the structure is less stiff for a non-associated flow rule, i.e. the displacements are larger for the same values of the vertical load. It is worth noting that calculations have converged for all values of θ greater than 20°, for smaller values convergence problems may occur as mentioned elsewhere in the literature [2, 4, 16].

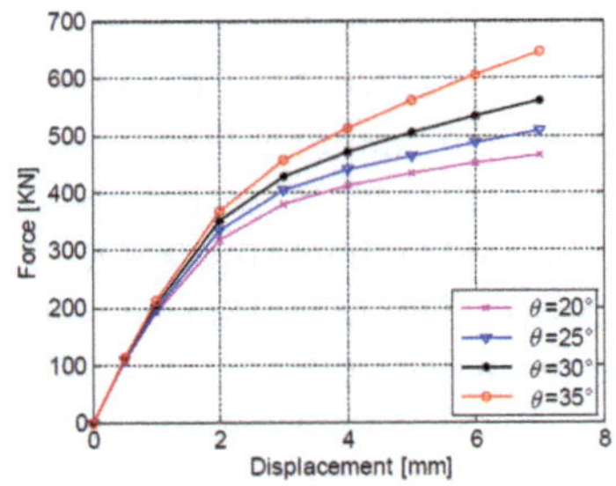

Fig. 3. Load-displacement curves

Plastic zones. A study of the plastic zone developed below the footing, the contour plots of the norm of the plastic strain at the end of the loading process are shown in the Fig. 4 for the coarse mesh. It shows that for smooth footings, all the soil immediately below the footing becomes plastic. For the non-associated flow rule case, the plastic zones are somewhat different from those for the associated flow rule case. They are not contiguous. As the dilatancy angle decreases, the shape of the failure surface for the non-associated model becomes significantly different from a log-spiral curve.

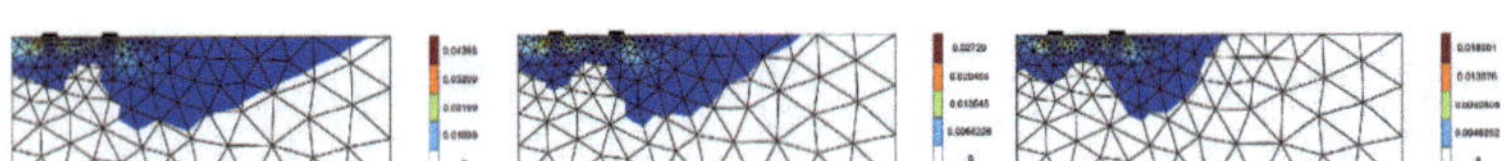

a. $\phi = \theta = 35°$ *b.* $\phi = 35°$ and $\theta = 30°$ *c.* $\phi = 35°$ and $\theta = 20°$

Fig. 4. Plastic zones (Mesh N1)

Error estimation. The evolution of the relative error with loading is depicted in Fig. 5a for the coarse mesh and for each value of the dilatancy angle (h = 100). It appear clearly that the relative error increases significantly with decreasing dilatancy angle. For instance, the relative error for θ = 20° is equal to 34.23% compared to 16.31% for $\theta = \varphi = 35°$. Let us define the degree of non-associativity ($\tan \varphi - \tan \theta$), this degree is always positive and is equal to zero if the flow rule is associated, we can say that the relative error increases with the degree of non-associativity even though the size of the plastic zone decreases with increasing values of degree of non-associativity.

Refinement strategies aims to highlight the effect of the space discretization error. For the problem under consideration, refinement is need it around the footing. Further the mesh pattern near the singularity at the edges of the footings must be kept. This mesh pattern permits the stress field to change rapidly and yield to accurate solution. The domain size and the associated finite element mesh are kept the same for all values of θ and it was found to be acceptable for all cases. The evolution of the relative global error with loading is depicted in Fig. 5b for the second mesh (Mesh N2) and Fig. 6 for the finest mesh (Mesh N3) and for each value of the dilatancy angle. The relative global error seems to evolve with the load in a similar manner for each mesh, only the magnitude of the relative error change with the mesh density. The relative global error at the end of the loading process for all meshes and all dilatancy is reported Table 1. For the associate case, the most refined mesh yield to a global relative error equal to 9.45%.

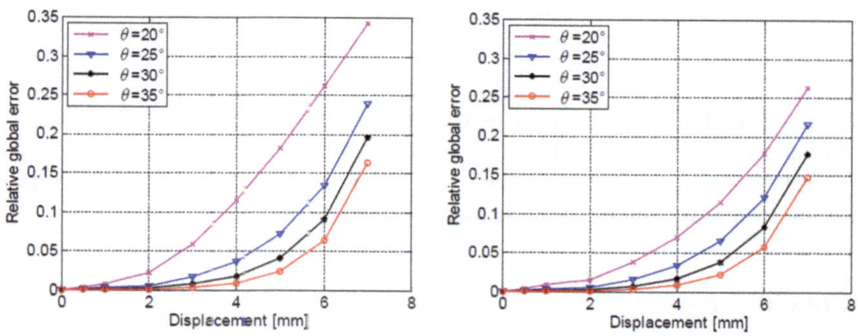

a. Mesh N1 b. Mesh N2

Fig. 5. Relative global error -displacement curves (Mesh N1 & N2, h = 100)

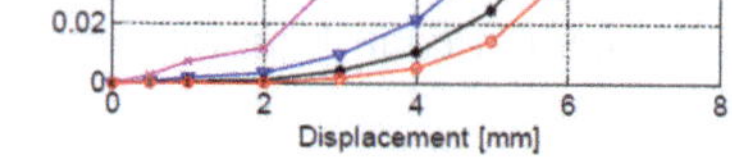

Fig. 6. Relative global error -displacement curves (Mesh N3, h = 100)

To this end numerical simulations have been carried out on given finite element mesh whereas the time step size Δt was changed (with h = 100). In general a reduction of the extended dissipation error is observed as the time step size Δt is reduced, though

Table 1. Relative error with different value of θ

Value of θ	Relative error at the end of loading		
	Mesh N1	Mesh N2	Mesh N3
$\theta = \varphi = 35^o$	16.31%	14.71%	9.45%
$\theta = 30^o$	19.60%	17.72%	11.36%
$\theta = 20^o$	34.23%	26.25%	15.91%

Table 2. Relative error with different numbers of steps (Mesh N1)

Number of steps	Relative error at the end of loading	
	Mesh N1: $\theta = \varphi = 35^o$	Mesh N2: $\theta = 20^o$
8	16.31%	34.23%
16	15.45%	31.97%
32	15.45%	31.57%

the reduction is not as pronounced as the one obtained by enrichment of the mesh. Dividing the size of step by four doesn't reduce significantly the error (see Table 2), it means that 8 steps is more than enough. We observe a decrease of the relative global error with decreasing values of the hardening modulus h.

6 Conclusions

In this paper, the assessment of the global quality of displacement finite element solutions for Drucker-Prager plasticity model, non-standard elastoplastic problems discretized in time has been presented.

On the introduction of a unique potential, function of the rates and the associated force, which is used by Ladevèze et al. [10–12] to extend the notion of dissipation error also to this class of material models, whereas applications of the new measure of error have been given in Hjiaj and Dao [6, 8]. This is a measure of the error in the constitutive equations, the extended dissipation error capability to capture the effects of time and space discretization has been shown in the assessment of the quality of finite element solutions of elastoplastic problems using Drucker-Prager plasticity model with linear hardening and illustrated with a numerical example. This has shown that all trends on the state law and dissipation contribution to the error were meaningful.

References

1. Ainsworth, M., Oden, J.T.: *A posteriori* error estimation in finite element analysis. John Wiley & Son and B G Teubner, Chichester (2000)
2. Halphen, B., Nguyen, Q.S.: Sur les matériaux standards généralisés. J. Méc. **14**, 39–63 (1975)

3. Babuska, I., Rheinboldt, W.C.: *A posteriori* error estimates for the finite element method. Int. J. Numer. Methods Eng. **12**, 1597–1615 (1978)
4. Drucker, D.C.: On the postulate of stability of material in the mechanics of continua. J. Méc. **3**(2), 235–249 (1964)
5. Lam, D.D.: *A posteriori* error estimation for non-associated plasticity problems, Ph. D. thesis, INSA de Rennes (2009)
6. Lam, D.D.: A bi-potential update algorithm for the non-associated plasticity model with hardening. In: Proceedings of CIGOS-2015: Innovations in Construction, Paris, France (2015)
7. de Saxcé, G.: Une généralisation de l'inégalité de Fenchel et ses applications aux lois constitutives. Comptes Rendus de l'Académie des Sciences de Paris (1992)
8. Hjiaj, M.: Sur la classe des matériaux standard implicites: Concept, Aspects discrétisés et Estimation de l'erreur a posteriori. Ph. D. thesis, Polytechnic Faculty of Mons. (1999)
9. Hjiaj, M., Lam, D.D., de Saxcé, G.: A family of bi-potentials describing the non-associated flow rule of pressure-dependant plastic models. Acta Mech. **220**, 237–246 (2011)
10. Ladevèze, P., Maunder, E.A.W.: A general procedure for recovering equilibrating element tractions. Comput. Methods Appl. Mech. Eng. **137**, 111–151 (1996)
11. Ladevèze, P., Moës, N., Douchin, B.: Constitutive relation error estimators for (visco) plastic finite element analysis with softening. Comput. Methods Appl. Mech. Eng. **176**, 247–264 (1999)
12. Ladevèze, P., Moës, N.: A new *a posteriori* error estimation for nonlinear time dependent finite element analysis. Comput. Methods Appl. Mech. Eng. **157**, 45–68 (1997)
13. Ladevèze, P., Pelle, J.P.: Mastering Calculations in Linear and Nonlinear Mechanics. Springer, New York (2004)
14. Ladevèze, P., Passieux, J.-C., Néron, D.: The LATIN multiscale computational method and the proper generalized decomposition. Comput. Methods Appl. Mech. Eng. **199**, 1287–1296 (2009)
15. Maunder, E.A.W., Moitinho de Almeida, J.P.: The stability of stars of triangular equilibrium plate elements. Int. J. Numer. Methods Eng. **77**, 922–968 (2009)
16. Moreau, J.: Evolution problem associated with a moving convex set in a hilbert space. J. Diff. Equ. **26**, 347–374 (1977)
17. Orlando, A., Peri´c, D.: Analysis of transfer procedures in elastoplasticity based on the error in the constitutive equations: Theory and numerical illustration. Int. J. Numer. Methods Eng. **60**, 1595–1631 (2004)
18. Simo, J.C., Hughes, T.J : Computational Inelasticity. Springer, New York (2000)
19. Zienkiewicz, O.C., Liu, Y.C , Huang, G.C.: Error estimation and adaptivity in flow formulation for forming problems. Int. J. Numer. Methods Eng. **25**, 23–42 (1988)
20. MATLAB Release: The MathWorks, Inc., Natick, Massachusetts, United States (2008)

Analysis of Structural Behaviour of Thick Composite Laminates on an Elastic Foundation Using Efficient Higher-Order Theory

Mokhtar Bouazza[1,2(✉)], Tawfiq Becheri[1], and Abderrahmane Boucheta[1]

[1] Department of Civil Engineering, University of Bechar, 08000 Bechar, Algeria
bouazza_mokhtar@yahoo.fr
[2] Laboratory of Materials and Hydrology (LMH), University of Sidi Bel Abbes, 2200 Sidi Bel Abbes, Algeria

Abstract. The present study is an attempt to analysis of structural behaviour of thick composite laminates on an elastic foundation using efficient higher-order theory for bending analysis of laminated composite plates subjected to uniformly distributed loading. This theory satisfies zero transverse shear stresses conditions at the top and bottom surfaces of the plate. Analytical solutions for deflections and stresses are obtained for simply supported rectangular plates. In the analysis, the two-parameter Pasternak and Winkler foundations are considered. Numerical examples are presented to verify the accuracy of the present theories.

Keywords: Laminated plates · Bending · Efficient higher-order theory · Winkler foundations · Pasternak foundation

1 Introduction

Plates supported by elastic foundations represent a very common model in structural engineering; as a consequence the number of technical problems connected with this model, is very large.

Generally the analysis is developed on the assumption that the reaction forces of foundation are proportional at every point to the deflection at that point (Winkler model). However, to represent the characteristic behavior of practical foundations, the Winkler model is not adequate and can be substituted by other models such as the Filonenko-Borodhic, Pasternak or Vlasov models [1–4]. In many cases the two-parameter Pasternak model has proved to be the most effective.

In the classical plate theory (CPT), it is assumed that the plane cross sections initially normal to the plate midsurface before deformation remain plane and normal to that surface during deformation. This is the result of neglecting the transverse shear strains. However, in thick and moderately thick laminated plates, significant transverse shear strains occur, and the theory gives inaccurate results for the plates. So, it is obvious that the shear strains have to be taken into account. There are numerous theories of plates and laminated plates that include the transverse shear strains [5–9].

H. Shehata and Y. Rashed (eds.), *Numerical Analysis of Nonlinear Coupled Problems*,
Sustainable Civil Infrastructures, DOI 10.1007/978-3-319-61905-7_10

The FSDT proposed by Reissner [10] and Mindlin [11] accounts for shear deformation effects by the way of linear variation of in-plane displacements through the thickness. Since the FSDT violates the equilibrium conditions on the top and bottom surfaces of the plate, a shear correction factor is required to compensate for the difference between actual stress state and assumed constant stress state [12–18].

Higher-order shear deformation theories (HSDTs) involving parabolic transverse shear distribution was proposed by Levinson [19], Lo et al. [20], and Reddy [21]. Over the last few decades, the rapid advancements in the use of composite structures have motivated researchers to develop rigorous plate the- ories accounting for accurate structural kinematics especially shear deformation. Notably among them are due to Kant and Pandya [22], Touratier [23], Soldatos [24], Karama et al. [25], Aydogdu [26], Bouazza and Bensedd:q [27], Grover et al. [28], Bouazza and his co-workers [29–35] and Mantari et al. [36]. In recent years, researchers have focused on inter-laminar continuity (IC) and Zig-Zag (ZZ) requirement [37] in addition to shear deformation. Nonlinear static and dynamic analyses of plates and shells of various shapes have been carried out by various researchers by Civalek et al. [38–46] and Sofiyev et al. [47–52].

In this paper, the behaviour of thick composite laminates on an elastic foundation using efficient higher-order theory. Pasternak's model is used here to describe the two-parameter elastic foundation, and getting a special case of Winkler's foundation model by considering a one-parameter elastic foundation. Analytical solutions for bending deflections and stresses of symmetric cross-ply laminates are found by using this theory. The results obtained are in a good agreement with those given in the literature. The effect of bending stiffness of laminated plates was also investigated.

2 Efficient Higher-Order Theory

Consider a rectangular plate of length a, width b, and thickness h and resting on elastic foundations (see Fig. 1). The mid-plane of the plate is taken as the xy plane, and the x- and y-axes are directed along the edges. The z-axis is taken perpendicular to the mid-plane. Let the plate be subjected to a distributed transverse load q(x, y), and the load-displacement relation between the plate and the supporting foundations follows the two-parameter Pasternak's model:

$$\Re = k_1 w - k_2 \nabla^2 w \tag{1}$$

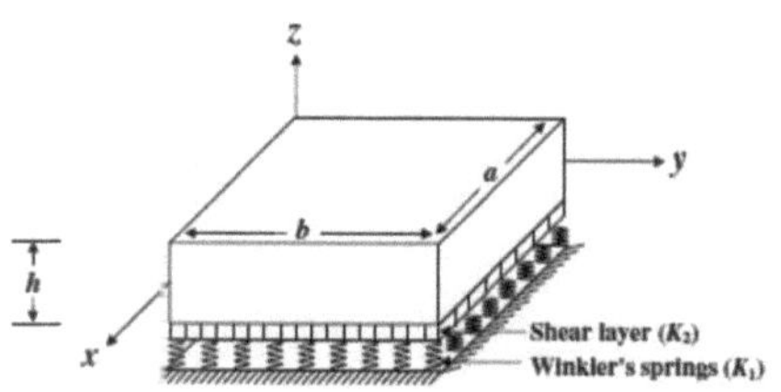

Fig. 1. Geometry and coordinate system for an orthotropic rectangular plate resting on two-parameter elastic foundations

where $\Re$ is the foundation reaction per unit area, k_1 and k_2 are the Winkler's and Pasternak's foundation stiffnesses, respectively, w is the plate deflection, and ∇^2 is the Laplace operator in x and y. This model is simply known as Winkler's type when $k_2 = 0$.

For the two-dimensional plate theory, the efficient higher-order theory has been widely studied in the elastic plate. Based on the efficient higher-order theory, the displacement field can be expressed as

$$\begin{aligned}
u_1(x,y,z) &= u(x,y) - z\frac{\partial w_b}{\partial x} - \frac{4z^3}{3h^2}\frac{\partial w_s}{\partial x} \\
u_2(x,y,z) &= v(x,y) - z\frac{\partial w_b}{\partial y} - \frac{4z^3}{3h^2}\frac{\partial w_s}{\partial y} \\
u_3(x,y,z) &= w_b(x,y) + w_s(x,y)
\end{aligned} \tag{2}$$

It should be noted that unlike the first-order shear deformation theory, this theory does not require shear correction factors. The kinematic relations can be obtained as follows:

$$\begin{aligned}
\varepsilon_x &= \varepsilon_x^0 + zk_x^b + f(z)k_x^s \\
\varepsilon_y &= \varepsilon_y^0 + zk_y^b + f(z)k_y^s \\
\gamma_{xy} &= \gamma_{xy}^0 + zk_{xy}^b + f(z)k_{xy}^s \\
\gamma_{yz} &= g(z)\gamma_{yz}^s \\
\gamma_{xz} &= g(z)\gamma_{xz}^s \\
\varepsilon_z &= 0
\end{aligned} \tag{3}$$

where

$$\begin{aligned}
&\varepsilon_x^0 = \frac{\partial u}{\partial x}, \quad k_x^b = -\frac{\partial^2 w_b}{\partial x^2}, \quad k_x^s = -\frac{\partial^2 w_s}{\partial x^2} \\
&\varepsilon_y^0 = \frac{\partial v}{\partial y}, \quad k_y^b = -\frac{\partial^2 w_b}{\partial y^2}, \quad k_y^s = -\frac{\partial^2 w_s}{\partial y^2} \\
&\gamma_{xy}^0 = \frac{\partial u}{\partial y} + \frac{\partial v}{\partial x}, \quad k_{xy}^b = -2\frac{\partial^2 w_b}{\partial x \partial y}, \quad k_{xy}^s = -2\frac{\partial^2 w_s}{\partial x \partial y} \\
&\gamma_{yz}^s = \frac{\partial w_s}{\partial y}, \quad \gamma_{xz}^s = \frac{\partial w_s}{\partial x} \\
&g(z) = 1 - f'(z), \quad f'(z) = \frac{df(z)}{dz}, \quad f(z) = -\frac{4z^3}{3h^2}
\end{aligned} \tag{4}$$

The linear constitutive relations of a orthotropic plate can be written as

$$\begin{Bmatrix} \sigma_x \\ \sigma_y \\ \sigma_{xy} \\ \sigma_{yz} \\ \sigma_{xz} \end{Bmatrix} = \begin{bmatrix} Q_{11} & Q_{12} & 0 & 0 & 0 \\ Q_{12} & Q_{22} & 0 & 0 & 0 \\ 0 & 0 & Q_{66} & 0 & 0 \\ 0 & 0 & 0 & Q_{44} & 0 \\ 0 & 0 & 0 & 0 & Q_{55} \end{bmatrix} \begin{Bmatrix} \varepsilon_x \\ \varepsilon_y \\ \gamma_{xy} \\ \gamma_{yz} \\ \gamma_{xz} \end{Bmatrix} \tag{5}$$

where

$$Q_{11} = \frac{E_1}{1 - \nu_{12}\nu_{21}}, \; Q_{12} = \frac{\nu_{12}E_2}{1 - \nu_{12}\nu_{21}}, \; Q_{22} = \frac{E_2}{1 - \nu_{12}\nu_{21}}$$
$$Q_{66} = G_{12}, \; Q_{44} = G_{23}\,, \; Q_{55} = G_{13} \tag{6}$$

Transforming the above equations of an arbitrary k layer in local coordinate system into the global coordinate system, the laminate constitutive equations can be expressed as

$$\begin{Bmatrix} \sigma_x \\ \sigma_y \\ \sigma_{xy} \\ \sigma_{yz} \\ \sigma_{xz} \end{Bmatrix}^{(k)} = \begin{bmatrix} \bar{Q}_{11} & \bar{Q}_{12} & \bar{Q}_{16} & 0 & 0 \\ \bar{Q}_{12} & \bar{Q}_{22} & \bar{Q}_{26} & 0 & 0 \\ \bar{Q}_{16} & \bar{Q}_{26} & \bar{Q}_{66} & 0 & 0 \\ 0 & 0 & 0 & \bar{Q}_{44} & \bar{Q}_{45} \\ 0 & 0 & 0 & \bar{Q}_{45} & \bar{Q}_{55} \end{bmatrix}^{(k)} \begin{Bmatrix} \varepsilon_x \\ \varepsilon_y \\ \gamma_{xy} \\ \gamma_{yz} \\ \gamma_{xz} \end{Bmatrix}^{(k)} \tag{7}$$

Where $\bar{Q}_{ij}$ are the transformed material constants given as

$$\begin{aligned}
\bar{Q}_{11} &= Q_{11}\cos^4\theta + 2(Q_{12} + 2Q_{66})\sin^2\theta\,\cos^2\theta + +Q_{22}\sin^4\theta \\
\bar{Q}_{12} &= (Q_{11} + Q_{22} - 4Q_{66})\sin^2\theta\,\cos^2\theta + Q_{12}(\sin^4\theta + \cos^4\theta) \\
\bar{Q}_{22} &= Q_{11}\sin^4\theta + 2(Q_{12} + 2Q_{66})\sin^2\theta\,\cos^2\theta + Q_{22}\cos^4\theta \\
\bar{Q}_{16} &= (Q_{11} - Q_{12} - 2Q_{66})\sin\theta\,\cos^3\theta + (Q_{12} - Q_{22} + 2Q_{66})\sin^3\theta\,\cos\theta \\
\bar{Q}_{26} &= (Q_{11} - Q_{12} - 2Q_{66})\sin^3\theta\,\cos\theta + (Q_{12} - Q_{22} + 2Q_{66})\sin\theta\,\cos^3\theta \\
\bar{Q}_{66} &= (Q_{11} + Q_{22} - 2Q_{12} - 2Q_{66})\sin^2\theta\,\cos^2\theta + Q_{66}(\sin^4\theta + \cos^4\theta) \\
\bar{Q}_{44} &= Q_{44}\cos^2\theta + Q_{55}\sin^2\theta \\
\bar{Q}_{45} &= (Q_{55} - Q_{44})\cos\theta\,\sin\theta \\
\bar{Q}_{55} &= Q_{55}\cos^2\theta + Q_{44}\sin^2\theta
\end{aligned} \tag{8}$$

The strain energy of the plate can be written as

$$U_P = \frac{1}{2}\int_V \sigma_{ij}\varepsilon_{ij} dV = \frac{1}{2}\int_V \left(\sigma_x\varepsilon_x + \sigma_y\varepsilon_y + \sigma_{xy}\gamma_{xy} + \sigma_{yz}\gamma_{yz} + \sigma_{xz}\gamma_{xz}\right) dV \tag{9}$$

The principle of virtual work for the present problem may be expressed as follows:

$$\begin{aligned} \iint \Big[& N_x \delta\varepsilon_x^0 + N_y \delta\varepsilon_y^0 + N_{xy}\delta\gamma_{xy}^0 + M_x^b \delta k_x^b + M_y^b \delta k_y^b + M_{xy}^b k_{xy}^b + M_x^s k_x^s \\ & + M_y^s k_y^s + M_{xy}^s k_{xy}^s + Q_{yz}\gamma_{yz} + Q_{xz}\gamma_{xz} \Big] dxdy = 0 \end{aligned} \tag{10}$$

where (N_x, N_y, N_{xy}) denote the total in-plane force resultants, (M_x^b, M_y^b, M_{xy}^b), (M_x^s, M_y^s, M_{xy}^s) denote the total moment resultants and (Q_{xz}, Q_{yz}) are transverse shear stress resultants and they are defined as

$$\begin{aligned} (N_x, N_y, N_{xy}) &= \int_{-h/2}^{h/2} (\sigma_x, \sigma_y, \sigma_{xy}) dz \\ (M_x^b, M_y^b, M_{xy}^b) &= \int_{-h/2}^{h/2} (\sigma_x, \sigma_y, \sigma_{xy}) z dz \\ (M_x^s, M_y^s, M_{xy}^s) &= \int_{-h/2}^{h/2} (\sigma_x, \sigma_y, \sigma_{xy}) f(z) dz \\ (Q_{xz}, Q_{yz}) &= \int_{-h/2}^{h/2} (\sigma_{xz}, \sigma_{yz})\, dz \end{aligned} \tag{11}$$

Substituting Eq. (7) into Eq. (11) and integrating through the thickness of the plate, the stress resultants are given as

$$\left\{ \begin{array}{c} \left\{ \begin{array}{c} N_x \\ N_y \\ N_{xy} \end{array} \right\} \\ \left\{ \begin{array}{c} M_x^b \\ M_y^b \\ M_{xy}^b \end{array} \right\} \\ \left\{ \begin{array}{c} M_x^s \\ M_y^s \\ M_{xy}^s \end{array} \right\} \end{array} \right\} = \left[\begin{array}{ccc} \begin{bmatrix} A_{11} & A_{12} & A_{16} \\ A_{12} & A_{22} & A_{26} \\ A_{16} & A_{26} & A_{66} \end{bmatrix} & \begin{bmatrix} B_{11} & B_{12} & B_{16} \\ B_{12} & B_{22} & B_{26} \\ B_{16} & B_{26} & B_{66} \end{bmatrix} & \begin{bmatrix} B_{11}^s & B_{12}^s & B_{16}^s \\ B_{12}^s & B_{22}^s & B_{26}^s \\ B_{16}^s & B_{26}^s & B_{66}^s \end{bmatrix} \\ \begin{bmatrix} B_{11} & B_{12} & B_{16} \\ B_{12} & B_{22} & B_{26} \\ B_{16} & B_{26} & B_{66} \end{bmatrix} & \begin{bmatrix} D_{11} & D_{12} & D_{16} \\ D_{12} & D_{22} & D_{26} \\ D_{16} & D_{26} & D_{66} \end{bmatrix} & \begin{bmatrix} D_{11}^s & D_{12}^s & D_{16}^s \\ D_{12}^s & D_{22}^s & D_{26}^s \\ D_{16}^s & D_{26}^s & D_{66}^s \end{bmatrix} \\ \begin{bmatrix} B_{11}^s & B_{12}^s & B_{16}^s \\ B_{12}^s & B_{22}^s & B_{26}^s \\ B_{16}^s & B_{26}^s & B_{66}^s \end{bmatrix} & \begin{bmatrix} D_{11}^s & D_{12}^s & D_{16}^s \\ D_{12}^s & D_{22}^s & D_{26}^s \\ D_{16}^s & D_{26}^s & D_{66}^s \end{bmatrix} & \begin{bmatrix} H_{11}^s & H_{12}^s & H_{16}^s \\ H_{12}^s & H_{22}^s & H_{26}^s \\ H_{16}^s & H_{26}^s & H_{66}^s \end{bmatrix} \end{array} \right] \left\{ \begin{array}{c} \left\{ \begin{array}{c} \varepsilon_x^0 \\ \varepsilon_y^0 \\ \gamma_{xy}^0 \end{array} \right\} \\ \left\{ \begin{array}{c} k_x^b \\ k_y^b \\ k_{xy}^b \end{array} \right\} \\ \left\{ \begin{array}{c} k_x^s \\ k_y^s \\ k_{xy}^s \end{array} \right\} \end{array} \right\} \tag{12}$$

$$\left\{ \begin{array}{c} Q_{yz}^s \\ Q_{xz}^s \end{array} \right\} = \begin{bmatrix} A_{44} & A_{45} \\ A_{45} & A_{55} \end{bmatrix} \left\{ \begin{array}{c} \gamma_{yz}^s \\ \gamma_{xz}^s \end{array} \right\} \tag{13}$$

Where $(A_{ij}, B_{ij}, D_{ij}, B_{ij}^s, D_{ij}^s, H_{ij}^s, A_{ij}^s)$ are the plate stiffness, defined by

$$(A_{ij}, B_{ij}, D_{ij}, B_{ij}^s, D_{ij}^s, H_{ij}^s) = \int_{-h/2}^{h/2} \bar{Q}_{ij}(1, z, z^2, f(z), zf(z), (f(z))^2)dz \quad (i,j = 1,2,6)$$

$$A_{ij}^s = \int_{-h/2}^{h/2} \bar{Q}_{ij}(g(z))^2\, dz \quad (i,j = 4,5)$$

(14)

The strain energy of the foundation can be expressed as

$$U_F = \frac{1}{2}\int_A \left\{ k_1 (w_b + w_s)^2 + k_2 \left[\left(\frac{\partial (w_b + w_s)}{\partial y} \right)^2 + \left(\frac{\partial (w_b + w_s)}{\partial y} \right)^2 \right] \right\} dxdy \quad (15)$$

The work done of the plate by applied forces can be written as

$$\delta V = -\int_0^L q\delta(w_b + w_s)dx + \frac{1}{2}\int_0^L \left[N_x^0 \frac{\partial^2 (w_b + w_s)}{\partial x^2} + N_y^0 \frac{\partial^2 (w_b + w_s)}{\partial y^2} + 2N_{xy}^0 \frac{\partial^2 (w_b + w_s)}{\partial x\, \partial y} \right] dx$$

(16)

where q and N_x^0, N_y^0, N_{xy}^0 are transverse and in-plane distributed forces, respectively.

Hamilton's principle is used herein to derive the equations of motion appropriate to the displacement field and the constitutive equation. The principle can be stated in analytical form as

$$0 = \int_0^t \delta(U_P + U_F + V)dt \quad (17)$$

Où δ indique une variation par rapport à x et y respectivement.

The governing stability equations are obtained for efficient higher-order theory as

$$\frac{\partial N_x}{\partial x} + \frac{\partial N_{xy}}{\partial y} = 0$$

$$\frac{\partial N_{xy}}{\partial x} + \frac{\partial N_y}{\partial y} = 0$$

$$\frac{\partial^2 M_x^b}{\partial x^2} + 2\frac{\partial^2 M_{xy}^b}{\partial x \partial y} + \frac{\partial^2 M_y^b}{\partial y^2} - k_w(w_b + w_s) + k_s \nabla^2 (w_b + w_s) + q + \bar{N} = 0$$

$$\frac{\partial^2 M_x^s}{\partial x^2} + 2\frac{\partial^2 M_{xy}^s}{\partial x \partial y} + \frac{\partial^2 M_y^s}{\partial y^2} + \frac{\partial Q_{xz}^s}{\partial x} + \frac{\partial Q_{yz}^s}{\partial y} - k_w(w_b + w_s) + k_s \nabla^2 (w_b + w_s) + q + \bar{N} = 0$$

(18)

3 Results and Discussion

In this section, numerical examples are presented and discussed for verifying the accuracy of the present theory in predicting the bending responses of plates. The numerical results of deflections and stresses are presented for plates with four edges simply supported resting on elastic foundations. The various non-dimensional parameters used are:

$$k_1 = \frac{a^4}{h^3}K_1,\ k_2 = \frac{a^2}{h^3}K_2,\ \bar{w} = \frac{10^2 h^3}{q_0 a^4} w\left(\frac{a}{2}, \frac{b}{2}\right)$$

The results for isotropic square plates resting on elastic foundations using refined efficient higher-order plate theory are compared with the corresponding results available in the literature [35–38]. The dimensionless deflection and foundations for this case are given by

$$\bar{k}_1 = \frac{a^4}{D}K_1,\ \bar{k}_2 = \frac{a^2}{D}K_2,\ w^* = \frac{10^2 D}{q_0 a^4} w\left(\frac{a}{2}, \frac{b}{2}\right)$$

Table 1 shows the comparison of nondimensional deflection w^* between the results of the present work and those of Buczkowski and Torbacki [53], Timoshenko and Woinowsky-Krieger [54], and Zenkour et al. [56]. The plate is subjected to uniformly distributed load and resting on Winkler's one-parameter elastic foundation. For all values of foundation parameters $\bar{k}_1$, the comparisons are well justified. Table 2 gives the nondimensional deflection w^* of an isotropic square plate on two-parameter elastic foundations. It should be noted that the present theory involves only four independent variables as against five in the case of Refs [55, 56]. Also, the present theory does not required shear correction factors.

Table 1. Comparison of the deflection w^* of an isotropic square plate on one-parameter elastic foundations under uniformly distributed load ($\nu = 0.3$, $\bar{k}_2 = 0$, h/a = 0.05).

$\bar{k}_1$	Ref. [53]	Ref. [54]	Ref. [56][a]	Ref [56][b]	Present
0	0.40624	0.41197	0.41150	0.41150	0.5388
1^4	0.40517	0.41088	0.41040	0.41040	0.5374
3^4	0.33472	0.33855	0.33814	0.33814	0.4458
5^4	0.15060	0.15114	0.15094	0.15094	0.2061
10^4	0.01115	0.01096	0.01108	0.01109	0.0210

[a] Simple first-order transverse shear deformation plate theory (SFPT),

[b] Mixed first-order transverse shear deformation plate theory (MFPT).

Table 2. Comparison of the deflection w^* of an isotropic square plate on two-parameter elastic foundations under uniformly distributed load ($\nu = 0.3$, h/a = 0.01)

$\bar{k}_1$	$\bar{k}_2$	Ref. [55]	Ref. [56][a]	Ref. [56][b]	Present
1	1	0.3853	0.38550	0.38550	0.5013
	3^4	0.0763	0.07630	0.07629	0.0996
	5^4	0.0115	0.01153	0.01153	0.0154
3^4	1	0.3210	0.32108	0.32108	0.4177
	3^4	0.0732	0.07317	0.07317	0.0955
	5^4	0.0115	0.01145	0.01145	0.0153
5^4	1	0.1476	0.14765	0.14765	0.1924
	3^4	0.0570	0.05704	0.05704	0.0746
	5^4	0.0109	0.01095	0.01095	0.0146

([a]:SFPT, [b]: MFPT)

4 Conclusions

A refined efficient higher-order plate theory is successfully developed for laminated composite plates on elastic foundation. Using the Navier solution method, the differential equations have been solved analytically for bending loads presented solutions. From this numerical study, the following conclusions may be drawn:

1. This theory is seen to behave well and the results of the sample problem show good agreement with the literature values as seen from the validation checks.
2. The theory accounts for a quadratic variation of the transverse shear strains across the thickness, and satisfies the zero traction boundary conditions on the top and bottom surfaces of the plate without using shear correction factors.
3. The present theory has strong similarity with the CPT in many aspects such as equations of motion, boundary conditions, and stress resultant expressions.
4. The results of present theory with four independent variables are comparable with those generated by other shear deformation plate theories containing more number of independent variables.

References

1. Xiang, Y., Kitipornchai, S., Liew, K.M.: Buckling and vibration of thick laminates on Pasternak foundation. J. Eng. Mech. **122**, 54–63 (1996)
2. Turvey, G.J.: Uniformly loaded, simply supported, antisymmetritally laminated, rectangular plate on a Winkler-Pastemak foundation. Int. J. Solids Struct. **13**, 43–44 (1977)
3. Jones, R., Xenophontos. J.: The Vlasov foundation model. Int. J. Mech. Sci. **19**, 317–323 (1977)
4. Shen, H.-S.: Postbuckling of orthotropic plates on two-parameter elastic foundation. J. Eng. Mech. **121**, 50–56 (1995)
5. Reissner, E.: The effect of transverse shear deformation on the bending of elastic plates. J. Appl. Mech. **12**, 69–77 (1945)

6. Midlin, R.D.: Influence of rotatory inertia and shear on flexural motions of isotropic, elastic plates. J. Appl. Mech. **18**, 31–38 (1951)
7. Reissner, E.: On the theory of bending of elastic plates. J. Math. Phys. **23**, 184–191 (1944)
8. Whitney, J.M., Pagano, N.J.: Shear deformation in heterogeneous anisotropic plates. J. Appl. Mech. **37**, 1031–1036 (1970)
9. Murty, M.V.V.: An improved transverse shear deformation theory for laminated anisotropic plates. NASA Technical Paper, No. 1903 (1981)
10. Reissner, E.: The effect of transverse shear deformation on the bending of elastic plates. J. Appl. Mech. **12**(2), 69–72 (1945)
11. Mindlin, R.D.: Influence of rotatory inertia and shear on flexural motions of isotropic, elastic plates. J. Appl. Mech. **18**(1), 31–38 (1951)
12. Castellazzi, G., Krysl, P., Bartoli, I.: A displacement-based finite element formulation for the analysis of laminated composite plates. Compos. Struct. **95**, 518–527 (2013)
13. Cui, X.Y., Liu, G.R., Li, G.Y.: Bending and vibration responses of laminated composite plates using an edge-based smoothing technique. Eng. Anal. Bound. Elem. **35**(6), 818–826 (2011)
14. Ferreira, A.J.M.: A formulation of the multiquadric radial basis function method for the analysis of laminated composite plates. Compos. Struct. **59**(3), 385–392 (2003)
15. Ferreira, A.J.M., Castro, L., Bertoluzza, S.: A high order collocation method for the static and vibration analysis of composite plates using a first-order theory. Compos. Struct. **89**(3), 424–432 (2009)
16. Ferreira, A.J.M., Fasshauer, G.E.: Analysis of natural frequencies of composite plates by an RBF-pseudospectral method. Compos. Struct. **79**(2), 202–210 (2007)
17. Bouazza, M., Tounsi, A., Adda-Bedia, E.A., Megueni, A.: Thermoelastic stability analysis of functionally graded plates: an analytical approach. Comput. Mat. Sci. **49**, 865–870 (2010). Elsevier
18. Bouazza, M., Tounsi, A., Adda-Bedia, E.A., Megueni, A.: Stability analysis of functionally graded plates subject to thermal loads. In: Altenbach, H., Eremeyev, V.A. (eds.) Shell-like Structures-Advanced Structured Materials, vol. 15, pp. 669–680. Springer, Heidelberg (2011)
19. Levinson, M.: An accurate simple theory of statics and dynamics of elastic plates. Mech. Res. Commun. **7**(6), 343–350 (1980)
20. Lo, K.H., Chiristensen, R.M., Wu, E.M.: A higher-order theory of plate deformation, part2: laminated plates. J. Appl. Mech. **44**, 669–676 (1977)
21. Reddy, J.N.: A simple higher order shear deformation theory for laminated composite plates. J. Appl. Mech. **51**, 745–752 (1984)
22. Pandya, B.N., Kant, T.: Higher-order shear deformable theories for flexure of sandwich plates Finite element evaluations. Int. J. Solids Struct. **24**(12), 1267–1286 (1988)
23. Touratier, M.: An efficient standard plate theory. Int. J. Eng. Sci. **29**(8), 745–752 (1991)
24. Soldatos, K.P.: A transverse shear deformation theory for homogenous monoclinic plates. Acta Mech. **94**, 195–220 (1992)
25. Karama, M., Afaq, K.S., Mistou, S.: A new theory for laminated composite plates. Proc. Inst. Mech. Eng. PartL: J. Mat. Des. Appl. **223**, 53–62 (2009)
26. Aydogdu, M.: A new shear deformation theory for laminated composite plates. Compos. Struct. **89**, 94–101 (2009)
27. Bouazza, M., Benseddiq, N.: Analytical modeling for the thermoelastic buckling behavior of functionally graded rectangular plates using hyperbolic shear deformation theory under thermal loadings. Multidiscipline Model. Mat. Struct. **11**(4), 558–578 (2015). doi:10.1108/MMMS-02-2015-0008

28. Grover, N., Singh, B.N., Maiti, D.K.: New non polynomial shear-deformation theories for structural behaviour of laminated composite and sandwich plate. AIAA J. **51**(8), 1861–1871 (2013)
29. Bouazza, M., Lairedj, A., Benseddiq, N., Khalki, S.: A refined hyperbolic shear deformation theory for thermal buckling analysis of cross-ply laminated plates. Mech. Res. Commun. **73**, 117–126 (2016)
30. Tounsi, A., Bouazza, M., Meftah, S.A., Adda-bedia, E.: On the transient hygroscopic stresses in polymer matrix laminated composites plates with cyclic and unsymmetric environmental conditions. Polym. Polym. Compos. **13**(5), 489–504 (2005)
31. Tounsi, A., Bouazza, M., Adda-bedia, E.: Computation of transient hygroscopic stresses in unidirectional laminated composite plates with cyclic and unsymmetric environmental conditions. Int. J. Mech. Mater. Des. **1**, 271–286 (2004)
32. Bouazza, M., Tounsi, A., Benzair, A., Adda-bedia, E.A.: Effect of transverse cracking on stiffness reduction of hygrothermal aged cross-ply laminates. Mater. Des. **28**, 1116–1123 (2007)
33. Adda-bedia, E.A., Bouazza, M., Tounsi, A., Benzair, A., Maachouc, M.: Prediction of stiffness degradation in hygrothermal aged [θm/90n]S composite laminates with transverse cracking. J. Mat. Process. Technol. **199**, 199–205 (2008)
34. Boucheta, A., Bouazza, M., Becheri, T., Benseddiq, N.: Hyperbolic four variable refined shear deformation theory for mechanical buckling analysis of functionally graded plates. U. P.B. Sci. Bull. Ser. D **78**(4), 65–78 (2016)
35. Becheri, T., Amara, K., Bouazza, M., Benseddiq, N.: Buckling of symmetrically laminated plates using nth-order shear deformation theory with curvature effects. Steel Compos. Struct. **21**(6), 1347–1368 (2016)
36. Mantari, J.L., Oktem, A.S., Soares, C.G.: A new trigonometric shear deformation theory for isotropic, laminated composite and sandwich plates. Int. J. Solids Struct. **49**, 43–53 (2012)
37. Carrera, E.: Theories and finite elements for multilayered, anisotropic, composite plates and shells. Arch. Comput. Methods Eng. **9**(2), 87–140 (2002)
38. Civalek, Ö.: An efficient method for free vibration analysis of rotating truncated conical shells. Int. J. Press. Vessels Piping **83**, 1–12 (2006)
39. Civalek, Ö.: Free vibration analysis of composite conical shells using the discrete singular convolution algorithm. Steel Comp. Struct. **6**(4), 353–366 (2006)
40. Civalek, Ö.: Nonlinear analysis of thin rectangular plates on Winkler-Pasternak elastic foundations by DSC-HDQ methods. Appl. Math. Model. **31**, 606–624 (2007)
41. Civalek, Ö.: Three-dimensional vibration, buckling and bending analyses of thick rectangular plates based on discrete singular convolution method. Int. J. Mech. Sci. **49**, 752–765 (2007)
42. Civalek, Ö.: A parametric study of the free vibration analysis of rotating laminated cylindrical shells using the method of discrete singular convolution. Thin-Walled Struct. **45**, 692–698 (2007)
43. Akgöz, B., Civalek, Ö.: Longitudinal vibration analysis for microbars based on strain gradient elasticity theory. J. Vib. Control (2014). doi:10.1177/1077546312463752
44. Civalek, Ö.: Geometrically nonlinear dynamic analysis of doubly curved isotropic shells resting on elastic foundation by a combination of HDQ-FD methods. Int. J. Pres. Ves. Pip. **82**(6), 470–479 (2005)
45. Akgöz, B., Civalek, Ö.: Strain gradient elasticity and modified couple stress models for buckling analysis of axially loaded micro-scaled beams. Int. J. Eng. Sci. **49**(11), 1268–1280 (2011)

46. Civalek, Ö.: Harmonic differential quadrature-finite differences coupled approaches for geometrically nonlinear static and dynamic analysis of rectangular plates on elastic foundation. J. Sound Vib. **294**, 966–980 (2006)
47. Sofiyev, A.H.: The buckling of FGM truncated conical shells subjected to combined axial tension and hydrostatic pressure. Compos. Struct. **92**, 488–498 (2010)
48. Sofiyev, A.H.: Thermal buckling of FGM shells resting on a two-parameter elastic foundation. Thin-Walled Struct. **49**, 1304–1311 (2011)
49. Sofiyev, A.H.: The buckling of FGM truncated conical shells subjected to axial compressive load and resting on Winkler-Pasternak foundations. Int. J. Pres. Ves. Pip. **87**, 753–761 (2010)
50. Sofiyev, A.H.: Buckling analysis of FGM circular shells under combined loads and resting on the Pasternak type elastic foundation. Mech. Res. Commun. **37**, 539–544 (2010)
51. Sofiyev, A.H., Alizada, A.N., Akin, O., Valiyev, A., Avcar, M., Adiguzel, S.: On the stability of FGM shells subjected to combined loads with different edge conditions and resting on elastic foundations. Acta Mech. **223**, 189–204 (2012)
52. Sofiyev, A.H.: Winkler-Pasternak elastic foundations under combined axial compression and external pressure. Compos. Struct. **113**, 208–215 (2014)
53. Buczkowski, R., Torbacki, W.: Finite element modeling of thick plates on two-parameter elastic foundation. Int. J. Numer. Anal. Methods Geomech. **25**, 1409–1427 (2001)
54. Timoshenko, S.P., Woinowsky-Krieger, W.: Theory of Plates and Shells. McGraw-Hill, New-York (1970)
55. Lam, K.Y., Wang, C.M., He, X.Q.: Canonical exact solution for Levy plates on two parameter foundation using Green's functions. Eng. Struct. **22**, 364–378 (2000)
56. Zenkour, A.M., Allam, M.N.M., Shaker, M.O., Radwan, A.F.: On the simple and mixed first-order theories for plates resting on elastic foundations. Acta Mech. **220**, 33–46 (2011)

3D Numerical Simulation of the Goaf Due to Large-Scale Longwall Mining

Samar S. Ahmed[1,3](✉), Marwan AlHeib[2], Yann Gunzburger[1], and Vincent Renaud[2]

[1] GeoRessources, Ecole des Mines de Nancy, Université de Lorraine, Nancy 54042, France
samar.ahmed@univ-lorraine.fr
[2] INERIS, Ecole des Mines de Nancy, Nancy 54042, France
[3] Mining Department, Faculty of Engineering, Cairo University, Cairo, Egypt

Abstract. Due to longwall excavations, the upper strata disturb, the roof and the floor of the opening become in contact. This disturbed area is commonly known as the "goaf area". The challenge of simulating numerically the goaf area is to identify its geometry and its equivalent mechanical properties. The main objective of this study is to improve the 3D numerical simulation of longwall mining and its accompanying goaf area, which will permit us to observe the stress changes due to longwall excavations. The Provence coal mine in the South of France has been chosen to be the case study of the current research, where the mined coal seam has 2.5 m thickness and the average depth of the mine is 1000 m. The longwall panels have a regular width of 200 m and various lengths from 400 up to 1400 m. A large-scale finite difference numerical model of the mine was constructed by using FLAC^{3D}. The numerical modeling of the goaf area was performed in two steps. The first step was to identify the geometry of the goaf area above longwall panels. The second step was to calibrate its equivalent mechanical properties with the total convergence between the roof and the floor. The goaf geometry and the mechanical properties were then calibrated with the in-situ surface subsidence. The results show that applying a linearly varying elastic modulus within the goaf area is a very effective method to express its heterogeneity, which also gave rational surface subsidence values for panels width less than 1000 m. In addition, in terms of stress redistribution, the induced vertical stress increases progressively with panel width, and it becomes close to the initial values at the center of the goaf for panel width larger than 1000 m.

1 Introduction

Longwall mining method, which involves removal of large rectangular panels, is widely used especially in coal mines. Due to excavation, the roof and floor of panel become in contact and high subsidence values appear at the earth's surface. Peng and Chiang (1984) considered that after the disturbance of longwall caving excavation, the above volume will divide into three different zones as shown in Fig. 1. The caved zone is the most influenced zone where the roof totally collapses into the floor of the excavation. Kenny (1969) performed in-situ measurements in order to determine the

H. Shehata and Y. Rashed (eds.), *Numerical Analysis of Nonlinear Coupled Problems*, Sustainable Civil Infrastructures, DOI 10.1007/978-3-319-61905-7_11

height of the caved zone. He found that, for a coal seam with height (*t*), the caved zone height (h_c) ranges between 2*t* and 4*t*. But, Hasenfus *et al.* (1998) found that h_c ranges between 4*t* and 6*t*. Recently, Shabanimashcool (2012) proposed that $h_c = 4t$ based on numerical modeling results.

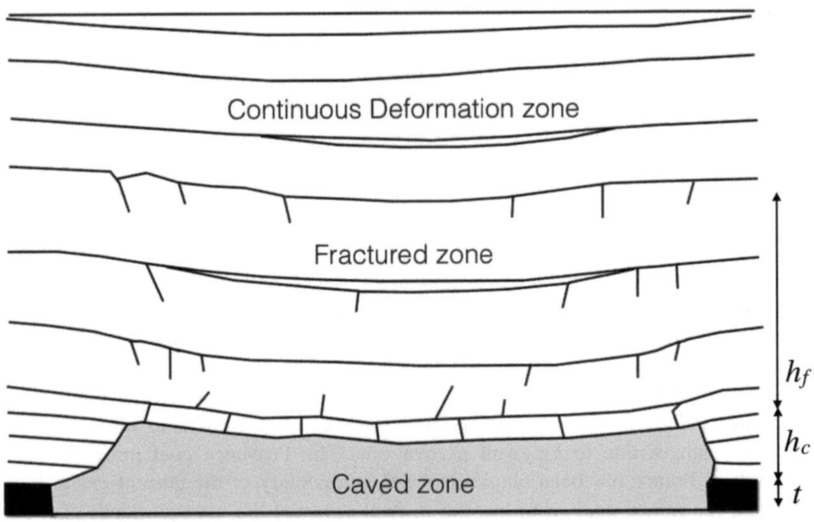

Fig. 1. Three zones of disturbance due to longwall caving mining method (Peng and Chiang 1984).

The second influenced zone is the fractured zone, which lies directly above the caved zoned, where the strata are broken into blocks associated with major horizontal and vertical crakes and bed separation. Peng and Chiang (1984) proposed that the fractured zone height (h_f) ranges between 28*t* and 42*t*. The last zone is the continuous deformation zone, where the rockmass behaves essentially as a continuous medium.

The principal issue of simulating numerically such damaged zones (caved zone + fractured zone) is the assessment of its equivalent mechanical properties. Finding an equivalent mechanical properties able to express the heterogeneity of those materials is very difficult because of the inaccessibility to this damaged area in the mine. Various authors have already tackled the issue of assessing the equivalent mechanical properties of the goaf area. For example, Kose and Cebi (1988) proposed a wide interval of elastic modulus within the goaf area. He suggested that the elastic modulus within the goaf area ranges between 15 MPa and 3500 MPa. Tajdus (2009) applied a back analysis method for determining the values of rockmass parameters for many areas disturbed by mining. He found that the elastic modulus in horizontal direction and in vertical direction are very low and range between 50 MPa and 150 MPa. Cheng *et al.* (2010) and Jiang *et al.* (2012) assumed that the values of elastic modulus and Poisson ratio in the goaf area are equal to 190 MPa and 0.25, respectively Salamon (1990) defined the stress strain relationship of the goaf material as:

$$\sigma = \frac{E_0 \varepsilon}{1 - (\varepsilon/\varepsilon_m)} \quad (1)$$

where, ε and σ are the vertical strain and stress respectively and E_O is the initial elastic modulus of the goaf material. ε_m is given by Eq. (2) using the buckling factor BF:

$$\varepsilon_m = \frac{BF - 1}{BF} \quad (2)$$

E_O (MPa) can be calculated as a function of the unconfined compressive strength of the intact rock, σ_c, and the buckling factor, (Pappas and Mark 1993) and (Yavuz 2004):

$$E_0 = \frac{10.39\sigma_c^{1.042}}{BF^{7.7}} \quad (3)$$

Salamon's model is valid for cave-in materials under hardening condition with non-elastic behavior. E_O and ε_m values must be determined firstly, then the hardening table could be estimated by using Eq. (1).

Wilson (1980) suggested that, after consolidation of the goaf, the vertical stress within the goaf increases linearly from zero at the ribside to the pre-mining vertical stress at a distance from the ribside equal to (0.3 – 0.4) times H, where H is the mining depth. Wilson (1982) suggested that the peak vertical stress on the ribside (the "abutment pressure") might be as high as six times the initial one. The generally accepted stress re-distribution developed by (Wilson 1982) is as shown in Fig. 2. However, Wilson proposed a 2D estimation and he did not consider the effect of the

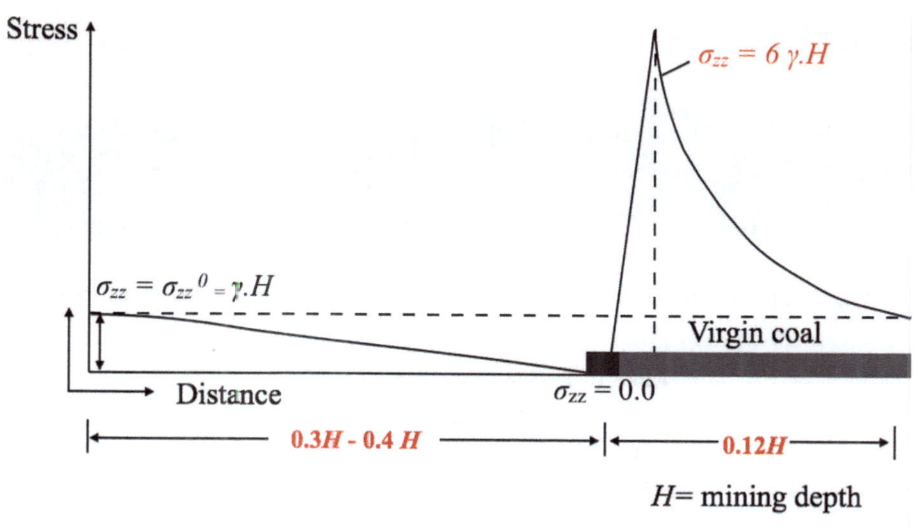

Fig. 2. Vertical stress distribution within the goaf and the ribside (Wilson 1982).

third dimension, which may play an important role. Also, he did not refer to the goaf mechanical properties and its effect on the vertical stress redistribution.

The aim objective of the current research is to simulate numerically the goaf area above excavated panels by using the elastic mechanical model. Also, we aim to assess the stress redistribution due to large-scale longwall mining. In the current research, the geometry of the goaf area was determined based on the previous studies. We assumed that goaf area behaves elastically, where the elastic modulus various linearly with the goaf height. Geometry and mechanical properties of the goaf area were calibrated with the convergence between the roof and the floor as well as the in-situ measurements of surface subsidence. The case study of our research is presented in the next section.

2 Case Study

The current case study is the Provence coalmine. It is located in the South of France. It had been exploited between 1984 and 2004 using the longwall mining method. The excavated panels have regular width of 200 m with various lengths. The width of the exploited panels reached up to 1400 m (7 panels). The average thickness of the exploited coal seam is $t = 2.5$ m, which lies at average depth of 1000 m. The overburden composes of Fuvelian limestone and Begudian-Rognacian limestone and marl as shown in Fig. 3. The stiffness of the Rognacian layer is low compared with the adjacent Fuvelian layer because it contains a high percentage of marl and clayey limestone, (Gaviglio 1985). The mechanical properties of the different layer within the rockmass are given in Table 1 as reported by Gaviglio *et al.* The estimation of the mechanical properties takes into account the rockmass quality and the mechanical characterizations from laboratory tests.

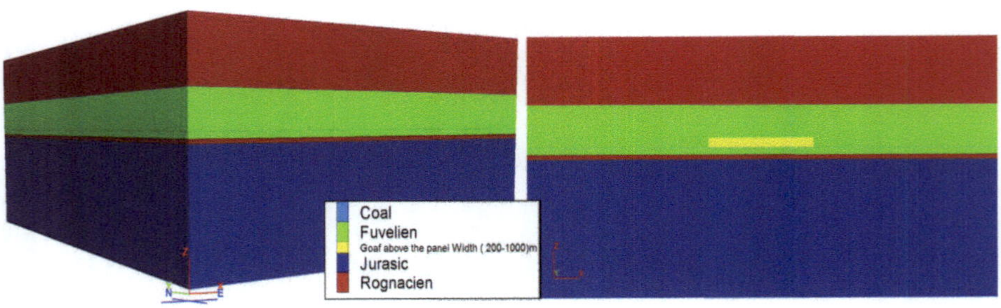

Fig. 3. 3D and 2D view of the model showing the mining panel and the goaf area.

Table 1. Rockmass mechanical properties.

Rock type	ρ(kg/m^3)	E (GPa)	v
Begudian-Rognacian	2400	1	0.25
Fuvelian	2400	8.4	0.24
Lignite coal	1500	3	0.32
Jurassic	2400	17	0.25

3 Numerical Model

A 3D numerical model of the mine was constructed using the finite difference code FLAC^{3D} (Fig. 3). Four different rock types are specified: the coal seam (2.5 m thickness), Fuvelian formation (400 m thickness) and Rognacian formation (600 m thickness) above the coal seam, and Jurassic limestone formation below the coal seam. The overall dimensions of the model are 4600 m in the x-direction, 6020 m in the y-direction and 2270 m in the vertical direction (z). The top of the model coincides with the ground surface at level z = 0.0. The excavated panels lie at depth (*H*) of 1000 m from the earth's surface. This model simulates the excavation of 7 parallel longwall panels. Each panel has 200 width and maximum length of 1400 m. This model contains 2.5 million mesh with a very high precision near to the excavated zone in order to overcome the mesh effect. The mechanical properties in Table 1 were used as input data in the numerical model. The initial vertical stress is assumed to equal to the overburden weight (ρ = 2400 kg/m^3, g = 9.81 m/s^2, H = 1000 m).

4 Goaf Simulation Methodology

4.1 Goaf Geometry

In the current study, the goaf simulation composes of two different steps. The first step is to estimate the goaf geometry (caved zone height and fractured zone height). From the previous studies, we found that the height of the caved zone equals to 4*t* (i.e., $h_{\text{caved-zone}} = 4t$) according to Shabanimashcool *et al.* (2012) and the height of the fractured zone equals to 28*t* (i.e., $h_{\text{fractured-zone}} = 28t$) according to Peng and Chaing (1984). For that, the total height of the goaf area was assumed to be equal to 32*t* (i.e., $h_{\text{goaf}} = 32t$), where *t* is the coal seam thickness = 2.5 m. The second step of simulating the goaf, which is the main concern of the current study, is to estimate numerically its equivalent mechanical properties.

4.2 Goaf Mechanical Properties

Two principal criteria have to be fulfilled while simulating the longwall exploitation. The first criterion is the roof-floor convergence of panel. The second criterion is the surface subsidence, which is discussed later on. To satisfy the first criterion, in the current numerical model, we assumed that the elastic modulus (*E*) of the goaf area varies linearly with the goaf height (32*t*) as shown in Fig. 4. It begins from a certain value $E_{immediate\text{-}roof}$ directly above the opening and increases linearly within the goaf up to $E_{hostrock}$ at height of 32*t*. Equation (4) was fitted to estimate E_{goaf} at any point within the goaf, by assuming that the Poisson ratio is $\upsilon_{goaf} = \upsilon_{hostrock}$ and the immediate roof above the excavation has $E_{immediate\text{-}roof}$. The only value that could be changed in this model is the $E_{immediate\text{-}roof}$.

$$E_{goaf(hg.t)} = \left(\frac{E_{hostrock} - E_{immediate-roof}}{x.t}.h_g.t\right) + E_{immediateroof} \quad (MPa) \qquad (4)$$

Where ($x.t$ = 32*2.5 = 80 m) is the maximum height of the goaf that corresponds $E_{goaf(32.t)}$, ($h_g.t$) is the height corresponds to $E_{goaf(hg.t)}$, h_g ranges between (0–32) and t is the coal seam thickness.

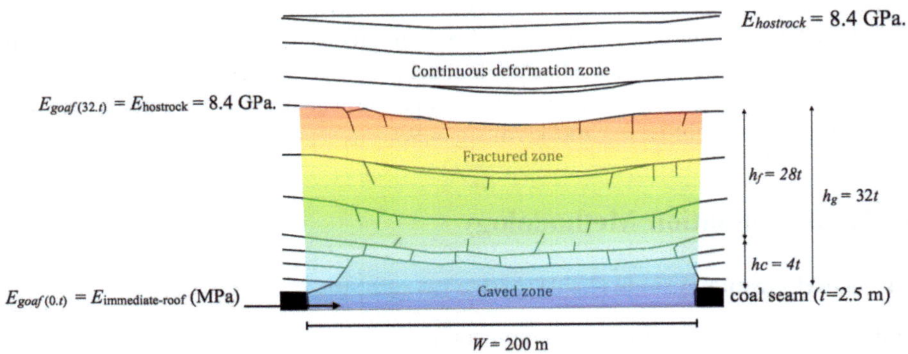

Fig. 4. Linear variation of elastic modulus within the goaf area.

The value of $E_{immediate\text{-}roof}$ was calibrated with the total convergence between the roof and floor of one isolated panel (200 m width and various lengths). Many iterations were performed by using different values of $E_{immediate\text{-}roof}$ as shown in Fig. 5. In order to satisfy the roof-floor convergence (i.e., *convergence = mining seam thickness (t)*) of one panel of 200 m width, we found that the $E_{immediate\text{-}roof}$ should not exceed 180 MPa (i.e., $E_{immediate\text{-}roof} = 0.021\ E_{hostrock}$) as shown in Fig. 5. We found also that the third dimension of a panel (panel length *L*) has a little influence in the convergence values (see Fig. 5). For that, Eq. (4) will be modified to be as follow:

$$E_{goaf(hg.t)} = \left(\frac{\mathrm{E_{hostrock}} - 0.021\ \mathrm{E_{hostrock}}}{\mathrm{x.t}}.\mathrm{h_g.t}\right) + 0.021\,E_{hostrock} \quad (\mathrm{MPa}) \qquad (5)$$

As mentioned above, the Provence coal mine contains multi-panels up to 7. The excavation progress was performed in the numerical model in one step. The sequence of panels exploitation is as shown in Fig. 6. The elastic modulus was changed in the goaf area after each excavation step to obey Eq. (5). Figure 6 shows the variation of the elastic modulus within the goaf area above the excavated panels from step 1 (excavation of one panel 200 m width (W/H = 0.2)) until step 7 (excavation of multi-panels 1400 m width (W/H = 1.4)). In order to calibrate the geometry and the mechanical properties of the goaf area in the case of multi-panels excavation (W/H > 0.2), the numerical subsidence values were calibrated with the subsidence in-situ measurements.

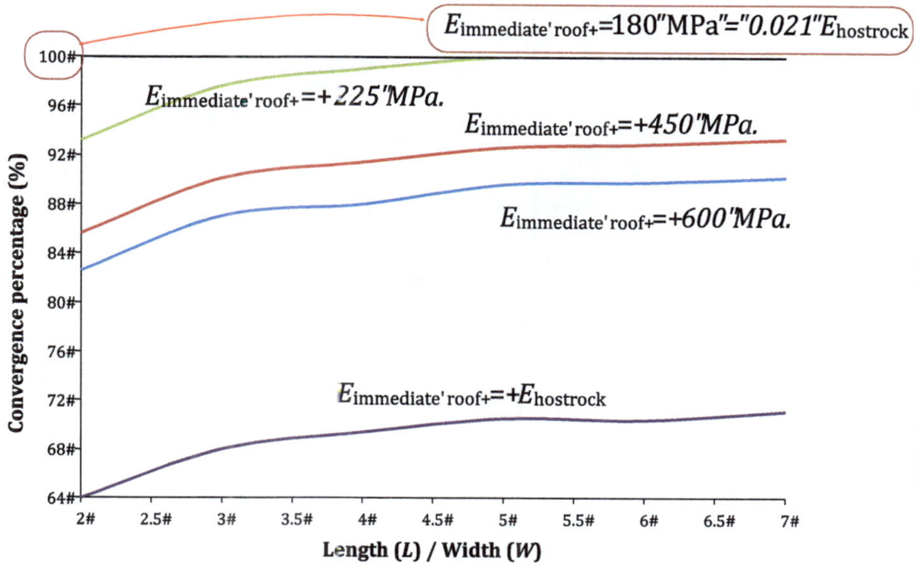

Fig. 5. Total convergence between the roof and floor for different panel lengths.

To calibrated the numerical model with the given in-situ values, a general flowsheet of the modeling process was generated and followed as shown in Fig. 7.

5 Results and Discussion

After fulfilling the roof-floor convergence criterion, in case of one-isolated panel, the subsidence values were determined from the numerical model to be then calibrated with the in-situ measurements. The linearly varying elastic modulus reported in Eq. (5) was implemented in the goaf area after each panel excavation step. We found that the surface subsidence increased with panel width to mining depth (W/H) ratio. The numerical subsidence values lie between the maximum and the minimum in-situ subsidence curves (for $W/H < 1$) as shown in Fig. 8a. However, for panels with high width to mining depth ratio ($W/H \geq 1$), the surface subsidence values lie above the maximum in-situ subsidence curve (see Fig. 8a).

In order to fulfill the surface subsidence for large panels (panel width to mining depth larger than 1 ($W/H \geq 1$)), the goaf geometry was decreased, which would consequently decrease the surface subsidence values to be compatible with the in-situ subsidence curves. New goaf height (h_g) values were implemented in the numerical model. These new values were calibrated with the surface subsidence. We found that the height of the goaf (h_g) decreased with mining width (W). In another word, the influenced zone due to excavation become smaller with mining advance, which could be the effect of the consolidation of fragmented rocks. We found that, to fulfill the surface subsidence criterion for large panels as shown Fig. 8b, the goaf should be simulated with height equals to $30t$ for ($W/H = 1$), $28t$ for ($W/H = 1.2$) and $26t$ for

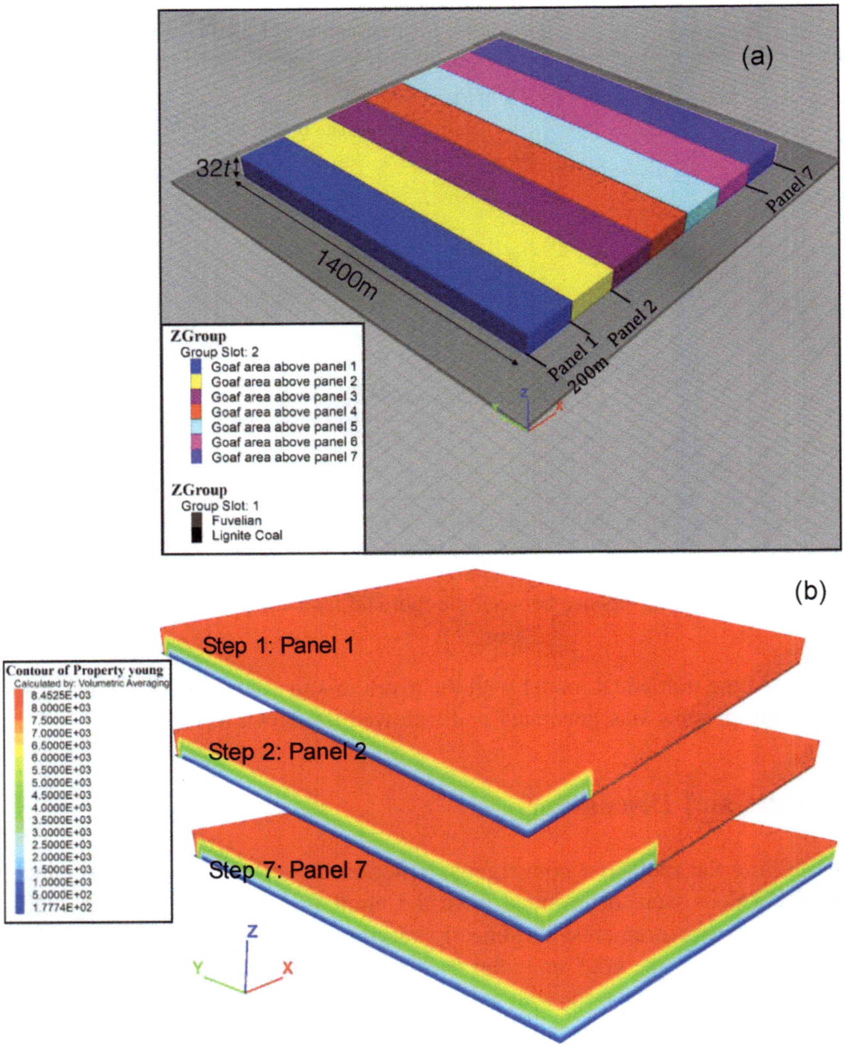

Fig. 6. Excavation sequence of the coal panels and elastic modulus variation within the goaf area (the elastic modulus values are in MPa).

(W/H = 1.4). The x factor in Eq. (6) could be classified in function of panel width to mining depth as follows:

$$E_{goaf(hg.t)} = \left(\frac{E_{hostrock} - 0.021 E_{hostrock}}{x.t} .h_g.t \right) + 0.021\, E_{hostrock} \quad \text{(MPa)} \tag{6}$$

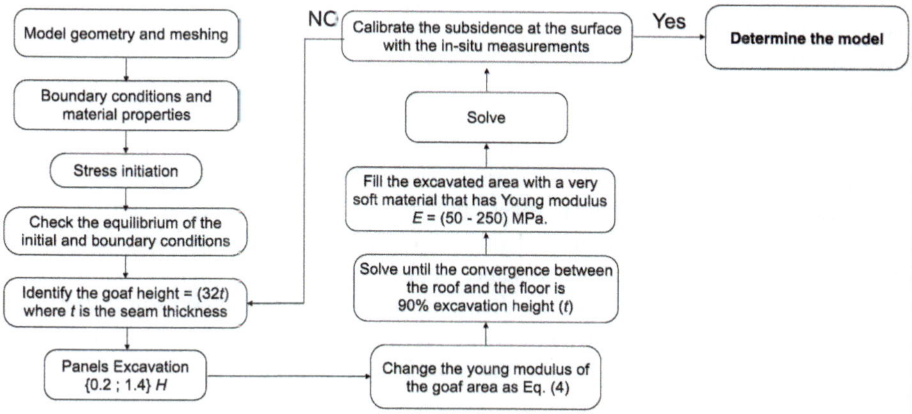

Fig. 7. General flowsheet of modeling process.

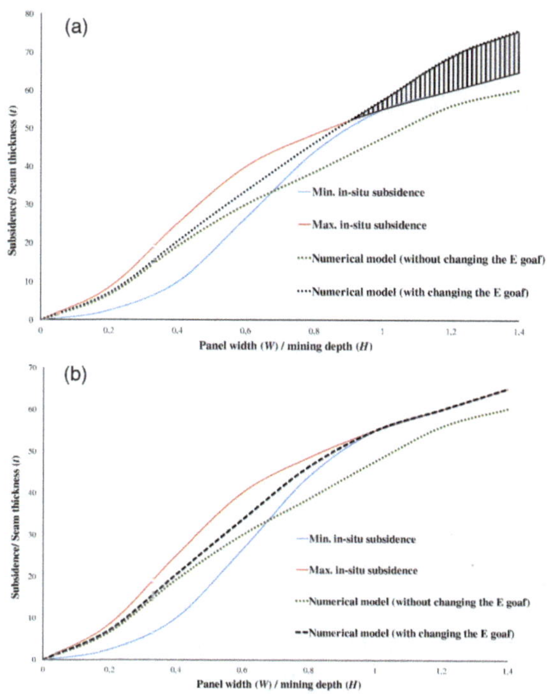

Fig. 8. Numerical and in-situ measurement of the subsidence-seam thickness ratio for various panel width-mining depth ratio (a) goaf height 32*t* (b) goaf height vary with the panel width.

(i) $W/H = 0.2\text{–}0.8$, $x = 32, h_g = (0\text{–}32)$
(ii) $W/H = 1$, $x = 30, h_g = (0\text{–}30)$
(iii) $W/H = 1.2$, $x = 28, h_g = (0\text{–}28)$
(iv) $W/H = 1.4$, $x = 26, h_g = (0\text{–}26)$

After simulating the goaf geometry and its equivalent mechanical properties, the next step is to obtain the induced vertical stress due to longwall mining. Figure 9 illustrates the normalized vertical stress (induced vertical stress/initial vertical stress) for different panel width to mining depth ratios. We noticed that the induced vertical stress reached its initial value at distance ranges between 0.6*H* and 0.7*H* from the mining face. These values are higher than the values proposed by Wilson (1982) who found from his 2D model that the induced stress reached its initial value at distance ranges between 0.3*H* and 0.4*H* from the mining face (Fig. 2).

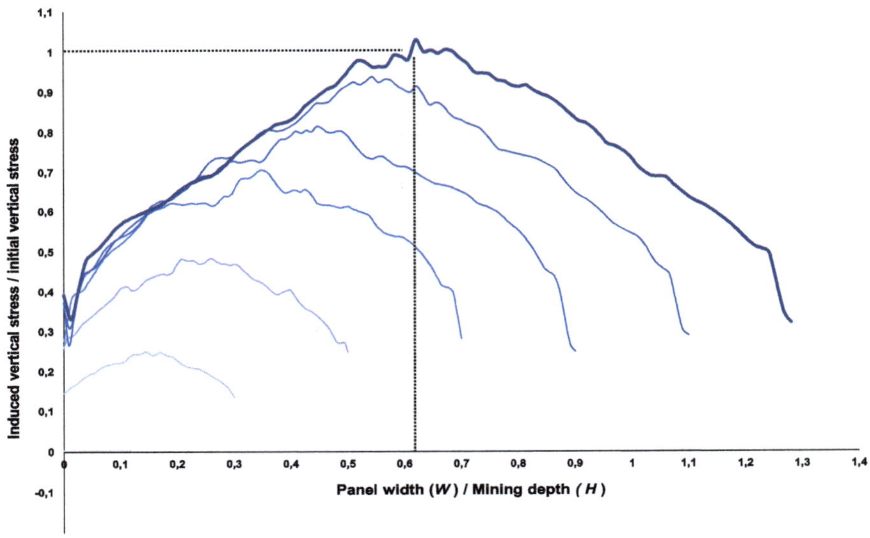

Fig. 9. Ratio between induced vertical stress and initial vertical stress.

6 Conclusion

This research concerns the 3D numerical simulation of the goaf area associated with longwall caving panels. Two principal issues are discussed in this paper. The first problem is how we can determine the goaf geometry. The second problem is how we can determine the equivalent mechanical properties of this part of mine (the goaf area) that should satisfy two principal criteria (the roof-floor convergence and the surface subsidence).

The height of the goaf area is found to be equal to 32*t*, where *t* is the coal seam. The proposed model with linearly varying elastic modulus within the goaf area from $E_{immediate\text{-}roof}$ (directly above the excavation), up to $E_{hostrock}$ (at the end of the goaf influence) is able to express the heterogeneity within this part of mine. To fulfill the total roof-and floor convergence criterion, the $E_{immediate\text{-}roof}$ should not exceed 180 MPa (i.e., $E_{immediate\text{-}roof} = 0.021\ E_{hostrock}$). To fulfill the surface subsidence criterion, we found that the goaf height decreases with panel width, which could be due to consolidation behavior of the fragmented rocks. And, the vertical stress at the center of the excavated zone increases with panel width. The vertical stress reaches its initial value at panel width to mining depth ratio equals to 1.4 ($W/H = 1.4$).

References

Peng, S.S., Chaing, H.S.: Longwall Mining, pp. 17–73. John Wiley & Sons Inc., New York (1984)

Kenny, P.: The caving of the waste on longwall faces. Int. J. Rock Mech. Mining Sci. Abstr. **6**, 541–555 (1969)

Hasenfus, G.J., et al.: A hydrogeomechanical study of overburden aquifer response to longwall mining. In: Peng Syd, S. (ed.) Proceedings of the 7th International Conference on Ground Control in Mining, pp. 149–162. West Virginia University, COMER, Department of Mining Engineering, Morgantown (1993)

Shabanimashcool, M., Charlie, C.L.: Numerical modelling of longwall mining and stability analysis of the gates in a coal mine. Int. J. Rock Mech. Mining Sci. **51**, 24–34 (2012)

Kose, H., Cebi, Y.: Investigation the stresses forming during production of the thick coal seam. In: 6th Coal Congress of Turkey, Zonguldak, pp. 371–383 (1988)

Tajdus, K.: New method for determining the elastic parameters of rock mass layers in the region of underground mining influence. Int. J. Rock Mech. Mining Sci. **46**, 1296–1305 (2009)

Cheng, Y.M., et al.: Three-dimensional analysis of coal barrier pillars in tailgate area adjacent to the fully mechanized top caving mining face. Int. J. Rock Mech. Mining Sci. **47**, 1372–1383 (2010)

Jiang, Y., et al.: Assessment of mitigation of coal bump risk during extraction of an island longwall panel. Int. J. Coal Geol. **95**, 20–33 (2012)

Salamon, M.D.G.: Mechanism of caving in longwall mining. In: Proceedings of the 31st US Rock Mechanical Symposium, Golden, Colorado, pp. 161–168. Balkema, Rotterdam (1990)

Wilson A.H.: Pillar stability in longwall mining. In: State of the Art of Ground Control in Longwall Mining and Mining Science, pp. 85–95. SME, New York (1982)

Wilson, A.H.: The stability of underground workings in the soft rocks of coal measures. Unpublished Ph.D Thesis, Univ. Nottingham (1980)

Yavuz, H.: An estimation method for cover pressure re-establishment distance and pressure distribution in the goaf of longwall coal mines. Int. J. Rock Mech. Mining Sci. **41**, 193–205 (2004)

Gaviglio, P.: La deformation cassante dans les calcaires Fuvéliens du basin de l'arc (Provence). PhD thesis. University of Provence, Marseille, France (1985)

Pappas, D.M., Mark, C.: Behaviour of simulated longwall gob material. Report of investigation no. 9458, US Department of the Interior, Bureau of Mines (1993)

A Suggested Model Using Quantitive and Qualitative Parameters for Cost Engineering of Mechanically Stabilised Earth Walls in Egypt

Joseph Meadows[1(✉)] and John Erian[2]

[1] Maccaferri Africa, New Germany, South Africa
[2] Strata Soil Systems, Cairo, Egypt

Abstract. To investigate current trends, existing international options and proposed new Technology in the Egyptian market, a model is suggested to define a method to assess by quantitative and qualitative methods, a cost Engineering approach to analyse suitable solutions. The numerous options of Mechanically Stabilised Earth Wall (MSEW) Technologies available internationally creates a difficult exercise to assess and establish accurate comparisons and evaluation criteria to arrive at a suitable choice of Technology, to be embraced in a developing market such as Egypt post 2011. The investigation into historically advanced and new MSEW technology begs the need to simplify the assessment process with a model, to clarify, best practice and economic criteria, in a method that demonstrates the least bias. A method has been proposed that allows Technology leaders and enterprising new comers to contribute to the understanding by Professionals and the layman on the merits of MSEW Technology, to maintain and promote the body of knowledge and state of the art of MSEW technology. The specific examples of geogrid soil reinforcement, concrete cladding panels, concrete block and steel mesh baskets used in MSEW projects in Egypt has been used from current contracts to demonstrate the assessment criteria and proposed model.

Keywords: Mechanically stabilised earth walls · Geogrid · Concrete block walls · Concrete panels · Cost engineering · Qualitative analysis · Quantitative analysis

1 Introduction

Egypt, as in many global areas that are in the process of rapid recent development, (UAE, Angola, West and East Africa, South Africa), all face a question to effectively assess, improvements and new technology, for the benefit of new construction and solutions to engineering projects. Egypt has the advantage to tap into proven systems that have been successful in the middle east regions that have similar conditions and challenges. Additionally, it can source the latest technology from Europe as a neighbour but also the economic benefits of the production advancements from the Far East. The crux of an assessment is to assess which system or solutions are appropriate and

H. Shehata and Y. Rashed (eds.), *Numerical Analysis of Nonlinear Coupled Problems*,
Sustainable Civil Infrastructures, DOI 10.1007/978-3-319-61905-7_12

which provides better guarantees of performance in the context of an efficient economic comparison. Several MSEW projects have been executed in Egypt, using a range of different systems in the past 5 years. Some structures stand as monuments to excellence while others are showing wear and tear or ageing and may soon need repair. Most of the Egyptian projects have small interference from rain and consequential damage, however Projects close to the coast would need to consider the effects of drainage and the need to design for the management thereof. Accordingly, deterioration of MSEW structures are expected to arise mainly from choosing low durable materials, poor consideration of foundation materials and low compliance with construction criteria.

The method and process: to list, analyse and compare products and solutions to arrive at a best choice needs a structured approach. Strata Soil Systems faced this question when it investigated the MSEW options available and concluded by engaging with the Italian company Maccaferri to become its Agent in Egypt.

2 Current Trends of MSEW Available in the Egyptian Market

1. Masonry (rip rap) walls- used extensively on slopes and for retaining walls
2. Block retaining walls- used on many projects particularly for road approach ramps
3. Wrap around slopes- Geotextiles/Geosynthetics used as soil reinforcement
4. Concrete panels in mechanically stabilised earth walls(MSEW)- introduced initially to the road sector
5. Eco friendly vegetated solutions- bio degradable materials to bind the soil on steep slopes
6. Reinforced concrete- (reinforced or unreinforced) traditional technique for walls and slopes
7. Wire mesh retaining systems- used in combination with geotextiles and geosynthetics
8. Shotcrete and nails- traditional technique using mechanically advanced application methods
9. Anchoring systems- traditional techniques reliant on rock mechanics designed by Engineering Geologists or Geotechnical Engineers
10. Sheet piling- traditional technique for shoring, lateral support to retain ground and soil.

3 Mechanically Stabilised Earth Wall (MSEW) Options Introduced by International Companies

1 Freyssinet- Most commonly known as Reinforced Earth operating in many countries using mainly steel reinforcement but also alternative Geosynthetics
2 Tensar- well established soil reinforcement solutions and products
3 Tencate- market lead in production of geosynthetic materials across a broad range of applications that compliments different facing designs for retaining walls

4 Keystone- originally an American retaining wall system well established to cover solutions using geosynthetic and steel ladder soil reinforcement
5 Canadian system- Lock and Load- a block wall system covering GRS, Gravity, MSEW Anchor and soil nail walls
6 Maccaferri- traditional Gabion manufacturer with associated MSEW systems, soil nail and anchor solutions covering a wide range of solutions.

4 New Technology

New is a relative word depending on the date reference and the person to whom it applies. To be more accurate to stabilise earth, is not new technology (the Ziggurats in Iraq are testimony to this historical fact 3300 years ago,). Several techniques have been tried tested and failed or faded from use over the past 300 years. What is relatively new is the development of Geosynthetics which is derived due to an industry explosion of synthetic materials that have migrated to the engineering sector, which are used specifically in construction. Currently there are many products available from more than 100 factories to choose from, yet to be fair, many of this number are similar or close copies of each other.

There are also many steel products in a variety of designs generating a mechanical soil interaction, and high shear resistance. These are sometimes coupled with: concrete, mesh, grids, geocomposites and geotextile materials.

Which increases the combinations even further to suit a vast array of ideas, that can be conjured up by engineers to solve geotechnical and structural problems and challenges.

To keep abreast of the options and solutions available requires dedicated assessment and comparative criteria to enjoy the economic and engineering advantages it offers.

In this respect the document "Durability of Geosynthetics", by Greenwood, Schroeder and Voskamp has been a useful document as a guide in the assessment and understanding of using Geosynthetics in a variety of applications.

5 Current Known Representatives in the Egyptian Construction Sector with MSEW Solutions

1. GEOS
2. Tensar
3. Freysinnet
4. Huesker
5. Maccaferri
6. Lock and Load

The most common system observed structures in Egypt to stabilise slopes is Rip Rap- (Stone masonry walls).

Egypt has thousands of square metres of slopes and walls using stone masonry (Rip Rap) that have stood the test of time. The availability of strong durable rock has supported the use of this technique coupled with the high masonry skill set that has been maintained in the labour market. The public acceptance of this technique as a standard solution is high because the newer technologies have been used on mainly new construction projects involved with roads and bridges. The ongoing and continued availability of rock in specific areas seems to have given this technique longevity and a continuous workforce willing to learn the masonry trade will perpetuate this option and supply customers into the foreseeable future. What will further support and develop this technique is the design in combination with Geosynthetic soil reinforcement which will maintain its viability as a solution into the future.

6 Steps to Develop a Model to Assess MSEW Solutions in Egypt

1. **Qualitatively**
   a. Identify the specific geosynthetic product regarding its type of polymer, physical structure, its manufacture, mechanical and hydraulic properties and descriptive features (mass/unit area).
   b. Define its function and application to its working environment with an order of priority of operation and secondary benefits.
   c. Establish the design life and conditions in which it must cope with aggressive challenges and the risk to degradation or loss of integrity.
   d. Decide which controlling factors have priority to determine the reduction factors to be applied in the design parameters.
   e. Analyse the data to list the testing or verification required to confirm suitability of material for the specific task/application.
   f. Differentiate the assessment options and cross check compliance with prescribed guidelines for predicted service life to satisfy not only the clients performance/usage requirements but also best practice from authoritative literature sources.
   g. State the consequences of this predicted differentiation and the margin between the design life and the predicted end life whether it is acceptable.
   h. Decide on appropriate inspection dates as a suitable time to confirm the level of performance and how to check if any degradation has occurred and the method for replacement or compensatory remedy if this becomes necessary.
   i. The physical assessment and testing is a specialized process to be performed by recognised competent organizations. Recommendations are the basic feedback require to make sense from lab test results. Thus, the lab chosen should have a Professional available to explain the results clearly.
   j. Valid certification of the materials and proposed solution is necessary by an independent body to satisfy that due diligence has been performed (e.g. per ISO/EN norms) and for reference as a permanent record.

7 Quantitative Methods Used to Assess MSEW Solutions in Egypt

At any stage of the analysis the allocation of current values to the cost of materials, systems and production rates, will show trends and these can also be graphically presented to gauge cost effective comparisons that should follow international trends. When the trends deviate from expectation it requires and investigation for bias or conditions that need to be defined and explain what has caused the anomaly (general example of MSEW cost trends are show in Fig. 1. Anomalies can be that very advantage that is identified for prompting a decision between what seems an obvious choice versus the most cost effective choice after the qualitative assessment has narrowed the field down to two or maybe three options.

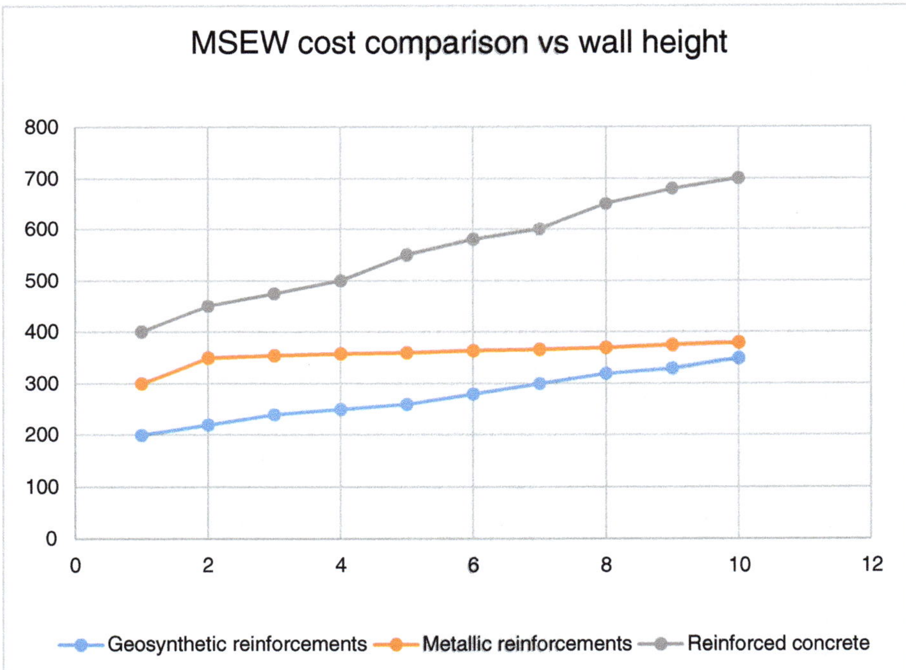

Fig. 1. General example of MSEW cost trends

8 Assessment Which Was Used to Identify Suitability of the Maccaferri Projects in Egypt

MacForce walls in Egypt (an MSEW system exclusive to Maccaferri using concrete panels and geosynthetic strips)

The set of Tables 1, 2, 3 and 4 in the Appendix show comparative information used in the assessment. The Project at the Regional Ring road is one of the most important

Table 1. Project details

No	Item	Unit rate	Unit
Materials			
1	Riprap including materials and execution fees	300	EGP/m^3
2	Sand	28	EGP/m^3
3	Sandy gravel	35	EGP/m^3
4	Granular	100	EGP/m^3
5	Concrete	560	EGP/m^3
6	Block	15	EGP/piece
7	Drainage gravel	100	EGP/m^3
8	Geotextiles	9	EGP/m^2
9	Labour	65	EGP/day
10	Technician	165	EGP/day
11	Site engineers	110	EGP/day
12	Consultant fees	2.50%	percentage
Installation rates			
13	Block wall	650	EGP/m^2
14	Freyssinet	550	EGP/m^2
15	Wraparound	690	EGP/m^2
16	TerraMesh	770	EGP/m^2
17	MacForce	650	EGP/m^2
Backfilling rates			
18	Backfilling	3	EGP/m^3
19	Backfilling compaction	7	EGP/m^3
Reinforced concrete			
20	Installation reinforced concrete wall	1800	EGP/m^2
Production rates			
21	Block wall	140	m^2/day
22	MacForce	200	m^2/day
23	TerraMesh	2	m^2/day/packer

infrastructure projects in Egypt due to its role in connecting several provinces outside the high traffic roads and opening new opportunities for development (Tables 5 and 6).

The 2nd sector is under construction by SAMCRETE – Engineers and Contractors which is connecting the 10th of Ramadan to Belbies. The First MacForce wall was to elevate the road to cross over the Ismylia Canal with a total Area of 1564 m2 with heights that vary from 1.6 m to 9.6 m.

Although it was the first MacForce wall built it wasn't an ordinary wall because an extra load of unreinforced soil embankment of 2 m height was an added feature. The Ismylia bridge abutment was the first of a set of MacForce walls in Regional Ring Road 2nd sector with total area of approximately 14,000 m^2.

Table 2. Comparative information used in the assessment

Item	Concrete panels	Block wall	TerraMesh	Masonry rock wall	Concrete retaining wall
Type of wall	MSEW	MSEW	MSEW	Gravity wall	retaining wall
Transport	Steel moulds, straps or geogrids and system accessories. Concrete panels casted at the site. Steel reinforcement + concrete at the site	The blocks, geogrids and systems accessories. The blocks are manufactured in a factory	The Terramesh units, rocks, geogrid + accessories. If rocks are available obvious method	Rocks, cement and sand	The steel bars, concrete and wooden sheathing and formwork
Components	Concrete, reinforcement bars, geotextile, connecting loops, reinforcing straps, lifting pins, rubber pads, plastic dowels, curing water, steel for fixing the straps in the soil, wooden clamps	Concrete blocks, geogrids, connectors, steel for fixing the geogrids in the soil and geotextile	TerraMesh units, rocks, lacing wire bracing wire, geogrid and geotextiles	Concrete and steel bars for the footing, rocks, cement, sand and tools to mix spread the mortar	Concrete, steel bars, wooden formwork, bitumen for insulation, steel threaded bolts, steel plates, steel nuts and water for curing
Labours skill level	Normal to moderate skilled labours are required	Normal to moderate skilled labours are required	Unskilled labours are required	Masons are required	Carpenter and steel bars team are required. High skilled labours
Labours number	Low	Moderate	Moderate	High	High
Equipment (crane, mixing trucks, loader, pump)	Moderate	Low	Low	Low	Moderate
Type of client	–	–	–	–	–

(*continued*)

Table 2. (*continued*)

Item	Concrete panels	Block wall	TerraMesh	Masonry rock wall	Concrete retaining wall
Location	–	–	–	–	–
Required area	Large casting yard Large stocking yard	Moderate stocking yard	Small stocking yard	Small stocking yard	Moderate stocking yard
Volume of concrete used	Small 0.14 to 0.18 m^3 concrete/m^2 facing	Small 0.16 to 0.18 m^3 concrete/m^2 facing	0	Moderate 0.25 to 0.35 m^3 concrete/m^2 facing	High 1 to 1.5 m^3 concrete/m^2 facing
Volume of mortar used	0	0	0	0.5 to 1.2 m^3 mortar/m^2 facing	0
Crushed stone filter	Required	Required	Not required	Required	Required
Max friction angle for backfilling soil	40	36	40	40	40
Engineering	–	–	–	–	–
Admin	Importing processes is required	Importing processes is required	Importing processes is required	No Importing processes is required	No Importing processes is required
Consulting fees	1.5 to 3%	1.5 to 3%	1.5 to 3%	1.5 to 3%	1.5 to 3%
Site visits	Moderate	Moderate	Moderate	Low	Moderate
Intensity to absorb execution errors	Moderate	Low	High	Low	Low
Intensity to absorb settlement	Moderate	Low	High	Low	Low

Table 3. Comparative information used in the assessment

Item	Unit	Reinforced concrete panels	Segmental block wall	Notes
Type of wall	–	MSEW	MSEW	
Area of facing	m^2	11	11	
Length	m	1	1	
Height	m	11	11	
Area of unit	m^2	3.2	0.08	
Units required/m^2 facing	unit/m^2	0.31	12.5	
Type of material used for facing	–	Reinforced concrete panels Concrete panels	Hollow concrete blocks	
Average volume of concrete/m^2 facing	m^3 concrete/m^2 facing	0.14–0.18	0	The thickness of the concrete if it is reinforced or not usually varies between 14 and 18 cm
Type of soil reinforcement used	–	Steel or polymeric straps	Monoaxial geogrids	
Linear meter of soil reinforcement/m2 facing	lm/m2	10.2	11.8	
Internal Friction angle for backfill soil	angle	33	33	Same soil for backfilling on both systems. Soil layer & compacting 25 cm
Geotextile area/m2 facing	m^2/m^2 facing	0.35	3	The geogrids are placed in case of the panels in the location of the gaps only. every layer (25 cm) and wrapped around in case of the block wall.
Crushed stone filter	m^3 crushed stone/m^2 facing	0.3	0.3	
PVC pipes for water drainage	m/linear meter of the wall	0	1	No PVC pipes is required to drain out the water. Countries were the rain is not so serious

(*continued*)

Table 3. (*continued*)

Item	Unit	Reinforced concrete panels	Segmental block wall	Notes
Connectors	no/m2 facing	1.5	8 for ruler connector 25 for connecting pins	
Required Team	–	3 labours 1 lifting crane (5 ton)	4 labours	
Rate of each team per day	m2/day/team	70–90	30–50	
Casting Yard	–	Required	Not required	
Curing	–	Required	Not required	
Wall with small area	–	Not preferable	Preferable	Concrete panels are not preferred in small area walls, casting & erection will be higher in case of small areas.

MacForce has been well received by the General Authority for the Roads and Bridges due to its quick installation, quick fixing of the Paraweb, and its attractive result.

Ismylia bridge abutment completed

Table 4. Comparative information used in the assessment

Item	Unit	Concrete retaining wall	TerraMesh	Notes
Type of wall	–	Concrete wall + Counterfort	MSEW	
Net Height	m	12.85	12.85	
Embedded depth	m	1.85	0.62	
Total height	m	14.7	13.47	
Length	m	60	60	
Area of facing	m2	882	808.2	The total area of TerraMesh is less embedment depth than the concrete retaining wall
Average volume of rocks/Area of facing	m^3 rock/m^2 facing	–	1	There is no rocks used in case of the concrete retaining wall
Plain concrete footing	m^3 concrete/m^2 facing	0.2	–	
Reinforced concrete footing	m^3 concrete/m^2 facing	0.7	–	A concrete footing level is required. TerraMesh wall no footing is required
Reinforced concrete for the wall	m^3 concrete/m^2 facing	0.45	–	The design of the concrete wall with thickness of 45 cm
Reinforced concrete for the counterfort	m^3 concrete/m^2 facing	0.32	–	The design of the concrete wall vs counterfort wall with thickness of 40 cm with a span of 5 m.
Total quantity for concrete/Area of facing	m^3 concrete/m^2 facing	1.43	–	
Volume of backfill at the back of the wall	m^3 backfill/m^2 facing	9.3	9	The volume of the backfill less for masonry rock wall volume of the rocks used is higher than TerraMesh wall.

(*continued*)

Table 4. (*continued*)

Item	Unit	Concrete retaining wall	TerraMesh	Notes
Internal friction angle for backfill soil	angle	33	33	Same soil for backfilling on both systems. Laying the soil and compacting layers of thickness of 25 cm
Crushed stone filter	m^3 crushed stone/m^2 facing	0.3	0	No filter is required for TerraMesh wall porosity 30% which acts as a filter.
PVC pipes for water drainage	m/linear meter of the wall	1.22	0	No PVC pipes is required to drain TerraMesh wall as the facing itself is acts like a filter.
Rate of construction for concrete wall - sheathing team	m^3 concrete/day/team	2	–	The Sheathing team includes one labour and one helper.
Rate of construction for concrete wall - reinforcement team	m^3 concrete/day/team	4	–	The reinforcement team includes one labour and one helper.
Construction (one builder + two helpers)	m^3 rock/day/team	–	6	Team for the TerraMesh wall one builder and two helpers.
Days to construct the wall/one team	days	328	135	The cost of the team required for TerraMesh is less than the cost of the concrete wall team

Table 5. Comparative information used in the assessment

Item	Unit	Masonry rock wall	TerraMesh	Notes
Type of wall	–	Gravity wall	MSEW	
Area of facing	m^2	9570	8662	The total area of TerraMesh has less embedded depth than the ordinary masonry rock wall
Length	m	979	979	
Average height	m	10	10	
Average volume of rocks/Area of facing	m^3 rock/m^2 facing	3.58	1	The quantity of rock for masonry rock wall is higher because the wall acts as a gravity wall. TerraMesh wall is a MSEW
Reinforced concrete footing	m^3 concrete/m^2 facing	0.25	0	The design of the masonry rock wall requires a concrete footing. TerraMesh wall no footing is required
Volume of backfill at the back of the wall	m^3 backfill/m^2 facing	3	5.65	The volume of the backfill is less in case of the masonry rock wall rocks used is higher than the TerraMesh wall.
Internal friction angle for backfill soil	angle	33	33	Same soil for backfilling on both systems. Laying the soil and compacting layers of 25 cm
Crushed stone filter	m^3 crushed stone/m^2 facing	0.3	0	No filter is required e TerraMesh wall porosity is 30% which acts as a filter.
PVC pipes for water drainage	m/linear meter of the wall	5.5	0	No PVC pipes is required to drain out the water in case of TerraMesh wall acts like a filter.
Weight of cement used in mortar	Kg/m^3 rock	100	0	No mortar is required in case of TerraMesh
Volume of sand used in mortar	m^3 sand/m^2 rock	0.333	0	No mortar is required in case of TerraMesh
Rate (one builder + four helpers)	m^3 rock/day/team	7	–	The construction team masonry rock wall one builder and four helpers due to the usage of mortar
Rate (one builder + two helpers)	m^3 rock/day/team	–	6	The construction team TerraMesh wall one builder and two helpers.
No of days to construct the wall/one team	days	4,894	1,444	The cost of the team for TerraMesh is less than Masonry rock wall team

Table 6. Comparative information used in the assessment

Item	Concrete panels	Block wall	TerraMesh	Lock + Load wall	Masonry rock wall	Concrete retaining wall
Type of wall	MSEW	MSEW	MSEW	MSEW	Gravity wall	retaining wall
Transport	Steel moulds, straps or geogrids accessories. panels casted at the site. reinforcement & concrete at the site	The blocks, geogrids and systems accessories. The blocks are manufactured in a factory	Terramesh, rocks, geogrids accessories. rocks at the site, perfect for TerraMesh system to be used	Steel moulds, geogrids, steel loops, facing panels, counterforts. The concrete is casted at site	Rocks, cement and sand	The steel bars, concrete and wooden sheathing and formwork
Components	Concrete, rebar, geotextile, connecting loops, reinforcing straps, lifting pins, rubber pads, plastic dowels, curing water, steel fixing the straps, wooden clamps	Concrete blocks, geogrids, connectors, steel for fixing the geogrids in the soil and geotextile	TerraMesh units, rocks, lacing wire bracing wire, geogrid and geotextiles	Concrete, reinforcement bars, geotextile, connecting loops, geogrids, concrete fibres, water for curing and steel for fixing the geogrids in the soil	Concrete and steel bars for the footing, rocks, cement, sand and tools to mix and manoeuvring the mortar	Concrete, steel bars, wooden formwork, bitumen for insulation, steel threaded bolts, steel plates, steel nuts and water for curing
Labours skill level	Normal to moderate skilled labours are required	Normal to moderate skilled labours are required	Unskilled labours are required	Normal to moderate skilled labours are required	Masons are required	Carpenter and steel bars team are required. High skilled labours
Labourers	Low	Moderate	Moderate	Moderate	High	High
Equipment (crane, mixing trucks, loader	Moderate	Low	Low	Low	Low	Moderate
Type of client	–	–	–	–	–	–
Location	–	–	–	–	–	–
Required area at the site	Large casting yard Moderate to large stocking yard	Moderate stocking yard	Small stocking yard	Large casting yard Moderate to large stocking yard	Small stocking yard	Moderate stocking yard
Volume of concrete used	Small 0.14 to 0.18 m^3 concrete/m^2 facing	Small 0.16 to 0.18 m^3 concrete/m^2 facing	0	Small 0.10 to 0.102 m^3 concrete/m^2 facing	Moderate 0.25 to 0.35 m^3 concrete/m^3 facing	High 1 to 1.5 m^3 concrete/m^3 facing
Mortar used	0	0	0	0	0.5 to 1.2 m^3 mortar/m^2 facing	0
Crushed stone filter	Required	Required	Not required	Required	Required	Required
Friction angle backfilling soil	40	36	40	36	40	40
Engineering	–	–	–	–	–	–
Administration	Importing processes is required	Importing processes is required	Importing processes is required	Importing processes is required	No Importing processes is required	No Importing processes is required
Consulting fees	1.5 to 3%	1.5 to 3%	1.5 to 3%	1.5 to 3%	1.5 to 3%	1.5 to 3%
Site visits	Moderate	Moderate	Moderate	Moderate	Low	Moderate
Absorb errors	Moderate	Low	High	Unknown	Low	Low
Absorb errors	Moderate	Low	High	Unknown	Low	Low

Regional Ring Road 2nd Sector- Ismylia Bridge abutment

Regional ring road-2nd sector Ramp2

Dawran Bridge under construction only a few kilometers away from Ismylia bridge

Stone Park in Egypt

The Petrified wood means “wood turned into stone”. A huge tree that was petrified through millions of years and solidified to stone was the reason for the name of Stone Park to a marvelous residential project of a total area of nearly about 450 Fadden. A solution was needed for the high slopes that separates the two phases between the Villas and residential buildings at the bottom of the slopes. However, the challenge was not only on stabilizing the slope but also providing a good-looking solution that satisfies

the luxurious nature of the project. A vegetated wrap around slope reinforcement system was proposed to reinforce the slopes that vary between 30 m down to 9 m for 1.5 km.

Petrified Rock feature at the entrance to the StonePark Compound east of Cairo after which the development has been named

Slopes from 9 to 30 m requiring stabilization and vegetation Terramesh next to Masonry wall

Telal El Ain Sokhana

In one of Egypt's luxurious resorts near the Red Sea, Telal El Ain Sokhna, which covers an area of 2.5 million square meters of real estate, the first Mechanically Stabilised Earth Wall using Terramesh was constructed. Its full area presents a 9000 m^2 Wall that ranges from 1 to 19 m in height, raising a claim to be the current highest full vertical MSEW structure in Egypt to date. The TerraMesh systems is consisting of a double twist mesh of 2.7 mm wire which forms a basket to be filled by rocks providing a facing with an average porosity of 25% which will act as a drain and soil reinforcement using Geosynthetics manufactured by the UK factory of Linear Composites. These products range from 200 kN to 500 kN in strength and was identified as a cost efficient, time saving and best practice engineered solution. The challenge to the designers was to satisfy all the Clients requirements namely; the 19-m height wall stability, accommodating a hard rock outcrop, durability in marine conditions and overall aesthetics in a prestigious Resort. The final analysis and comparison of suitable solutions versus the accurate cost engineering concluded with the Maccaferri Terramesh system. Additional to this assessment was the benefit to local labour force to learn and acquire new skills and support an introduction of new Technology into the Egyptian market.

The Design, material selection and construction methodology proposed by the supplier Maccaferri was verified and approved by a respected Geotechnical Consulting firm in Egypt. The successful execution of the current highest MSEW wall in Egypt using the TerraMesh system gives an indication to the development and confidence to introduce new Geotechnical applications in the Egyptian Market.

Excavation into dolomitic foundations

-villas on the ridge, MSEW- Red sea

close up view of the Terramesh system

Progress at the end of 2016

Precasting in moulds without a welded base

One of the effective approaches that were defined in the Egyptian market was using a four sided mould without a steel base in casting the concrete panels which started to grow rapidly in Egypt. Freyssinet was the pioneers in using this approach in Egypt. The approach depends on using a concrete base or an Epoxy fibre glass base fixed on a concrete base which has some texture. The four sides are manufactured from steel.

The benefit of using this approach:

- Low cost moulds.
- Using a textured base gives good looking panels to fulfil the architectural aspect.
- The manufacturing of the moulds will quicker, easier and low risk in having defects.
- The probability of having defects in the casted panels is less due to the strong fixation of the four sides and the base in the concrete base.
- More efficient handling on site and lower in the cost of transportation in comparison to the welded steel moulds is which requires a loader and takes more space Freyssinet as well is using a very good approach in lifting the panels which is a

temporary system for lifting. In other words, the lifting part is used more and it isn't a consumable item like in other panel systems which decreases the cost as well.

These moulds are easy to use for patterns

and efficient to store

The system has a novel lifting connector

A traditional system with a fixed base plate

<u>All MSEW systems face the question of the connection between the facing and the soil reinforcement</u>

<u>Block walls:</u>

1. Connecting pins: through attaching the geogrids with the block through a polymeric connector in the shape of pins or a L shape connector which is inserted into a grove in the block itself.
2. Connecting bar: through attaching the geogrid with a connecting bar which placed between the blocks.
3. Placing geogrids between the blocks: depending on the weight of the block the geogrid is attached to the blocks.

<u>Concrete Panels:</u>

1. Geosynthetic Loops: in precast phase half of a polymeric loop is placed inside the casted reinforced concrete panels and the other half is available for the soil reinforcement straps.

2. Loops and toggle: a steel loops are inserted into the concrete panels. The toggle is inserted into the casted loops forming the connection where the straps is attached to the panels.

TerraMesh:

A gravity and friction connection between the facing and the reinforcing straps through the placing of the geogrids underneath the tail of the TerraMesh which is formed of steel mesh.

Placing the geogrids inside concrete panels

A new approach was introduced in the Egyptian market by inserting geogrids inside the panels. Unfortunately, after casting the panels a lot of cracks were observed around all the layers of geogrids. Later, when the erection started, the panels started to move and the walls started to show poor performance which forced the contractor to build another wall in front of it to compensate for the bulging and the leaning wall. The problems are being resolved and this concept seems to have restarted so it remains to be seen what the outcome of this method will produce.

Some considerations to stimulate a better understanding of factors influencing MSEW structures

1. Assessment of available workforce capability and planning a method to upgrade and develop the labour skill set
2. Consistent and ongoing education of the Engineering participants on market status, research and changes
3. Motivating Construction leaders to engage with companies that offer new developments and techniques
4. Establishment of formal quality assurance and quality control standards for evaluation and acceptance of MSEW work and Projects.
5. Creating a platform to present well documented Case histories, articles, editorials, seminars and develop interest group forums

Contribution to the body of Knowledge with Specialised Interest groups-

1. International Geosynthetic Society
   The IGS Egyptian chapter is undergoing formal registration with the International body to establish its presence in Egypt. Dr. Fatma Baligh has been selected as the first President for the IGS Egyptian chapter and Dr. Rami Sherbiny is one of the founders.
2. Mechanically Stabilised Earth Interest Group
   Egyptian members have the option to be registered independently via the internet and await the emergence of a champion to rally the members to form an Egyptian interest group.
3. International Society for Soil mechanics and Geotechnical Engineering
   Egypt hosted the 17th International Conference in Alexandria in 2009
   With a keen interest by Egyptian participants who have become active in keeping up to date with this groups activities. Dr. Fatma Baligh from Helwan University is the vice President for Africa.

The Egyptian market has shown an intensive development in the last 2 years. The Egyptian market includes now Tenax, Tensar, Huesker, Freyssinet, Maccaferri, Bonar and NAUE.

The seven companies are competing to supply the client the most efficient solutions for their problems. In addition to the competition from the technical aspects, the commercial competition is very strong which serves the interest of the clients and eliminates the presence of any monopoly. Some companies were making more than 60% profit margin when they were alone in the market.

New systems which have recently appeared in the Egyptian market are the discrete concrete panels (MacForce/Reinforced Earth), the TerraMesh (double twist wire mesh) system, lock and load system (concrete blocks) and Geosynthetic wrap around (for vegetated front face).

Discrete concrete panels
Concrete panels which are reinforced or unreinforced are casted in special moulds at site. Then they are erected by a crane. Polymeric or steel straps are connected to the facing panels through special loops or boxes.

TerraMesh
Steel boxes which are manufactured from steel wire mesh coated with PVC are filled with rocks to form a rock facing walls. The TerraMesh units without soil reinforcement can be used up to 4 m height wall. and additional geogrids (Paragrid or Paralink) are used for walls above 4 m in height.

The modular steel – and polyfiber – Reinforced concrete units consist of a facing panel and an anchoring/reinforcing component known as counterfort, mechanically joined together during wall assembly to form an integral unit. Geosynthetic reinforcement is used to increase the retaining wall height beyond the maximum height limit of reinforced-soil wall created by counterforts alone.

Green facing wrap around system: Geogrids are placed on layers and wrapped around forming a slope with angles between 45 and 60 degrees. Geosynthetics are wrapped around with the geogrids forming a triangle that is filled by vegetation soil. Hydro seeding is done to plant the slope formed by the geogrid. The result is a marvellous green slope.

The presence of different solutions and different companies serves the interest of the client in having several solutions and with high efficiency technically and commercially

9 Conclusion

MSEW is an alternative to reinforced concrete retaining walls and its success in using geosynthetic soil reinforcement has made a great contribution to solve problems with weak soils. The potential to compliment the stone masonry/rip rap solution can extend this age-old technique into the future. With so many MSEW systems to choose from it is necessary to agree to use an assessment technique that can create a local suitable model to analyse geosynthetics which forms one of the critical items in determining the most cost effective MSEW solution. Investigation has identified the usefulness to develop a simple assessment model that satisfies all the qualitative and quantitative characteristics to arrive at a means to decide which product or solution is most suitable for the Egyptian market.

By approaching the soil reinforcement options as a first step to examine the appropriate material and associated cost to compliment an MSEW system, the investigator can use the suggested tables as a model to arrive at an unbiased collation of information and make an informed decision.

If the soil reinforcement for an MSEW system is a geosynthetic it should be assessed for its suitability and durability in a manner that satisfies the design and life time expectation of the owner of the structure. To perform due diligence, it needs a comparative study to ensure it is the most suitable option and the consequence of its use about durability in long term performance and how to resolve future changes that may need replacement/repair. The assessment model proposes methods of testing the material for verification and organizations for validation. It allows the collation of information in a systematic way to enable a prediction to be made as an estimate of performance. Numerous authors have demonstrated that MSEW solutions can be between 25 to 50% more cost effective than traditional Reinforced Concrete retaining walls. Therefore, the next step in the decision process is using a simple method of assessing which of the many options available are most suitable. A simple cost benefit to cost ratio analysis can be done on an excel spreadsheet once the information has been tabulated in a comparative manner by differentiating between the quantitative and qualitative aspects. The authors Koerner have raised the alarm of risk in an assessment in the USA which claims that 86% of MSEW structures less than 4 m in height show distress or may fail. For this single reason, it would be best practice to apply some logic in determining which system to use on the next project.

Acknowledgements. We as authors thank our friends and colleagues who supplied the information and photos to contribute to the final report

References

Greenwood, Hartmut, Schroeder: Durability of Geosynthetics (2015)

Sharma, M., Goliya, H.S.: Design and economic analysis of reinforced earth walls. Int. J. Emerg. Trends Eng. Devel. **6**(4) (2014)

Koerner, R.M., Koerner, G.R.: The importance of drainage control for geosynthetic reinforced mechanically stabilised earth walls. J. GeoEng. **6**(1), 3–13 (2011)

Schmidt, J.M., Harpstead D.L.: MSE wall Engineering-A new look at Contracting, Design and Construction

Singh, H., Aktar, S.: A review on economic analysis of Reinforced Earth wall with different types of Reinforcing materials. Ultemas **iv**(xii) (2015)

Geotechnical Engineering Manual. mechanically stabilised Earth System, Inspection manual GEM 16, New York Department of Transport

Anderson, P.L., Gladstone, R.A., Sankey, J.: (22222) State of Practice of MSE Wall design for Highway structures. www.reinforcedearth.com

Influence of Asphalt Mixture Ageing and Lowered Laboratory Compaction Rate on Stiffness and Cracking Behavior

Pavla Vacková(✉), Jan Valentin, and Adriana Kotoušová

Department of Road Structures, Faculty of Civil Engineering,
CTU in Prague, Thakurova 7, 166 29 Prague, Czech Republic
{pavla.vackova.1, jan.valentin}@fsv.cvut.cz,
adrianakotousova@gmail.com

Abstract. Ageing is one of the fundamental phenomena related to bitumen and asphalt mixtures. Traditional problem of asphalt testing is that most of the characterizations done presently reflect ageing only in limited manner. Nevertheless the behavior of a pavement structures is on the other hand dominated by natural effects like ageing. Similarly very important issue related to good job quality is in case of asphalt mixtures the compaction rate of asphalt layers. Compaction rate is often a factor which has very significant impact on the durability, but it is often neglected or the attention which is given to proper compaction is not sufficient enough.

In this research the asphalt mixtures with different compaction rates were exposed to artificial laboratory ageing to evaluate the effect on selected asphalt properties of both impacts. The assessed properties were stiffness modulus at three testing temperatures (0 °C, 15 °C and 30 °C) and crack propagation test at 0 °C. The properties were always determined for virgin (unaged) as well as aged test specimens. The ageing indexes were calculated for both of the characteristics discussed in this paper.

1 Introduction

Compaction of asphalt mixtures constitutes a very important, yet often ignored factor which has a significant impact on asphalt layer quality and lifespan. Insufficient compaction might reduce the life of an asphalt layer considerably, either due to structural deformations, easier oxidising (ageing) or more intense water immersion.

Asphalt compaction quality is influenced, first and foremost, by the mix design. The smoother the grading curve, the better the compaction. Mixtures with finer particles are usually easier to compact than coarse asphalt mixtures although their stability is markedly lower (Zajicek 2014). Higher bitumen content improves compactability to a certain degree and ensures a higher compaction rate. However, if there is too much bitumen in the mix, all of the air voids in the aggregate skeleton may be filled and the mix cannot be further compacted. This might cause a reduction in the bulk density of the mix. Another important aspects of compaction is the temperature under which the mechanical process is taking place. The compaction temperature influences workability of the asphalt mixtures (Hassan 2005). The lower the temperature the harder it is to

H. Shehata and Y. Rashed (eds.), *Numerical Analysis of Nonlinear Coupled Problems*,
Sustainable Civil Infrastructures, DOI 10.1007/978-3-319-61905-7_13

compact the mixture. The temperature of asphalt mix preparation should not exceed the maximum recommended level to avoid thermal degradation and so-called overburing of the bitumen. At the same time transport of asphalt mixtures should not exceed one hour (in extreme case of high summer temperatures up to 2 h may be accepted) from uploading the truck. In region like the Czech Republic the mixture cools down during transport, primarily due to the weather conditions (temperature, wind strength, possibly rain), distance of the site, speed of the truck (additional air flow) and the arrangement of protection from such factors. Unprotected trucks may be allowed to transport only some types of asphalt mixtures on very short distances and solely in windless weather (Hanzik 2015).

The insufficient compaction has influence on lower compaction rate/higher air voids content and leads to contribution of early failure of pavement (Kennedy 1984).

2 The Asphalt Mix

The study presented in this paper was conducted at the Faculty of Civil Engineering of the Czech Technical University in Prague on the basis of the initial findings from the diploma thesis (Sedlacek 2016). The study focuses on AC_{surf} 11 asphalt mixture with paving grade bitumen 50/70, which was produced at a mixing plant and then compacted in the laboratory under various temperatures and using various compacting energies (number of blows of Marshall compactor) to determine the impact of incorrect paving/insufficient compaction. The effect of the insufficient compaction rate is demonstrated on properties characterizing asphalt behaviour in the low temperature range.

The asphalt mixture was designed with 5.2 M% of bituminous binder. The test specimens were prepared according to standard CSN EN 12697-30 - Specimen preparation by impact compactor. The referential air voids content (air voids content according to initial type testing (ITT)) was determined on specimens compacted with 2 × 50 blows of Marshall compator at 150 °C. The referential air voids content was determined by the value of 2.8%-vol. The test specimens were gradually compacted at 150 °C, 120 °C and 90 °C. To complete the series, test specimens compacted at 135 °C were added later. The mix for set of test specimens compacted at 135 °C was prepared at later date by the same mix design in the same mixing plant; however, its maximum density is slightly lower than in the case of the former mixtures. The maximum density of the first set amounted to 2.686 g/cm^3 while, the set compacted at 135 °C only scored 2.637 g/cm^3. This means that although the mix was prepared in the same mixing plant under the same conditions and using the same input material, the results of the second set of test specimens must be interpreted in light of this circumstance.

3 Laboratory Testing of the Mixes

3.1 Air Voids Content

As has been mentioned earlier, the test specimens from the selected asphalt mix were compacted according to standard EN 12697-30 under various temperatures (150 °C;

135 °C; 120 °C; 90 °C) using various compacting energies of 2 × 50; 2 × 40; 2 × 25 blows of the Marshall impact compactor. (2 × 50 means 50 blows from each side of specimen). EN 13108-1 standard sets the compaction temperature for asphalt concretes with bitumen 50/70 from 140 °C to 180 °C.

Figures 1 and 2 show the effect of the paving and compaction discipline (including insufficient compaction setup) on voids content. The air voids content grows with decreasing blows of Marshall compactor and decreasing temperature.

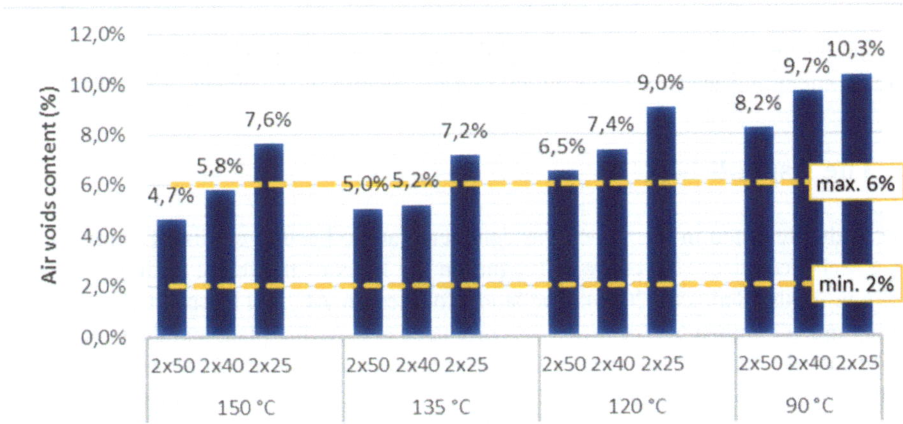

Fig. 1. Air voids content for AC_{surf} 11 mix – different temperatures and compaction energies

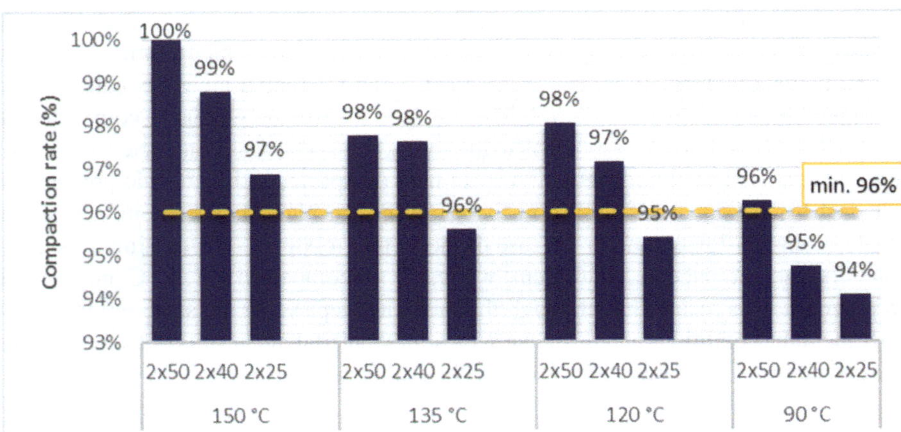

Fig. 2. Compaction rated for assessed variants of AC_{surf} 11 with different compaction efforts and temperatures

Figure 1 demonstrates the limit of 2.0%-vol. to 6.0%-vol., which is determined for control testing of voids content in case of AC_{surf} 11 mix type by EN 13108-1.

The limits were only met by mixtures compacted at 150 °C and at 135 °C by higher compacting energies. The mixtures compacted at 120 °C and less failed to meet the voids content requirements even with the standard number of blows (2 × 50) by the Marshall compactor.

The air voids content has influence on all of the asphalt mix properties. It affects the strength characterizations, rutting resistance to crack propagation etc.

The compaction rate was calculated from the bulk density of the compacted test specimens. The bulk density of test specimens compacted under the standard requirements at 150 °C and by 2 × 50 blows of the Marshall compactor was used as the referential value. The minimum permitted compaction rate is stipulated by CSN 73 6121 to be 96%. This limit was not met by all of the mixtures compacted by 50% of the required compacting energy defined by the standard, with the exception of the test specimens compacted under the highest temperature. It must be emphasised that even test specimens compacted at 90 °C with the standard compacting energy of 2 × 50 blows met the minimum compaction rate requirement although its voids content exceeds the limit by over 2.0%-vol. and it almost doubles the value of voids content of the test specimens compacted in the standard manner.

Although CSN 73 6121 stipulates the minimum compaction rate as 96%, it has been proven by multiple practical and experimental experience in the past that an asphalt layer compacted to 96% has a significantly lower lifespan than would be expected with such relatively small reduction in the optimum compaction level. An example might be the findings of (Zak 2012), describing asphalt mixtures compacted to 96% which only reach 25% of the lifespan of layers compacted to 100% level.

With respect to the fact that the standard fatigue resistance measurement method as one of the key lifespan indicators given by EN 12697-24 is rather demanding in the terms of both time and costs, a method of comparing the asphalt mix lifetime by the parameters of reference test specimens and the test specimens aged in the laboratory was selected. Therefore the test specimens were subject to a laboratory ageing cycle according to one of the methods defined in prEN 12697-52. Based on that, compacted test specimens are stored in a thermal chamber with air circulation under 85 °C for 5 days. Once the ageing simulation is finished, the specimens are left to cool down in the same position to prevent deformation or damage caused by handling of hot specimens. The test specimens are put in the thermal chamber with no other protective collars as has been verified e.g. in Valentova (2016). The aged test specimens have demonstrated no changes in voids content due to the ageing process. The specimens showed voids contents identical to those prior to the simulated laboratory ageing.

The stiffness modulus test under three temperatures (0 °C, 15 °C and 30 °C) and resistance to crack propagation in semi-cylindrical test specimens at 0 °C were chosen to compare the properties characterising mix behaviour.

3.2 Stiffness Modulus

The stiffness modulus was determined in compliance with EN 12697-26 by the IT-CY (indirect tensile stress on cylindrical specimens, non-destructive test) always with at least 4 specimens. The overall results of reference test specimens for AC_{surf} 11 are depicted in Figs. 3, 4 and 5 for temperatures 0, 15 and 27 °C. (There are marked out standard deviations for each set of tested specimens). The data clearly shows that the stiffness modules tend to decrease with increasing voids content in the mix, which is a natural and expected trend. The difference in stiffness modules measured at 15 °C for specimens compacted by 2 × 50 blows under 150 °C and 90 °C amounts to almost 45% (and is even greater for other compacting energies). It must be again emphasised that test specimens compacted under 90 °C met the minimum requirement for compaction rate as stipulated by the standard. However, it is obvious that the impact on deformation behaviour is much greater in such cases – not to mention the effects of water or cyclic water and frost impacts.

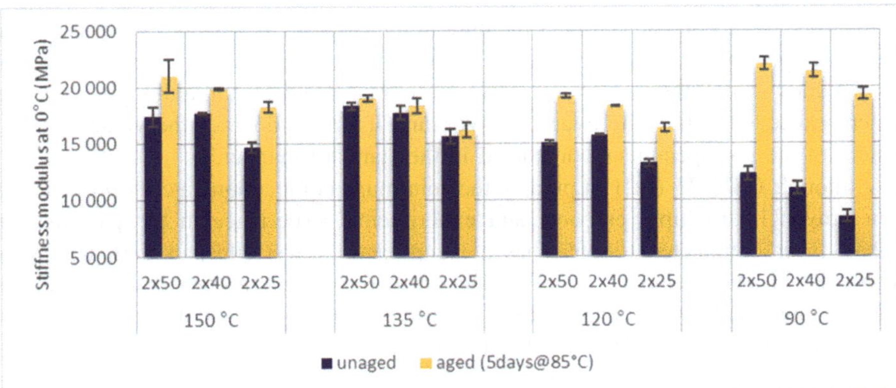

Fig. 3. Stiffness modulus at 0 °C of AC_{surf} 11 test specimens

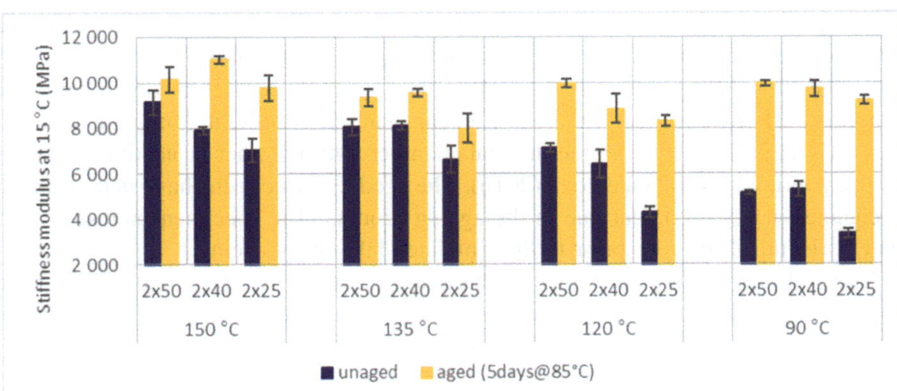

Fig. 4. Stiffness modulus at 15 °C of AC_{surf} 11 test specimens

The ageing index was calculated for the individual compaction variants and individual temperatures of stiffness modulus test as a ratio of the stiffness modulus of an

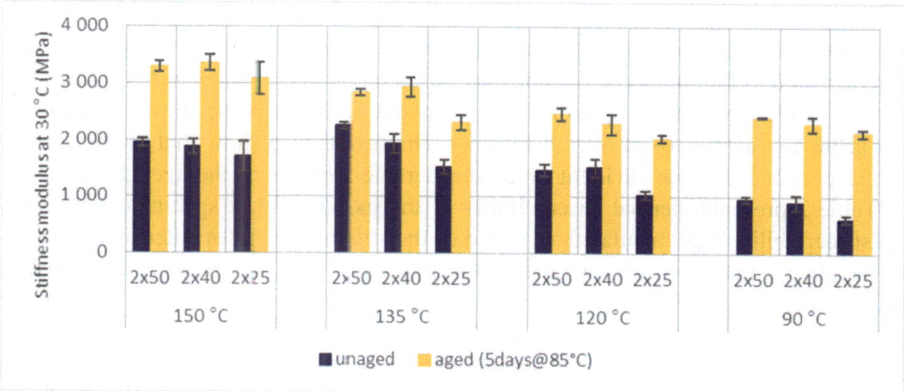

Fig. 5. Stiffness modulus at 30 °C of AC_{surf} 11 test specimens

aged test specimen to an unaged test specimen. The results of determination of the ageing index are summarized in Table 1. Analysing the results it can be seen that there is a distinctive effect of the compaction temperature, the ageing index grows with decreasing compaction rate. The only exception are test specimens compacted at 135 °C which deny the trend slightly. As has been mentioned above, the asphalt mix for this compaction temperature was prepared at a later date and, therefore, its properties may differ.

Table 1. Ageing indexes for stiffness modulus of AC_{surf} 11 test specimens

Temp. of compaction	Compaction energy	Ageing index *(ratio of stifness modulus of aged specimen to unaged specimen)*		
		0 °C	15 °C	30 °C
150 °C	2 × 50	121%	111%	168%
	2 × 40	112%	140%	178%
	2 × 25	116%	129%	165%
135 °C	2 × 50	120%	124%	106%
	2 × 40	121%	119%	118%
	2 × 25	123%	139%	140%
120 °C	2 × 50	127%	131%	193%
	2 × 40	116%	149%	193%
	2 × 25	122%	186%	225%
90 °C	2 × 50	156%	193%	256%
	2 × 40	154%	167%	237%
	2 × 25	192%	249%	334%

The lower the ageing index (the closer to the value of 100%) is, the less susceptible to ageing of the test specimen (asphalt mixture) is, and it can be potentially considered more resistant to fatigue. This hypothesis is based on the fact that the test specimen will demonstrate less brittleness and, therefore, it should resist the cyclic stress in the fatigue test, achieving an improved overall lifespan. The results also show that the decreasing compaction rate becomes evident on a considerable increase in the ageing index.

Table 2 presents thermal susceptibility of the individual test specimen sets. Thermal susceptibility was calculated as a ratio of the stiffness modules determined at 0 °C and at 30 °C. Paradoxically, ageing in the majority of tested specimens resulted in reduced thermal susceptibility. The expected trend was the opposite, i.e. ageing had been expected to increase this parameter slightly. The reason supporting the assumption was the fact that under higher temperatures, the asphalt mixtures will reach higher stiffness while their slightly deteriorated brittleness will be demonstrated under lower temperatures. In the case of the stiffness modules of assessed AC_{surf} 11, and the comparisons of thermal susceptibilities of unaged and laboratory aged test specimens, the lower value means an improved resistance to temperature changes which, to a certain degree, also means improved resistance to lower temperatures. As has already been mentioned, aged bituminous binder tends to be stiffer but more brittle which should make both the bituminous binder and the asphalt mix prepared therewith more susceptible to temperature changes. However, the presented results did not verify this assumption and, therefore, we claim this cannot be considered the right approach which proves improved asphalt mix behaviour following simulated ageing.

Table 2. Thermal susceptibility of assessed AC_{surf} 11 mixtures

Temp. of compaction	Compaction energy	Voids content	Thermal susceptibility	
			Unaged	Lab aged
150 °C	2 × 50	4.7%	8.9	6.4
	2 × 40	5.8%	9.4	5.9
	2 × 25	7.6%	8.6	6.0
135 °C	2 × 50	5.0%	8.1	9.1
	2 × 40	5.2%	9.1	9.3
	2 × 25	7.2%	10.2	9.0
120 °C	2 × 50	6.5%	10.2	6.7
	2 × 40	7.4%	10.4	6.2
	2 × 25	9.0%	12.8	7.0
90 °C	2 × 50	8.2%	12.8	7.8
	2 × 40	9.7%	12.2	8.0
	2 × 25	10.3%	13.9	8.0

Subsequently, Fig. 6 compares the values of the stiffness modules and air voids contents of the individual test specimen groups. It clearly shows the decreasing trend of the lines connecting the individual values when a distinctive drop in the stiffness modulus is recorded under lower temperature while the voids content increases (a higher slope).

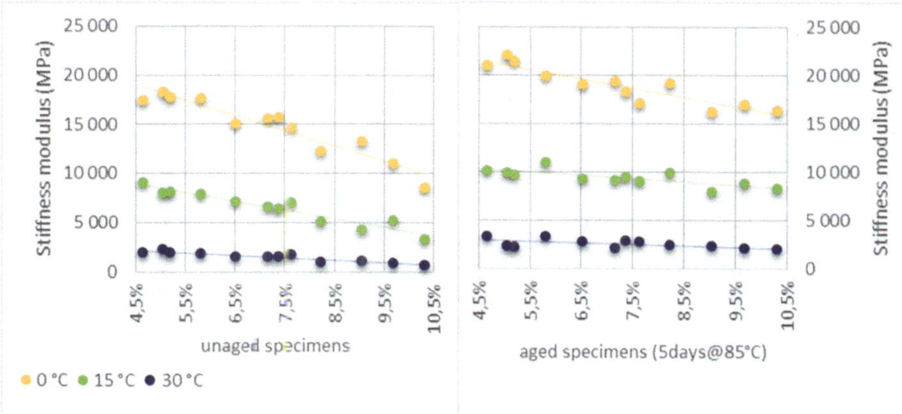

Fig. 6. Comparison of stiffness modules and air voids content for assessed AC_{surf} 11 mixtures

3.3 Resistance to Crack Propagation

The other test chosen to examine the effects of ageing in combination with the impact of optimum and insufficient compaction rate was the laboratory procedure for testing asphalt mix behaviour in the low temperature range; the asphalt mixtures were tested for resistance to low-temperature crack propagation in compliance with EN 12697-44. After the stiffness modulus testing, the specimens were cut down to the required height of 50 mm, and then halved. An artificial crack, 10 mm deep, was cut in the bottom of the semi-cylindrical specimens to weaken the cross-section – this helps the correct formation of a controlled low-temperature induced crack. The test was conducted at 0 ° C employing the standardised loading speed of 5 mm/min.

The initial assumption was that the aged test specimens would achieve lower fracture toughness values – as has already been mentioned, aged bituminous binder is stiffer on one hand and more brittle on the other, which makes it less resistant from the point of view of behaviour in the lower temperature range. This assumption was verified only for five test specimen sets (highlighted in green in Fig. 7); the trend was quite the opposite in the remaining sets – the aged specimens achieved higher values. A similar trend is demonstrated by thermal susceptibility results for stiffness modules (see Table 2).

From the perspective of the aforementioned assumption of bituminous binder brittleness, this trend is inexplicable but persists in most types of asphalt mixtures tested in the road laboratory of the Faculty of Civil Engineering, CTU Prague, and is undergoing further analysis preconditioned by research focused on actual effects impacting the structural layer in the pavement under low temperatures and traffic loads. The determining parameter of the test of asphalt mix resistance to crack propagation in semi-cylindrical test specimens is fracture toughness which might be affected by a number of factors, the most significant seeming to be the preparation, i.e. cutting, of the test specimens. It is very important for the specimens not to be damaged at all during cutting, and for the crack to be located exactly in the centre of the specimen in the

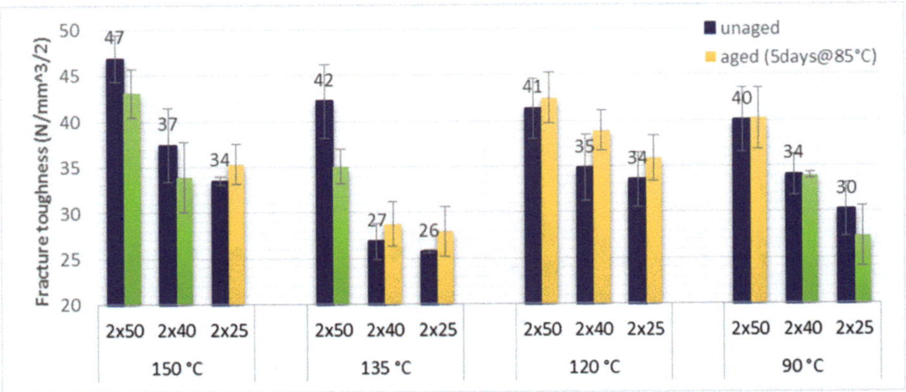

Fig. 7. Critical value of fracture toughness for test specimens of AC_{surf} 11

required thickness and depth. The inaccuracy of cutting negatively influence the test results.

Therefore, a comparison of aged and unaged test specimens seems to be slightly pointless in this regard, thanks to the unclear link between them. However, partial values can be compared to one another. The trend of the critical value of fracture toughness falling with the compaction rate is no longer as noticeable as in the case of the determination of stiffness modules; however, it is still quite obvious.

The decrease of fracture toughness with lower compacting energy (by 10 blows) is around 20%. Reducing the compacting energy to 50%, the decrease of fracture toughness drops by approximately 30% for specimens compacted at 150 °C. The decreases via the compaction energies are slightly lower at compaction temperature of 120 °C and 90 °C.

4 Conclusion

The laboratory measurements assessed on unaged and aged (5 days at 85 °C) asphalt test specimens clearly show the importance of professional discipline during asphalt mixture compaction. The specimens with insufficient compaction rate shows worsening of all characteristics in comparison to referential variant (compacted at 150 °C with standardized 2 × 50 blows of Marshall impact compactor for AC_{surf} with unmodified binder). An important role in this undoubtedly plays the temperature of asphalt mix production, the temperature of paving – which is affected by the transportation (insulation of vehicle's loading body, distance etc.) and weather conditions (possibly being able to reduce the temperature quicker if unfavourable), the temperature of compaction and, last but not least, the compacting equipment which should always ensure the highest possible compaction rate.

If there is complaint on quality of asphalt road, the very first checked data are usually parameters of input asphalt mixtures. The asphalt plants are pressurised to control regularly the quality of outgoing asphalt mixtures (granularity and air voids content). The quality of compaction on construction side is considered later, if ever. The compaction rate of asphalt layer should be checked every time, because even though the laboratory air void content is in standard limits, the laboratory compaction is performed at ideal temperature and compaction conditions, which are never this ideal on construction site. There are existing some non-destructive test methods (e.g. Troxler radiometric probe) for checking the compaction rate easily and quickly right on site. Checks made by such equipment are becoming increasingly popular; their availability might lead to improved quality of compacted layer application, thus helping improve the quality of pavement structures and extending the lifespan of the entire road network.

From the presented results, it is obvious how sensitive the asphalt layer quality might be. The air void content grows very steeply with decreasing compaction temperature and compaction energy. The air void content has very significant impact on all asphalt mixture properties. It influences mainly strength characteristics. If the air void content is too higher (higher than standard limits), the asphalt mixture has usually lower strength characteristics. With decreasing air voids content, the strength characteristics grows, of course only to certain level. If the air void content is below standard limits, it can have reverse effect Too dense mixtures could have high strength characteristics, but usually have problem with rutting or parameters measured at higher temperatures.

The problem of decreasing strength characteristics is even more significant for aged asphalt specimens. The specimen after the artificial ageing shows how the specimen might behave during the lifetime. The asphalt layer is exposed to oxidation, UV, climate change and so on and everything runs under the traffic load. The more constant the data stays after the artificial ageing, the more stable the asphalt layer will be over the time. If there is significant increase of stiffness modulus after the ageing process, it means, that the layer is very influenced by the ageing and it parameters will change. The bituminous binder in the asphalt layer will become stiffer and stiffer, but therefore increasingly brittle and the layer would degrade sooner.

Acknowledgement. This paper was realized within the research project SGS16/054/OHK1/1T/11.

References

Hanzik, V., et al.: Pokladka hutnenych asfaltovych smesi. Sdruzeni pro vystavbu silnic Praha, Praha (2015). ISBN 8090392563; 9788090392564. (in Czech)

Hassan, Y.A.: Methodology for determining most suitable compaction temperatures for hot mix asphalt. J. Eng. Sci. **33**, July 2005. Assiut University

Kennedy, T.W., Roberts, F.L., McGennis, R.B.: Effects of compaction temperature and effort on the engineering properties of asphalt concrete mixtures. In: Wagner, F.T. (ed.) Placement and Compaction of Asphalt Mixtures, pp. 48–66. ASTM Special Technical Publication 829. American Society for Testing and Materials, Philadelphia (1984)

Sedlacek, J.: Vliv technologické nekazne pri provadeni asfaltovych vozovek na vybrane materialove charakteristiky asfaltoveho betonu. Master Thesis. Faculty of Civil Engineering at Czech Technical University in Prague (2016). (in Czech)

Valentova, T., Atman, J., Valentin, J.: Impact of asphalt ageing on the activity of adhesion promoters and the moisture susceptibility. Transp. Res. Procedia. **14**, 768–777 (2016). http://www.sciencedirect.com/science/article/pii/S2352146516300667. ISSN 2352-1465

Zajicek, J.: Technologie Stavby Vozovek. ČKAIT, Praha (2014). ISBN 978-80-87438-59-6. (in Czech)

Zak, J., Luxemburk, F.: Effect of compaction and winter maintenance of asphalt mixtures on a pavement lifetime. In: Conference: GTZ Tuzla a GEO-EXPO 2012, Tuzla, Bosnia and Herzegovina (2012)

CSN 73 6121. Road building - Asphalt Pavement Courses - Construction and conformity assessment

EN 12697-26, Bituminous mixtures – Test methods for hot mix asphalt – Part 26: Stiffness

EN 12697-44, Bituminous mixtures – Test methods for hot mix asphalt – Part 44: Crack propagation by semi-circular bending test

EN 13108-1, Bituminous mixtures – Material specifications – Part 1: Asphalt Concrete

EN 13108-21, Bituminous mixtures – Material specifications – Part 21: Factory Production Control

prEN 12697-52 Bituminous Mixtures - Test Methods - Part 52 Conditioning to Address Oxidative Ageing

Numerical Study of the Failure Surface in Granular Soil Under Two Closely Spaced Strip Footings

Assma Benbouza[1(✉)], Liela Arabet[2], and Khelifa Abbeche[1]

[1] Laboratory of Applied Hydraulics (LRHYA), University of Batna, Batna, Algeria
bassma_gc@yahoo.fr
[2] Civil Engineering and Environment Laboratory LGCE, University of Jijel, Jijel, Algeria

Abstract. This work aims to present a numerical study using finite element analysis using the model of plane strain performed on sand with two closely spaced strip footings. The calculations are performed using the finite element code Plaxis. The soil is represented by the nonlinear model with hardening of soil (Hardening Soil Model) is an elasto-plastic and hyperbolic model. A parametric study revealed the role of the distance between footings on the interfering factor ($\zeta\gamma$) and failure surfaces of soil under footings. The numerical results agree with the theory of Stuart [3].

1 Introduction

The calculation of the bearing capacity of a foundation is usually done using a method that is similar to that developed by Terzaghi [4], for isolated footings. In practice, the footings are rarely isolated and they interfere with each other to some extent. A number of studies, dealing with the interference of a group of two strip footings, have been reported in literature (Stuart [3]; West and Stuart 1965; Saran and Agarwal 1974; Das and Larbi-Cherif 1983; Kumar and Saran 2003; Kumar and Ghosh 2007a, b; Kumar and Kouzer 2008). For this, the main objective of this work is to study the phenomenon. The finite element method is used in this study based on the Plaxis2D V.8 software. The results will be compared with those obtained by the theory of Stuart [3].

1.1 Theory of Stuart [3]

If the foundations are placed close to each other with the same soil conditions, the ultimate bearing capacity of each foundation may decrease due to the interference effect of the failure surface in the soil.

This theory has been studied by Stuart [3], for granular soils. The results of this study are summarized in Fig. 1.

H. Shehata and Y. Rashed (eds.), *Numerical Analysis of Nonlinear Coupled Problems*, Sustainable Civil Infrastructures, DOI 10.1007/978-3-319-61905-7_14

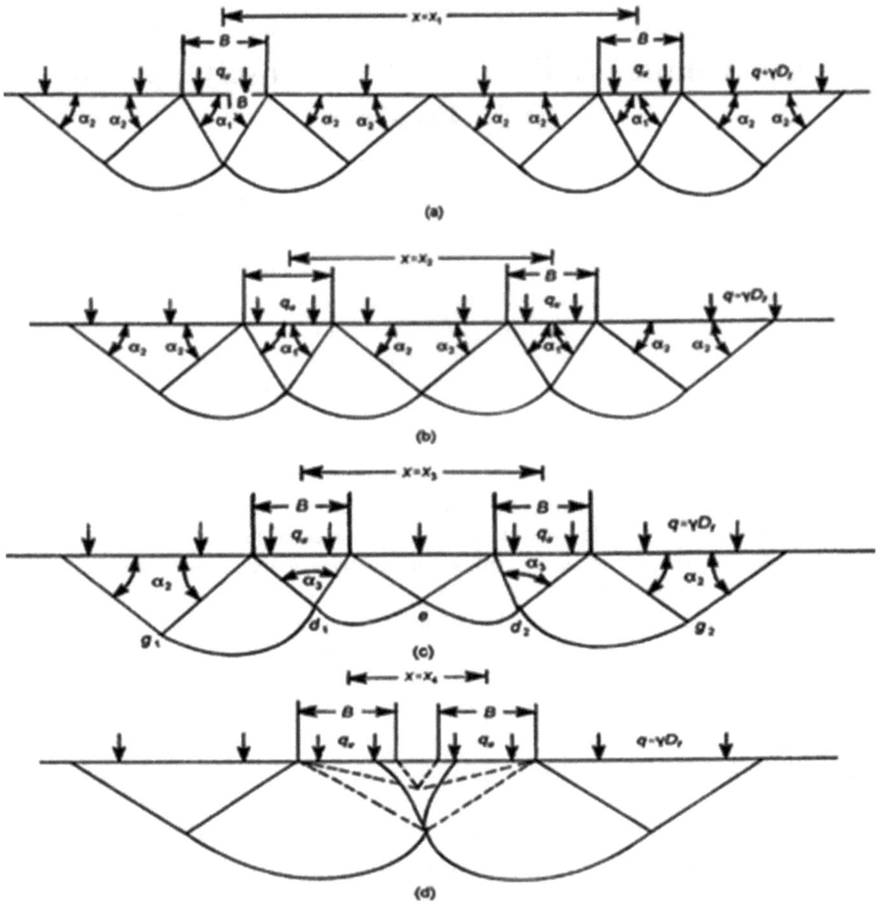

Fig. 1. Assumptions for the failure surface in granular soil under two closely spaced rough continuous foundations

Figure 1a: note that the rupture surface in the soil under each foundation will not overlap. So, the ultimate bearing capacity of each strip foundation can be given by the Terzaghi's equation (Eq. (1)). For c = 0

$$q_u = qN_q + \frac{1}{2}\gamma BN_\gamma \tag{1}$$

where Nq, N_γ = Terzaghi's bearing capacity factors.

Figure 1b: in this case the rupture surface in the soil under foundations will just overlap, then at q_u always be given by Eq. (1).

Figure 1c: This is the case where the rupture surface in the soil under each foundation will not overlap, the ultimate bearing capacity of each foundation can be given as (c = 0):

$$q_u = qN_q\zeta_q + \frac{1}{2}\gamma BN_\gamma\zeta_\gamma \tag{2}$$

where ζ_q, ζ_γ = efficiency ratios

The efficiency ratios are functions of *x*/*B* and soil friction angle φ.

The theoretical variations of ζ_q and ζ_γ are given in Figs. 2 and 3.

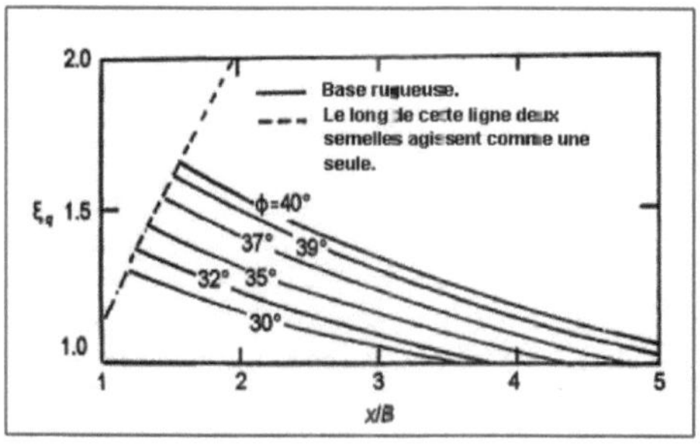

Fig. 2. Stuart's interference factor ζ_q

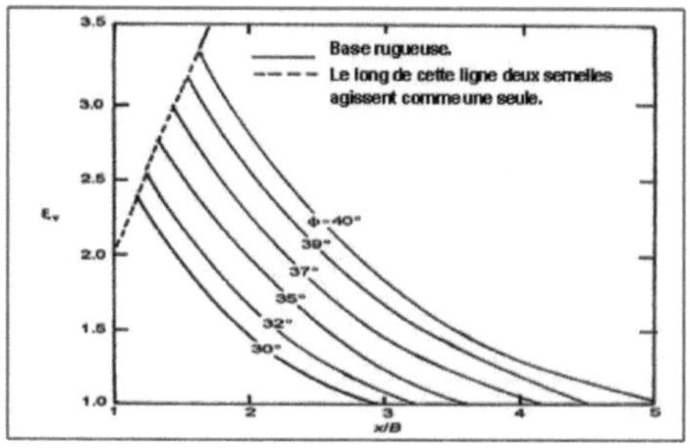

Fig. 3. Stuart's interference factor $\zeta\gamma$

Figure 1d: If the distance between the foundations is further reduced, a blockage occurs, and the pair of foundation acts as a single foundation. The soil between the two foundations will be an inverted arc that moves down with the foundation when the load is applied. When the two foundations touch, the arc zone disappears and the system behaves as a single foundation of a width equal to 2B. In this case the ultimate bearing capacity can be given by Eq. (2), with B replaced by 2B in the second term.

1.2 Numerical Model

Figure 4 shows a geometry of two closely spaced strip footings with B and Δ parameters that represent the foundation width and the spacing between the two foundations respectively.

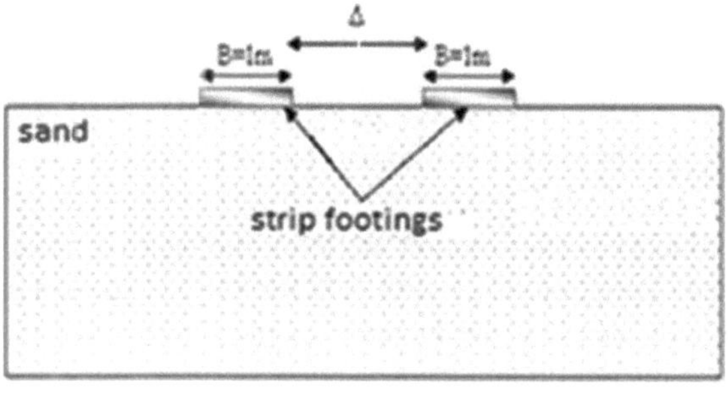

Fig. 4. Geometry of the problem

Considering drained sandy soil with the following geotechnical characteristics: $\gamma_{unsat} = 18.9$ KN/m^3; $\gamma_{sat} = 20.7$ KN/m^3, friction angle $\varphi = 35°$, cohesion c = 0 but we take c = 2 KN/m^2 for numeric computation. Using a plane strain model. the hardening soil model is used and a non-associative plastic flow rule.

To reduce the time required for each execution, only a symmetrical half-model structure was built in computer simulation according to Kumar and Kouzer (2008) [2] for example Fig. 5.

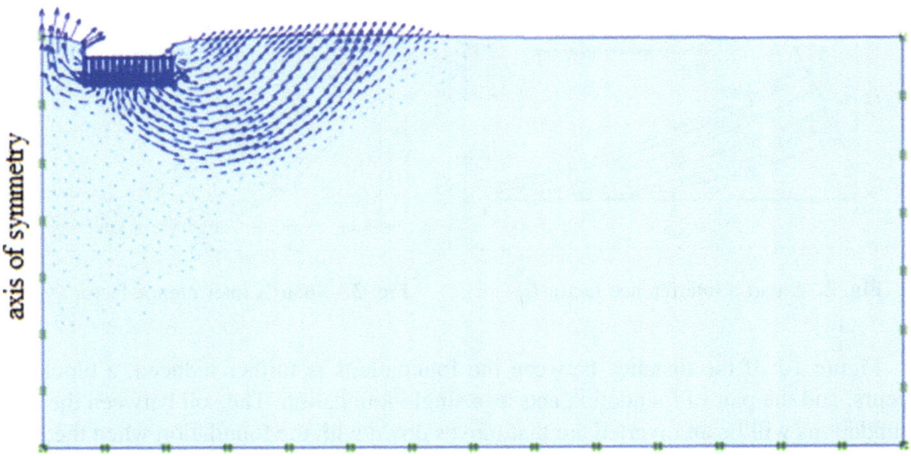

Fig. 5. Total increment displacement

In all analyzes performed in this study, we assume that footings are located on the surface of the soil and c = 0. Thus, both N_c and N_q coefficients in equation of Terzaghi are negligible. The rigidity of the footings was simulated by applying a vertical displacement on all nodes in the soi the beneath footing.

1.2.1 Efficiency Factor ($\zeta\gamma$) from the Present Analysis

According Ghazavi and Lavasan [1], to evaluate the bearing capacity of a foundation interfering in unreinforced soil, the *Efficiency factor* $\zeta\gamma$ is defined as:

$$\zeta\gamma = \frac{q_{u_{\text{int}}}}{q_{u_{\text{isolated}}}} \tag{3}$$

where $q_{u_{int}}$ and $q_{u_{isolated}}$ represents the ultimate bearing capacity for an interfering and isolate foundation respectively.

To study the influence of the friction angle on the bearing capacity, 40 numerical tests were performed. Variation of $\zeta\gamma$ with Δ/B for various friction angles is shown in Fig. 6. This figure indicates that the increase of the interference factor $\zeta\gamma$ is proportional to the increase in friction angles φ.

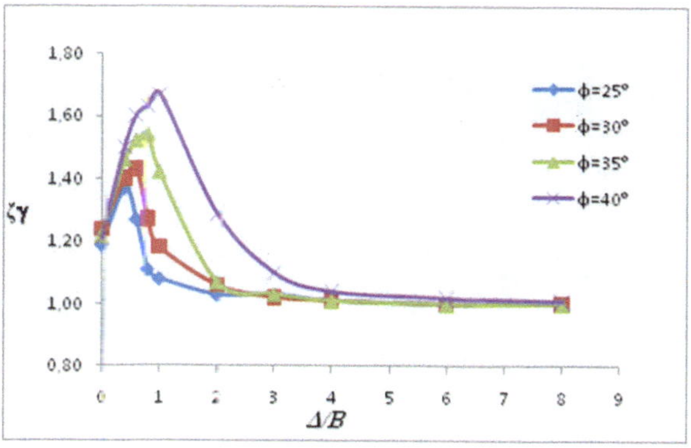

Fig. 6. Variation of $\zeta\gamma$ with Δ/B for variation of friction angle φ (25°–40°)

1.2.2 Rupture Surface

The rupture surface of the symmetrical half-model is shown in Figs. 7, 8, 9, and 10. This type of rupture surface is called generalized shear failure.

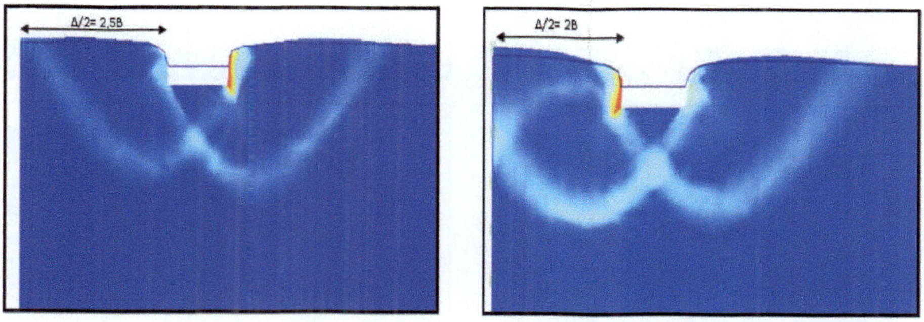

Fig. 7. Rupture surface for Δ/B = 5 et 4 respectively

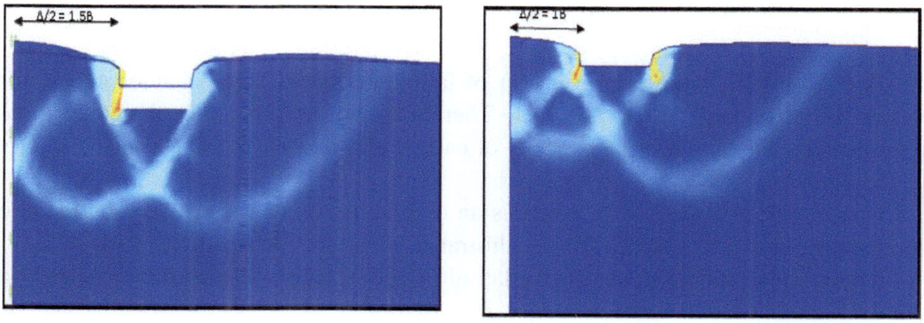

Fig. 8. Rupture surface for Δ/B = 3 et 2 respectively

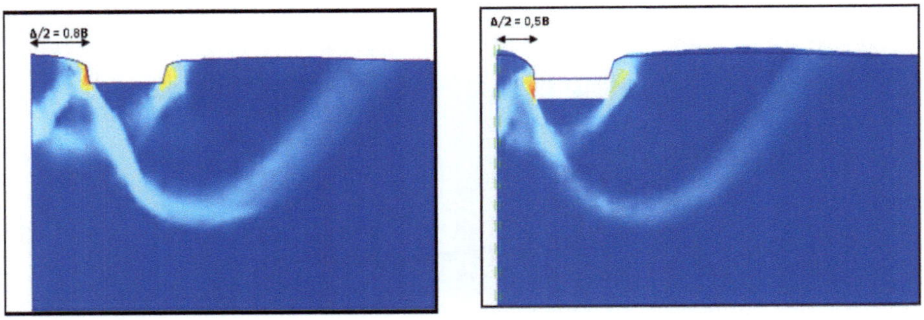

Fig. 9. Rupture surface for Δ/B = 1.6 et 1 respectively

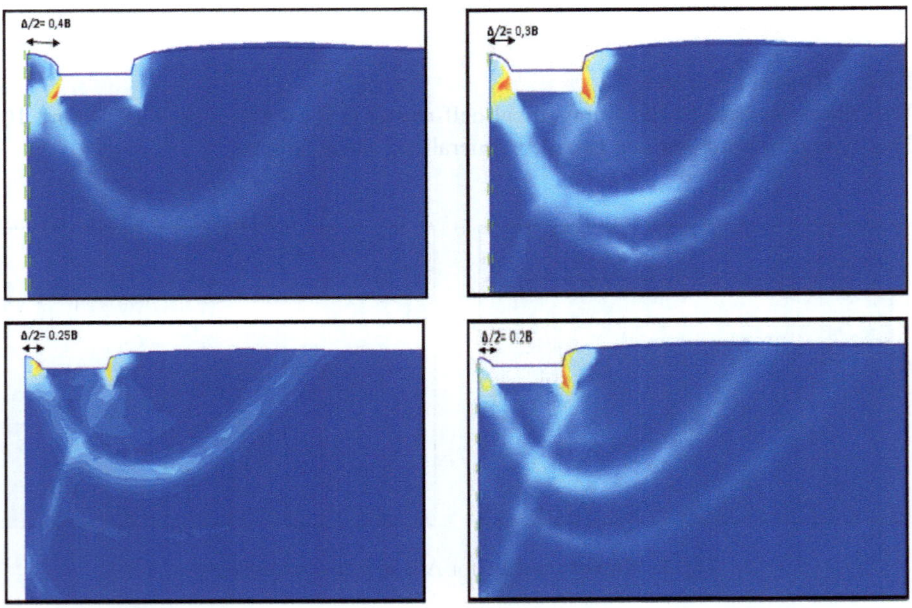

Fig. 10. Rupture surface for Δ/B = 0.8, 0.6, 0.5 et 0.4 respectively

Note in Fig. 7, there is no overlap of failure surfaces. This behavior occurs for values of Δ/B greater than $\Delta_1 = 4B$. Therefore, we can conclude that there's no interference effect and bearing capacity of each foundation is calculated independently of the other.

In case where $\Delta = \Delta_2 < \Delta_1$, there is an overlap of passive areas between the two foundations began to appear. This is illustrated in Fig. 8, for which $1.6 < \Delta_2 < 4$.

Figure 9 also shows a case of overlap of failure surfaces. This case corresponds to that of Fig. 1(c) and it's for: $\Delta = \Delta_3 < \Delta_2 = 2$.

As illustrated in Fig. 10 when the distance between the two foundations is very small or zero, the behavior is so similar to that of a single foundation with width of 2B + Δ.

1.3 Comparison and Validation of Results

To validate the results obtained by the Plaxis code, they were compared with the results of numerical analyzes (FLAC3D) which were obtained by Ghazavi and Lavasan (2008), the theoretical analysis of Stuart [3], and testing experimental Das and Larbi - Cherif (1983a, b), which are presented on Fig. 11.

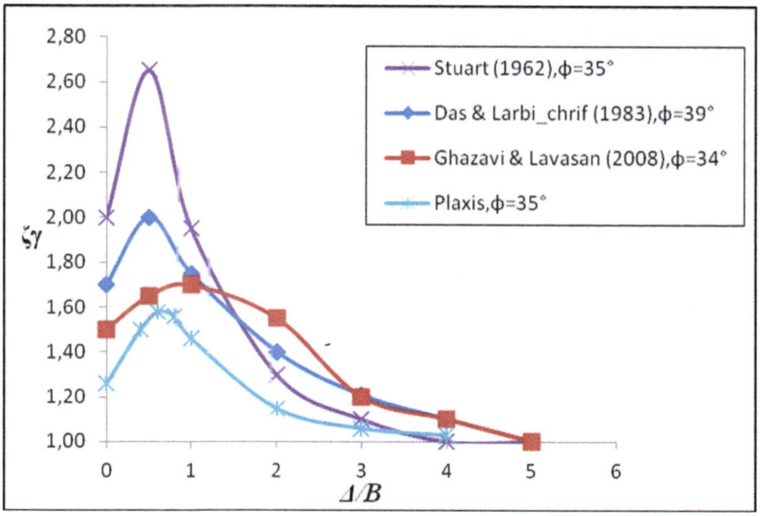

Fig. 11. Comparison of numerical, experimental and analytical results for interfering strip footing.

It appears that the general trend of interference factor variations found in this study is similar to those predicted by other studies, but there is a large variation in amplitudes between the theory and the experimental and numerical results.

From this figure, the numerical results agree very well with the results of experimental tests.

2 Conclusions

The numerical analysis showed that for $0 \leq \Delta/B \leq 1$, the ultimate bearing capacity of two closely spaced footings increases for $1 \leq \Delta/B \leq 4$, the interference factor will decrease with increasing ratio' spacing. Finally, for, $\Delta/B \geq 4$ bearing capacity remains constant. This means that for a ratio of spacing greater than 4B, no interference effects were observed and each foundation acted as an isolated foundation.

Failure mechanisms observed for granular soil are shown in Figs. 7, 8, 9 and 10, they conform to those presented by theory Stuart [3] Figs. 1(a, b, c, d) respectively.

It appears that the general trend of interference factor variations found in this study is similar to those predicted by other studies, but there is a large variation in amplitudes between the theory and the experimental and numerical results. But the numerical results agree very well with the results of experimental tests.

References

1. Ghazavi, M., Lavasan, A.A.: Interference effect of shallow foundations constructed on sand reinforced with geosynthetics. Geotext. Geomembr. **26**, 404–415 (2008)
2. Kumar, J., Kouzer, K.M.: Bearing capacity of two interfering footings. Int. J. Numer. Anal. Meth. Geomech. (2007). doi:10.1002/nag.625
3. Stuart, J.G.: Interference between foundations with special reference to surface footings in sand. Geotechnique **12**(1), 15–23 (1962)
4. Terzaghi, K.: Theoretical Soil Mechanics. Wiley, New York (1943)

Evaporation Rate Dependence with Saturation Degree

Houcem Trabelsi(✉)

Laboratoire de Génie Civil, Ecole Nationale D'ingénieurs De Tunis, Université Tunis El Manar, Tunis, Tunisia
houcemtr@yahoo.fr

Abstract. Evaporation in wet and particularly in saturated soils depends on several factors. Temperature and relative humidity are two essential parameters for determining potential evaporation (PE). Water retention is a soil characteristic that will directly influence actual evaporation (AE). The objective of this paper is to determine the evaporation rate as a function of the degree of saturation of the soil.

1 Introduction

In geotechnical and environmental engineering, increasing attention attributed to thermo-hydro-mechanical coupled problems of clayey soils regarding to the increased number of catastrophic landslides induced by degradation of soil strength caused by humidification/desiccation cycle [1–3]. In Tunisia, dryness phenomena caused considerable damage to structures built on initially quasi-saturated soils. Especially the dams and the clayey landfills may experience shrinkage with different intensities as function of the stress histories, initial hydraulic conditions, grain-size distributions of the used materials, permeability, etc. [4]. The shrinkage can also be affected by chemical interactions [5].

Two principal processes govern the exchange of water flux between soil and atmosphere. Water enters the soil surface as liquid and goes out from the soil surface as vapour through the evaporation process. Evaporation in porous media is an important process in geotechnics and environment. Many physical effects considered fluid flow, heat transfer in soil and transport of moisture. Prediction of the flux-boundary condition with respect to water flow across the soil atmosphere boundary is useful to predict the behaviour of soil during desiccation process. This paper describes how dry air flowing through a saturated soil can partially dry it and change the volume (shrinkage) and the mechanical behaviour [4]. Recent developments in the study of moisture and saturation degree in the soil are applied to the evaporation problem at soil surfaces. Desiccation cracking in drying soil is a common natural phenomenon, and it significantly affects the soil's mechanical and hydraulic behaviour [6, 7]. In this study, experimental desiccation tests were conducted on clayey soil. The preparation method considered consist of an initially saturated soil layer. Several aspects of the behaviour of the soil (water evaporation, permeability) were investigated. The results show a linear curve between the degree of saturation and evaporation rate for the different samples. Engineers have

H. Shehata and Y. Rashed (eds.), *Numerical Analysis of Nonlinear Coupled Problems*, Sustainable Civil Infrastructures, DOI 10.1007/978-3-319-61905-7_15

traditionally used a term defined as potential evaporation (PE) to estimate evaporation or evapotranspiration rates of water [8], and actual evaporation (AE) or evaporation rate (Re) to estimate the actual evaporation in unsaturated soil.

2 Material and Method

2.1 Material

Tibar clayey soil from the Tunisian area was used in this investigation. The grain size distribution of the clay is presented in Fig. 1(a). The physical properties of the soil are presented in Table 1. The water retention curve for initial slurry soil and dried at 22 °C (D) shown in Fig. 1(b) was determined following a drying path under unstressed conditions and starting from remoulded and slurry state (initial void ratio: $e_0 = 2.13$, initial dry density: $\rho_{d0} = 0.86$ g/cm^3).

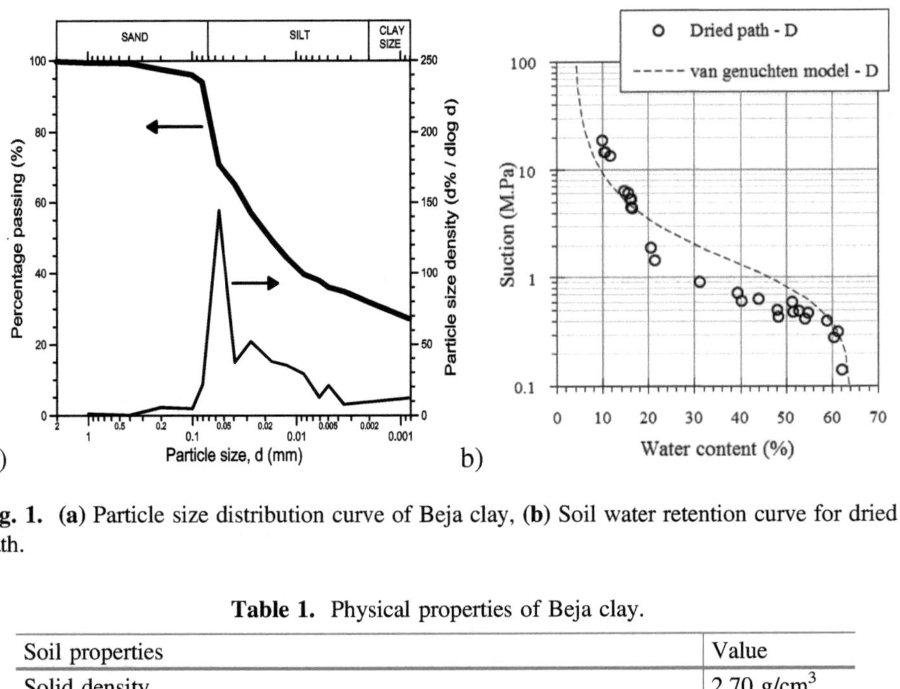

Fig. 1. **(a)** Particle size distribution curve of Beja clay, **(b)** Soil water retention curve for dried path.

Table 1. Physical properties of Beja clay.

Soil properties	Value
Solid density	2.70 g/cm^3
Liquid limit	62%
Plastic limit	30%
Shrinkage limit	around 15%
Plasticity index	32%
Fraction of fines (<75 μm)	98%
Clay-size fraction (<2 μm)	18%
Water content under hygroscopic conditions (relative humidity: 50%)	4.3%
Clay minerals (qualitative DRX)	Illite, smectite
Specific surface (mercury intrusion porosimetry)	24 m^2/g

The test material was saturated with distilled water, and sieved to 400 μm. Initial water content is approximately $1.5\omega_l$, and then it was poured into circular glass plates with 10 cm of diameter. The glass plates were placed in climatic chamber to impose fixed temperature (the humidity is quasi-constant about 60% ± 5%). When the sedimentation of the slurry soil starts a film of water increases to the surface. This film of water was not taken into account in the measure of thickness, but this quantity was taken into account in the measure of weight. This gives a saturated degree greater than one. By using the preceding procedure, some specimens were prepared for the desiccation tests. They were placed in a climatic chamber with controlled temperature and measured relative humidity (RH). The climatic chamber was designed in the laboratory of soil mechanics at the National school of Engineering of Tunis (ENIT). It was equipped with a box, digital cameras, balances (with an accuracy of 0.01 g), thermocouples and bulbs. The two cameras and the balances are connected to a computer. During drying, the specimens were weighed to an accuracy of 0.01 g, which enabled recording the water loss at varying intervals. Changes of water content ω and water evaporation rate Re (mm per unit hour) versus elapsed drying time for the tested specimens were then calculated. At the same time, a digital camera fixed directly above the specimens was used to monitor surface crack pattern evolution. Image were taken and used subsequently for images analysis technique.

2.2 Permeability and Porosity

The saturated permeability coefficient for most soils is considered constant measured. In this work, the saturated permeability coefficient is determined as a function of the porosity. Tests are carried out in an odometer suitably adapted for the determination of permeability and porosity. In a first stage, we increase the consolidation stress, and after 24 h, we determine the porosity and the void ratio. The Classical Method, developed by Terzaghi is used. After a hydraulic gradient is applied. The hydraulic head as a function of time is evaluated through Darcy law for a falling water head, and the permeability k [m/s] is determined. An exponential law has been fitted to express K_{sat} as a function of porosity given by Rodríguez Sánchez by Eq. (1) and comparing with proposed exponential law (Eq. 2).

$$K_{sat}(n) = K_0\, exp(b(\emptyset - \emptyset_0)) \tag{1}$$

$$K_{sat}(n) = K_0 \left(\frac{\emptyset}{\emptyset_0}\right)^n \tag{2}$$

Where b is a parameter of material, and K_0 is the saturated permeability of reference corresponding to the reference porosity $\emptyset_0$. Hydraulic conductivity of 5×10^{-8} m/s was measured for a porosity of $\emptyset_0 = 0.45$. A value of b = 40 was determined with experimental data shown in Fig. 2(a) and this law reproduces the measured data relatively well for the range of considered porosities. For the proposed law (Eq. 2) K_0 is the saturated permeability of reference corresponding to the reference porosity $\emptyset_0$. Hydraulic conductivity of 5×10^{-8} m/s was measured for a porosity of $\emptyset_0 = 0.45$.

A value of n = 14 was determined with experimental data shown in Fig. 2(a). This law fits quite well the experimental results. Figure 2(b) presents the variation of permeability and void ratio in a function of effective stress for loading and unloading path.

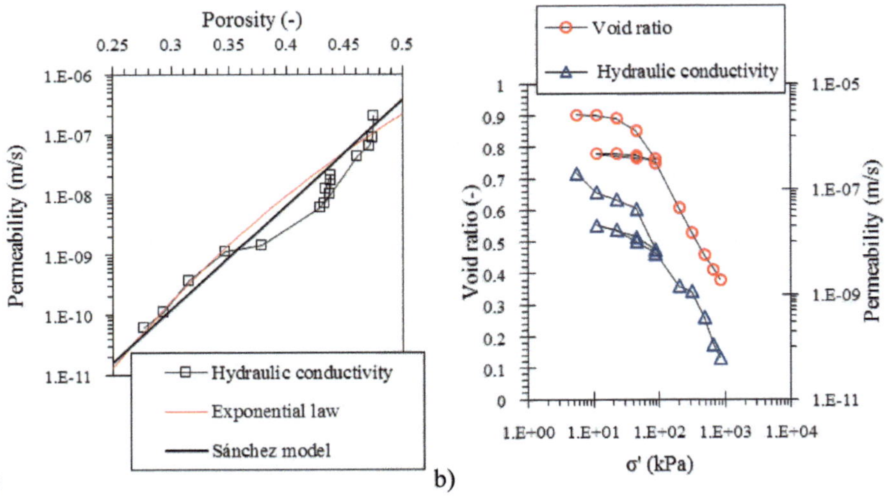

Fig. 2. **(a)** Change in permeability as a function of the porosity for the slurry sample, **(b)** Change in permeability as a function of the effective stress of the slurry sample.

With the permeability versus porosity model, it is possible to define the variation of the permeability during the desiccation. This part will be the objective of future research.

2.3 Preparation of Test Specimen and Experimental Device

Four specimens with different thickness were prepared at $1.5\omega_l$ in a cylindrical mould of a fixed diameter at the bottom and in surface grooved 80 mm. The bowl is a supple plastic. The initial water content, void ratio and thickness are presented in Table 2. The grooves of the mould intended to prevent the shrinkage during the drying. These specimens then were placed together in a climatic chamber. The air relative humidity and temperature of the neighbouring air were measured through hygrometer.

Table 2. Initials conditions of specimens.

	ω_i (Initial water content)	H_i (Initial thickness)	e_i (Initial void ratio)	PE (Potential evaporation)
E1	109%	1	2.97	2.85
E2	78.6	8	2.09	2.49
E3	54.8	14	1.42	3.91
E4	98.6	19	2.6	4.71

The specimen is placed in an air-conditioned room the temperature of which is of 35 °C ± 0,5 °C. This device allows the continuous measure of the mass, the height, and the surface of cracks determined by the analysis of the images obtained throughout the test for different time chosen manually.

3 Evaporation Rate for Slurry Clay

The used soil in this study is presented in Table 1. In Tunisia, several regions are concerned by deep or shallow landslides induced by rainfall infiltration, particular in Beja (Tibar). For this reason, we select the site of the used soil.

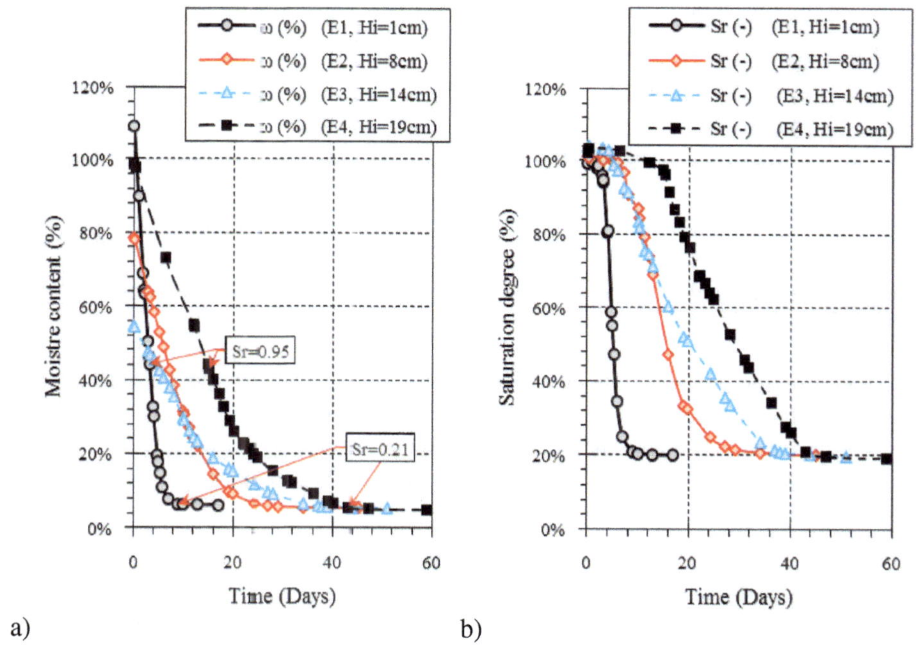

Fig. 3. **(a)** Moisture water content during desiccation process, **(b)** saturation degree during desiccation process.

Figure 3(a) Presents the variation of moisture content during the time. The moisture content decreases linearly initially in the first stage and reached the hydric equilibrium for different time. Comparing the specimens E1 and E4 the initial moisture content is similar and three stages will be presented:

- The first stage is characterised by quasi-constant evaporation (Fig. 4(a)) and quasi-saturated soil (1 < Sr < 0.95) Fig. 3(b). The decrease of water mass is linear and resulting in this stage a linear decrease of moisture content during time. The long period of this phase depends on the hydraulic conductivity, as well as the climatic conditions. This first stage is the pendular stage.

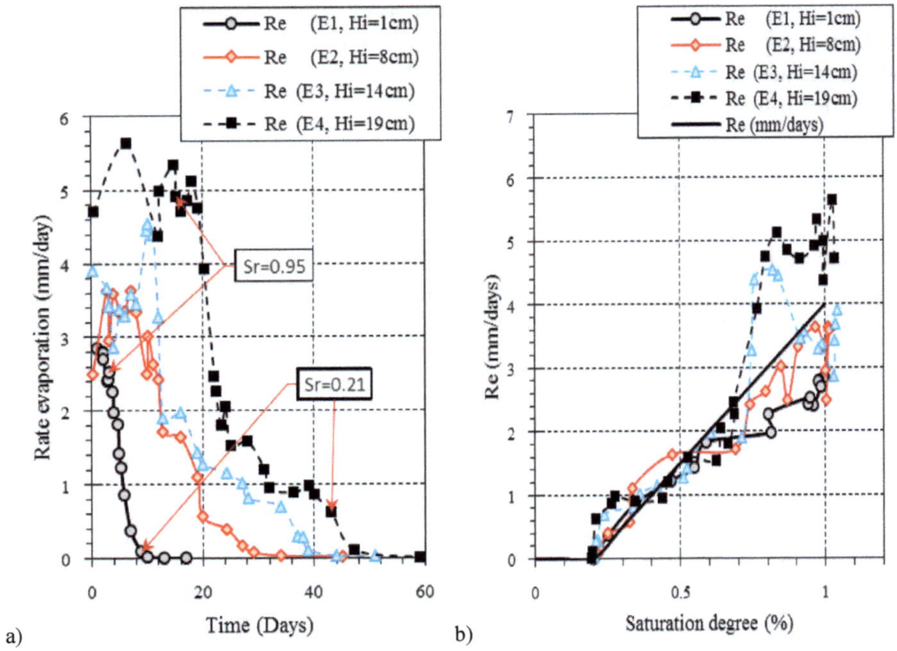

Fig. 4. Evaporation rate during desiccation process and relationship between evaporation rates versus saturation degree.

- The saturation degree and the rate evaporation decrease linearly for the second stage (Figs. 3(b) and 4(a)). In this stage, the saturation degree is less than one and the soil is unsaturated. This stage is the funicular stage (0.95 < Sr < 0.21).
- The last stage is the capillary regimes stage (Sr ≈ 0.21). In this stage the soil tend to the equilibrium regime.

Comparing E4 and E3 (Fig. 3b) in term of saturation degree we can notice a light gap between both curves and this gap increases when the thickness is smaller for E2 and E1. We vocalize our study in this part on the variation of the evaporation rate according to the saturation degree of the ground. Figure 4(a) presents the variation of the evaporation rate during the time for four specimens. The evaporation rate depends on relative humidity, temperature and the air velocity [8]. The relationship between evaporation rate and saturation degree is defined by the equation:

$$R_e = PE\left(\frac{S_r - S_r^{res}}{S_r^i - S_r^{res}}\right) \tag{3}$$

Where:

S_r^{res}: Residual saturation degree.

S_r^i: Initial saturation degree.

PE: Potential evaporation.

4 Conclusion

In this paper, we study a soil from Tunisian area particularly from Tibar city. This region concerned by deep or shallow landslides induced by rainfall infiltration. In general, the desiccation humidification phenomena induce degradation of soil hydro-mechanical characteristics. We are also interested to study the evaporation phenomena during desiccation (Wet period). Four specimens with same material and different thickness are studied. The evaporation rate behaviour depends on three stages. In the first stage, the soil is quasi-saturated and the evaporation rate is quasi-constant. In the second stage the evaporation rate and the saturation degree decrease linearly. Finally the capillary regimes stage is reached. A linear relationship between the rate evaporation versus saturation degree is proposed.

References

1. Philip, J.R.: Evaporation, and moisture and heat fields in the soil. J. Meteorol. **14**(4), 354–366 (1957)
2. Or, D.: Wetting-induced soil structural changes: the theory of liquid phase sintsintering. Water Resour. Res. **32**(10), 3041–3049 (1996)
3. Tang, C., Shi, B., Liu, C., Zhao, L., Wang, B.: Influencing factors of geometrical structure of surface shrinkage cracks in clayey soils. Eng. Geol. **101**, 204–217 (2008)
4. Trabelsi, H., Jamei, M., Zenzri, H., Olivella, S.: Crack patterns in clayey soils: experiments and modeling. Int. J. Numer. Anal. Meth. Geomech. **36**(11), 1410–1433 (2012)
5. Thyagaraj, T., Thomas, S.R., Das, A.P.: Physico-chemical effects on shrinkage behavior of compacted expansive clay. Int. J. Geomec. (2016). American Society of Civil Engineers
6. Louati, F., Houcem, T., Jamei, M.: Experimental and numerical investigation about permeability of clay subjected to humidification-desiccation cycles. In: 3ème colloque international sur les sols non satures (UNSAT Batna 2015), at Batna (2016)
7. Trabelsi, H., Jamei, M.: 3D desiccation crack pattern effects on ucs of clay. In: 3éme colloque international sur les sols non saturés, UNSAT Batna 2015, at Batna (2016)
8. Wilson, W.G., Fredlundan, D., Barbour, S.: Coupled soil-atmosphere modelling for soil evaporation. Can. Geotech. J. **31**, 151–161 (1994)

Comparative Analysis of a Deep Excavation in a Clays Sequence in Bogota City, with an Emphasis on FEM and Auscultation from the View Point of the Soil-Structure Interaction

Lucero Amparo Estevez Rey(✉)

Universidad Santo Tomas, Bogota, Colombia
luceroestevez@usantotomas.edu.co

Abstract. This document presents the results of a geotechnical analysis in a deep excavation (11.25 m) in a sequence of lacustrine clays in the city of Bogota, Colombia, where it can be found clays of high plasticity and compressibility with void ratio (e) of 3 to 5, compression index (Cc) of 1.2 to 2.0, and young's modulus E of elasticity under 1000 kPa with organic matter from medium to high. It will be analyzed the comparison of the projected measurement through the modeling of finite elements, as well as auscultated measurements of the deformation of the diaphragm walls, checked within the framework of the geological conditions studied for the execution of the project and the analysis of the soil - structure interaction used in the geotechnical design phase.

This study counts with the analysis of seven inclinometers installed in order to monitor the spatial displacements induced during the construction of three basements and piles of 40 m length to deep friction work under the structure of the building. Measurements lasted about sixteen months. In the research, it was found that the general behavior of predicted deformation is not consistent with the measured deformations. Mainly, due to the differences between the real contour conditions and the model, the wrong localization of the instrumentation, the measurement of the water table, the complex properties of rigidity of the materials, and the basic analysis of the soil - structure interaction of the project.

1 Introduction

Based on several characterizations that are made for the study of a soil profile through field and laboratory tests, the soil studies have evaluations by means of software modeling that allows to know in a theoretical way the behavior of the materials before the imposition of loads and/or removal of materials, giving the designer an idea of how the project will behave geotechnically in real scenarios.

Based on the geotechnical tests that will be carried out to study a type of soil and the determination of its bearing capacity and containment alternatives, the construction scenario should be modeled with its stages of material removal and melting of structural elements. As this allows the designer to have an idea of the in situ behavior of the soil submitted to a series of loads, and as in this case, to observe the deformation of the

H. Shehata and Y. Rashed (eds.), *Numerical Analysis of Nonlinear Coupled Problems*,
Sustainable Civil Infrastructures, DOI 10.1007/978-3-319-61905-7_16

diaphragm walls that for the present study support the subsoil that adjoins three basements of 11.25 m of total depth in a sequence of lacustrine clays.

Through the analysis of the installed instrumentation for the monitoring of a screen wall system, together with the comparison of finite element modeling results, the designer can understand the situations to which the structure is exposed, such as substantive failure, deformations, bending rupture, among others.

This paper will present a series of geotechnical analysis in different conditions of a construction project in lacustrine soils in the city of Bogotá, Colombia, along with observations that show the shortcomings when considering methodologies to carry out a containment system for deep basements, as well as from the point of view of soil - structure interaction. It was supported by research and studies by authors from different parts, including: Henri Cambefort "Geotechnics of the Engineer: Recognition of Soils" [Spanish Version], Julio Moya Barrios "Excavations in Complex Situations" [Spanish Version], Carola Sanhueza Plaza "Criteria and Design Parameters for Continuous screens in Madrid" [Spanish Version], Manuel Delgado Vargas "Soil - structure interaction (introduction to static soil interaction - foundations structure)" [Spanish version], Mariano Ruiz Vásquez "Geology applied to Civil Engineering"[Spanish Version], Proceedings of the International Geotechnical conference "Soil - Structure Interaction, Underground Structures and Retaining Walls", TC 207 ISSMGE.

2 Properties of Materials

Based on Colombian regulations for the construction of resistant earthquakes (NSR-10) and its constant updating through legal decrees, such as decree 523 of 2010, it is possible to identify the limits of the seismic microzone and adopt the design spectra in the city. The project studied is located in the "Lacustrine A" area, which consists mainly of deposits of soft clays normally consolidated reaching depths of 500 m with occasional presence of peat and/or sand of medium to low thickness. In addition, a pre-consolidated surface layer with a variable thickness of approximately 10 m is presented. Figure 1 shows the map of the seismic microzone in Bogotá.

The above information was verified and complemented by specific studies for the development of geotechnical designs. For this, five depth 50 m probes and a Down Hole test were performed; As well as the laboratory tests of a soil study. As a result, a stratigraphic profile of three strata was formed as follows: (1) Dark brown organic rubbish and limes with roots presence, from firm to hard consistency (0.0/0.5–1.5 m); (2) Greenish-gray and brown silt clay with presence of roots and rust, soft to medium consistency (1.5–4.0 m); (3) Clay in sandy matrix of olive-gray color, of fine grain with presence of sand (4.0–50.0 m). Figure 2 presents the location of the probes and the Down Hole test, as well as in Table 1 the index properties per layer.

The resistance properties of the materials were determined by the constitutive model "Mohr - Coulomb", speaking of Young's modulus, Poisson relation, Cohesion and Friction Angle. For this, a correlation between the Down Hole trial of a nearby project and this project was made. Figure 3 shows the graphical comparison of the values and in Table 2 are shown the resistance qualities.

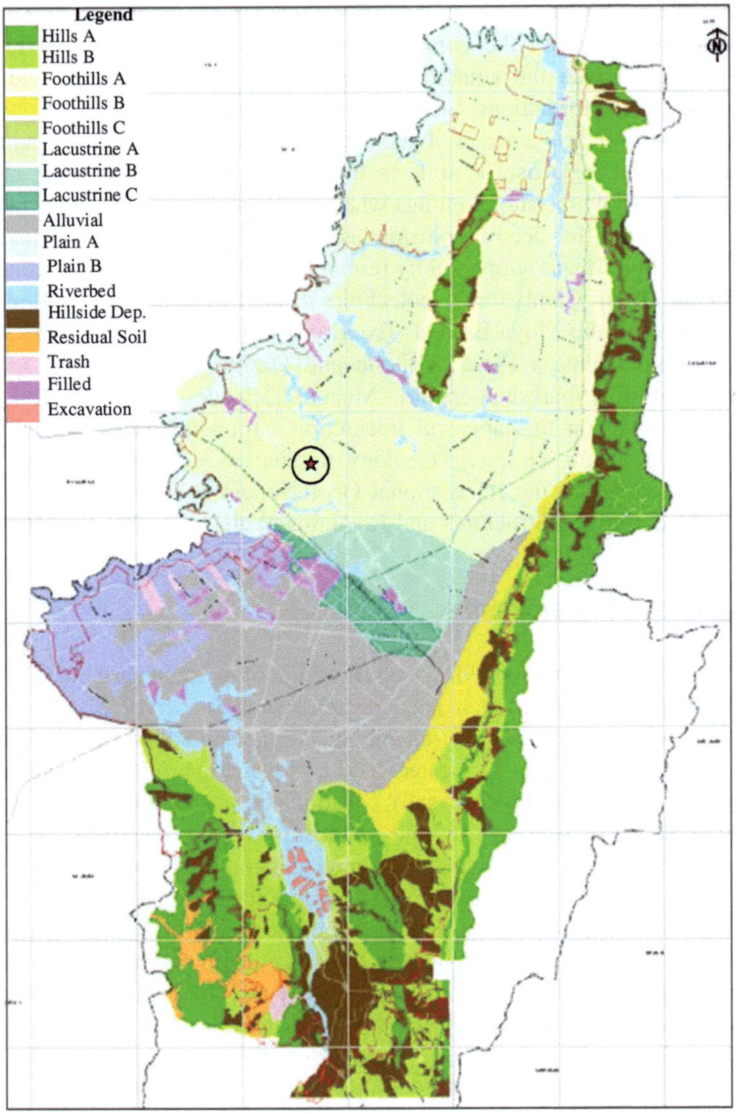

Fig. 1. Map of seismic microzoning of Bogotá and the project location

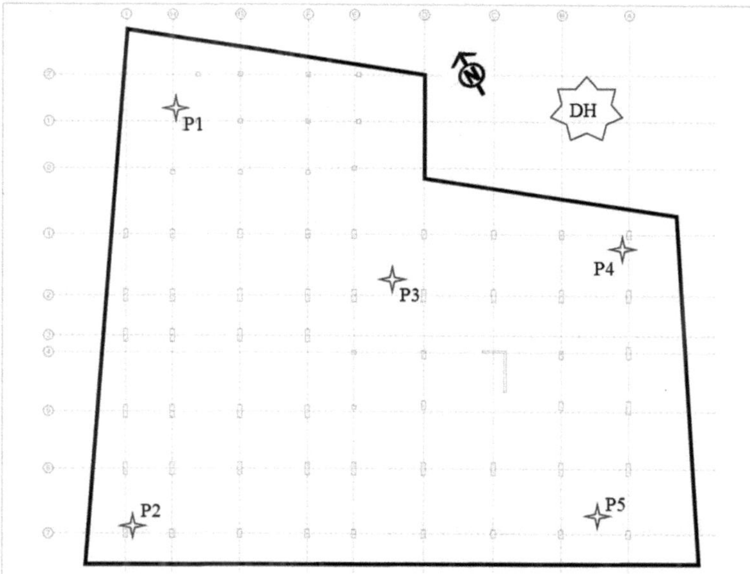

Fig. 2. Location map of probes and the Down Hole test

Table 1. Geotechnical properties varying in depth

Depth (m)	γ (T/m³)	LL (%)	LP (%)	IP (%)
0.00/0.50–1.50	1,5	155,95	57,08	98,87
1.50–4.00	1,5	169,5	56,73	112,77
4.00–50.00	1,5	141,75	71,16	121,06
Depth (m)	e	Gs	Qu (Kg/cm²)	C (Kg/cm²)
0.00/0.50–1.50	2,402	2,4	1	0
1.50–4.00	2,281	2,459	0,62	0,08
4.00–50.00	2,679	2,459	0,28	1,88
Depth (m)	υ	Friction Angle (°)	E (MPa)	N_{SPT}
0.00/0.50–1.50	0,4	26	145,34	2
1.50–4.00	0,4	22	213,34	3
4.00–50.00	0,4	19	82,45	3

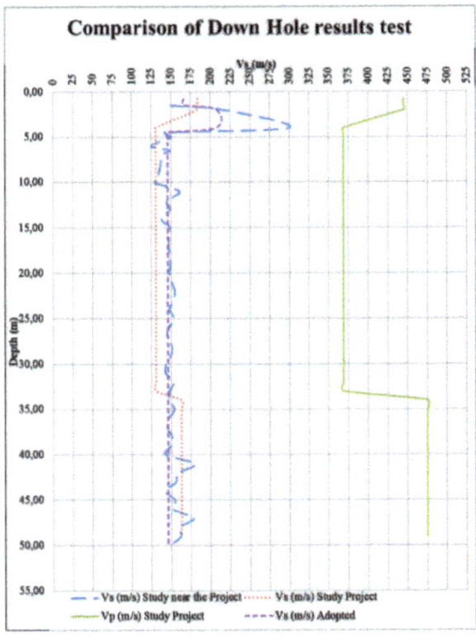

Fig. 3. Graphical comparison of results from the Down Hole test

Table 2. Summary table of resistance properties

Layer	Young's modulus (kPa)	Young's modulus (MPa)	Cohesion (kPa)	Cohesion (MPa)	Friction angle (°)	Poisson's relation
1	113329,14	113,32914	0,00	0,00	26	0,40
2	167172,95	167,17295	0,08	0,00008	22	0,40
3	79454,04	79,45404	1,88	0,00188	19	0,40

3 Modeling in Finite Elements

The project consists of a seven-story building and three basements in a structure of conventional porches with 670 T specific loads, for commercial use and with an area of 3648 m^2, which contains the material by pre-excavated diaphragm walls of depth of 16.15 m and thickness of 0.40 m. The basements have a depth of 11.25 m including underlayment plate, and it's for parking use. The deconfining was performed in 81 stages by an intervening sequence based on an octagonal geometry. The schematic plan of the project is shown in Figs. 4 and 5 the process of removal of the material.

3.1 Modeling in Phase2 Software

The finite element model was carried out using the Phase2 software of the Rockscience house, which was developed using the Mohr - Coulomb constitutive model in two dimensions. This software allows to relate all the structural scope of the project and to model through sequence of excavations. 13 constructive stages were modeled in which the behavior of the screens was observed:

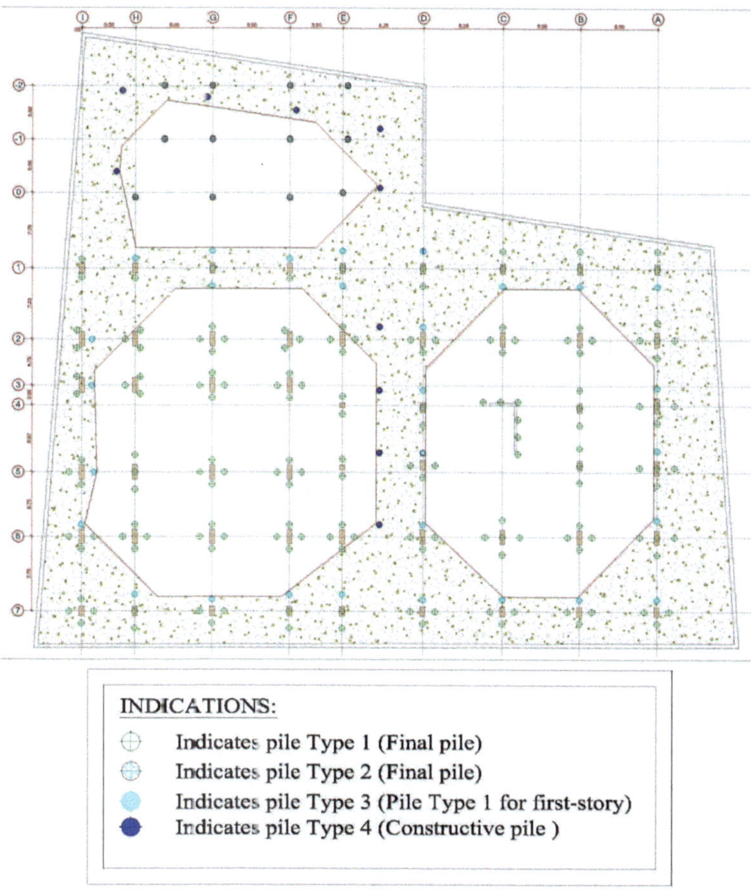

Fig. 4. Schema of shoring beams in basements, and constructive piles and final piles

1st Stage: Definition of land boundaries and soil characteristics; 2nd stage: conformation of pre-excavated screen walls, construction and final piles (01/31/2011); 3rd Stage: Establishment of the head beam (28/02/2011); 4th stage: first moment of deconfining the first basement in the evaluated profile (03/31/2011); 5th stage: deconfining of the entire basement in the assessed profile (04/28/2011); 6th Stage: melting of the belt beam (04/28/2011); 7th Stage: first moment of deconfining the second basement in the evaluated profile (06/30/2011); 8th Stage: deconfining of the entire second basement in the evaluated profile (28/07/2011); 9th Stage: melting of the belt beam (08/31/2011); 10th Stage: first moment of deconfining the third basement in the evaluated profile (09/30/2011); 11th Stage: deconfining of the entire third basement in the assessed profile (10/31/2011); 12th stage: melting of suppression plate (11/30/2011); 13th Stage: application of loads on piles due to superstructure (12/30/2011) The schematic conformation of the model through the software is shown in Fig. 6.

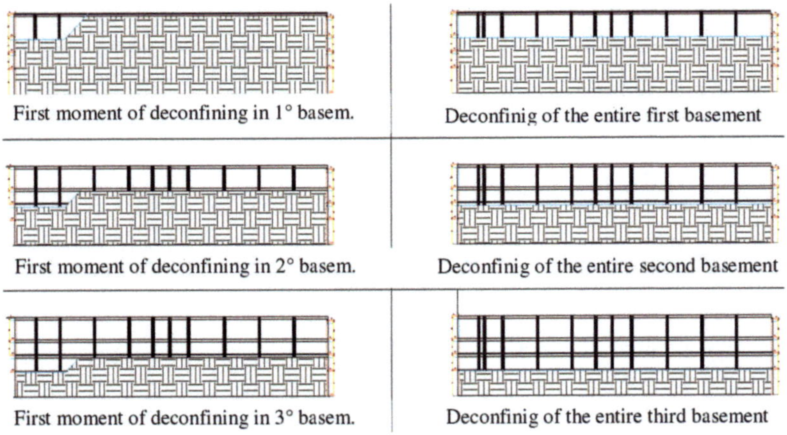

Fig. 5. Layout of material withdrawal in stages

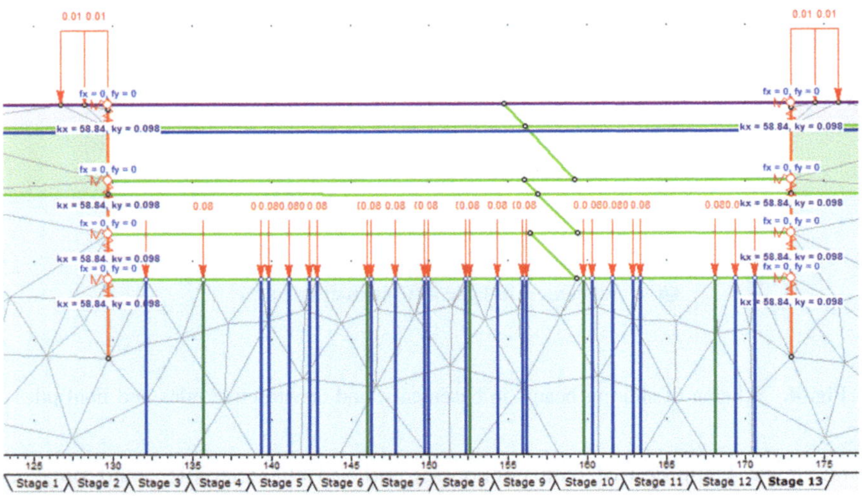

Fig. 6. Conformation of the model in the software

3.2 Results

It is observed in the results of the color range that in the stages where the head and shoring beams are added there are no significant changes in the pressures applied by the soil against the constructive element, an opposite result to when the partial and total deconfinements start. The material of the backfilling tries to move pretending to compensate the freed efforts. The screen wall is therefore prone to high deformations, and if necessary, to mass removal phenomena such as substantive failure. The underpressure slab and the action of the piles make up the first restriction to this phenomenon. The maximum displacements of the soils are in a range of 0.27 to 0.33 m. The movement of

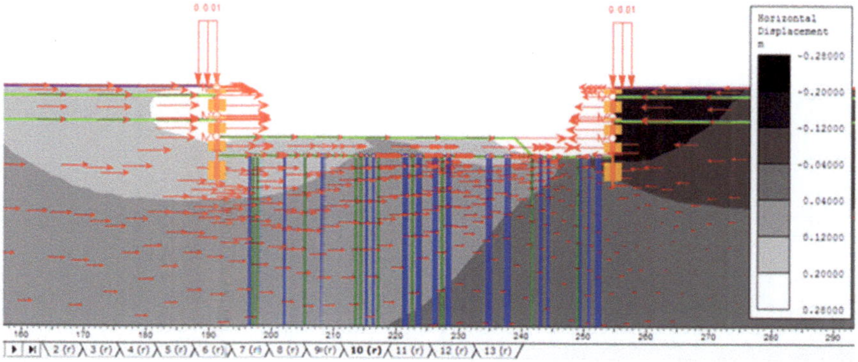

Fig. 7. Color range result for partial deconfining

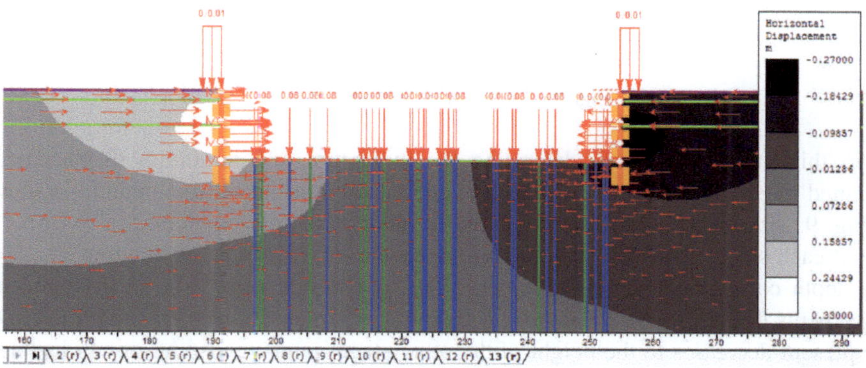

Fig. 8. Color range result for total deconfining

the soil for partial deconfining is shown in Fig. 7 in a color range, as well as in Fig. 8 in a color range the displacement for total deconfining.

In the graphical results, it is observed that the head and shoring beams do not affect the increase of the deformations but they are the structural control point of the wall, since they show the change of tendency or movement caused by the following deconfinement. This element acts elastically, because it allows deformation according to the elimination of the internal pressures of the excavation.

4 Instrumentation

Slope Indicator inclinometers and Casagrande piezometers were used for the monitoring of the screen walls in the project. The inclinometers were placed inside the wall reaching the total depth of the screen. Controls were recorded for sixteen months in total, including time before the excavation process began. Their results correspond to the sum of the incremental displacements starting from a reference point or initial date.

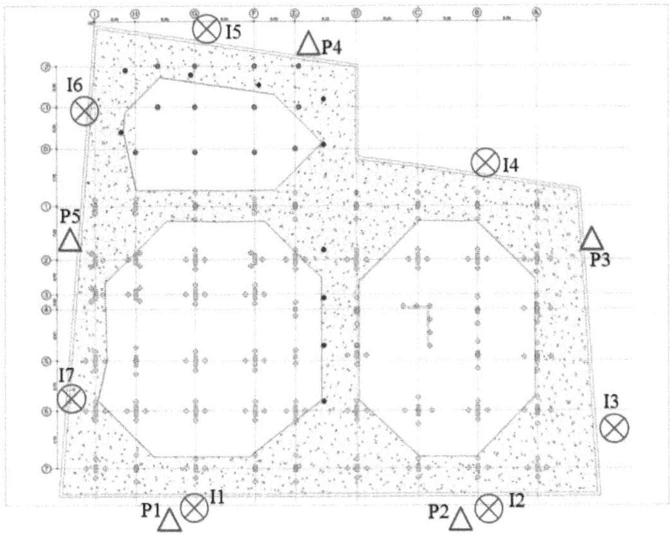

Fig. 9. Location of instrumentation

From this, it is possible to determine the maximum and minimum displacements obtained in each date or stage evaluated. The location of the instrumentation is shown in Fig. 9, distinguishing "I" as inclinometers and "P" as piezometers.

At each stage of analysis the measurements are increasing to the depth of 11.25 m, this depth corresponds to the location of the suppression plate. The most significant movements date from the phases of deconfining. There are movements that mostly do not present decreases to the height of the other stages, and a constant growth in stages like imposition of structural loads until the end of the monitoring.

5 Comparative Analysis of Inclinometry and Modeling

It is intended to give an interpretation between similarities and differences in the comparison of results of instrumentation and modeling. Since in the construction phases the strut beams did not show significant changes, significant comparisons were selected, as shown in Table 3.

Table 3. Guide for comparison of results

Results comparison	
Instrumentation (Date)	Modulation (Stage)
31/03/2011	4
28/04/2011	5
30/06/2011	7
28/07/2011	8
30/09/2011	10
31/10/2011	11

Trends in some comparisons were similar in movement, but distant in values. Also, totally different trend comparisons were found. As a representative sample of coincident and different cases, four examples will be presented, such as inclinometers 3 and 7 as cases of near results and inclinometers 4 and 5 as the different ones. Remember to observe their location in Fig. 9.

5.1 Quantitative Analysis

The dotted lines correspond to the inclinometry, the complete lines represent the results of the modeling, and the positive direction of the graphic shows the movement towards

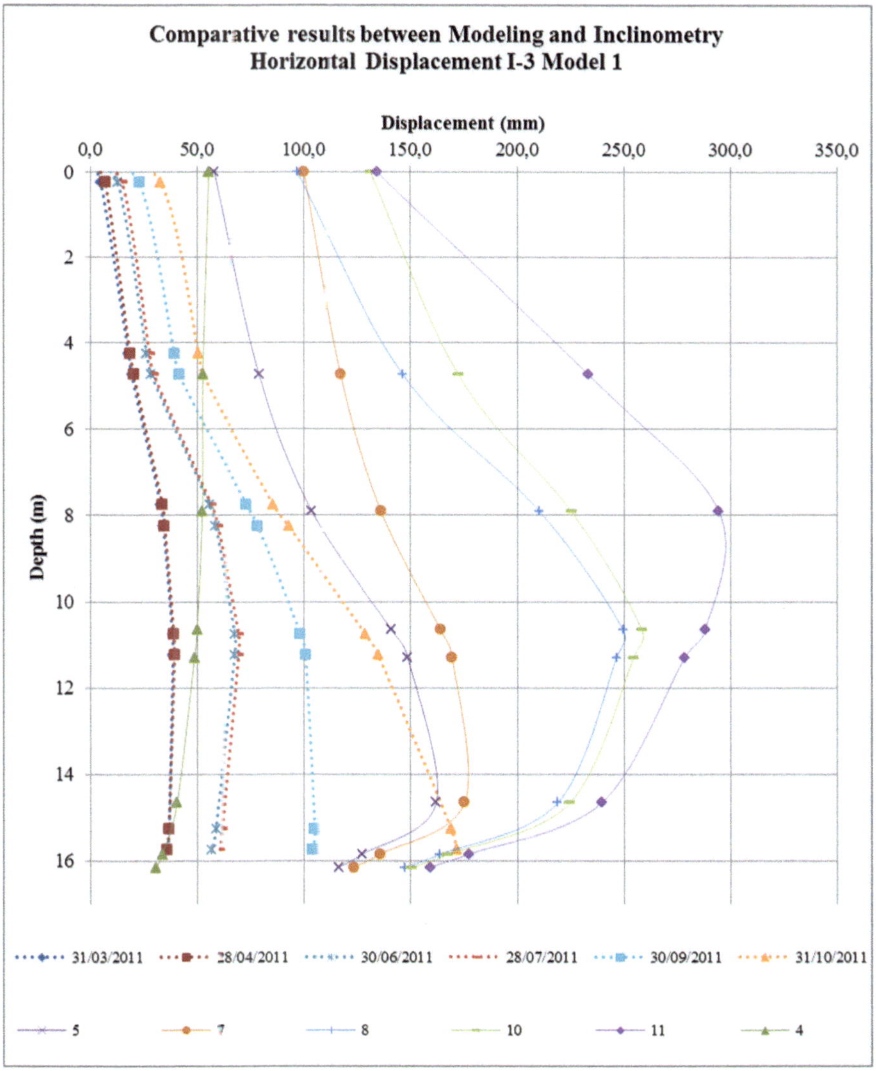

Fig. 10. Comparative graphic between modeling and inclinometry for Inclinometer 3.

the excavation. Figures 10 and 11 correspond to the coincident trend comparison of inclinometers 3 and 7 respectively, whereas Figs. 12 and 13 correspond to the comparison of different trends of inclinometers 4 and 5 respectively.

As it can be seen in Figs. 12 and 13, the instrumented tendency is always increasing in size as it descends on the screen wall, reaching the maximum displacements at the height of the subfloor slab, while the model indicates maximum deformations of 0.28 m to the depth of 8.0 m. The opposite happens in the comparison of Figs. 10 and 11, where both the instrumentation and the model have maximum displacements in the depths of 8.0 to 11.0 m corresponding to the second and third basements.

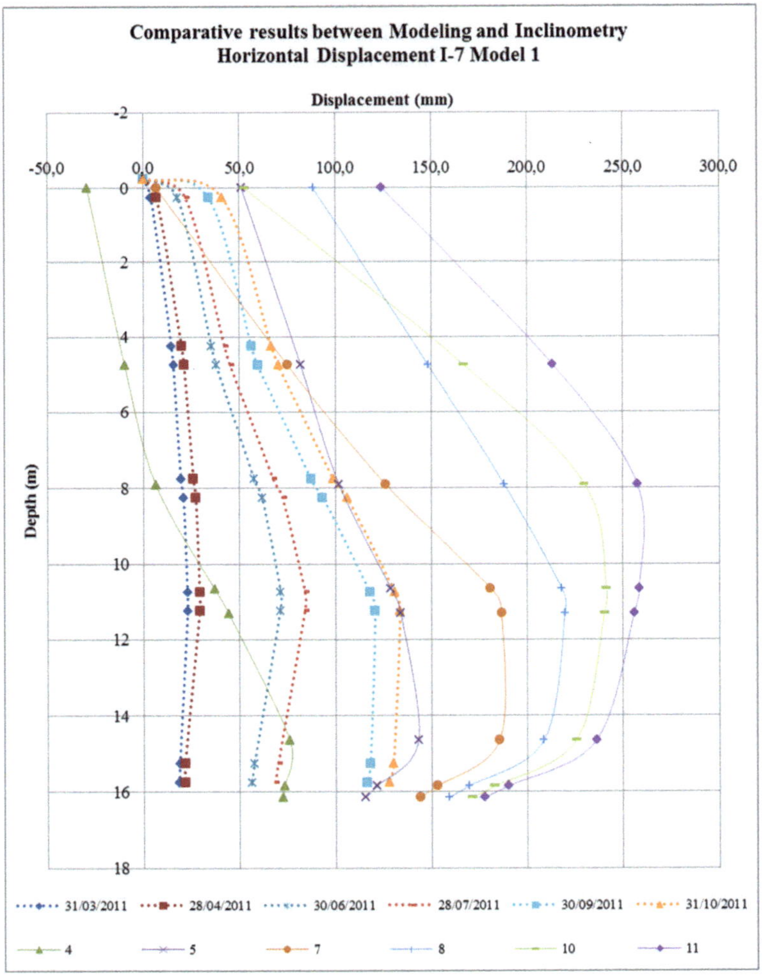

Fig. 11. Comparative graphic between modeling and inclinometry for Inclinometer 7.

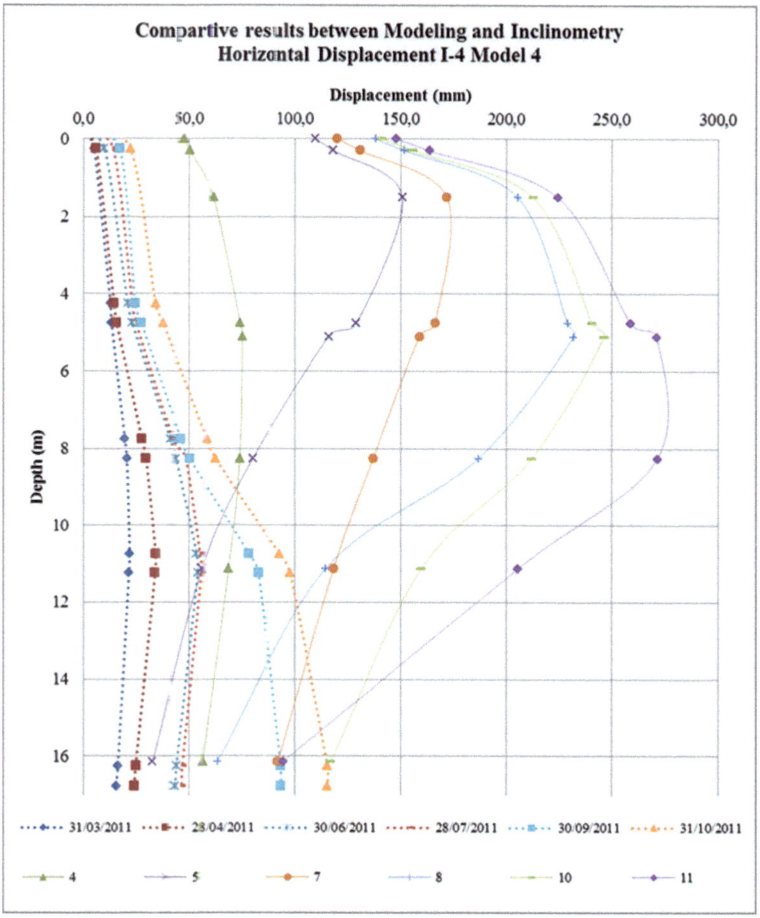

Fig. 12. Comparative graphic between modeling and inclinometry for Inclinometer 4.

5.2 Results Discussion

5.2.1 Instrumentation and Construction Process

A length of inclinometry greater than the length of the screens, located outside them and which in turn concludes their results with inclinometry in the piles, generates comprehensive results for the case study, because the type of soil (cohesive soil), needs it.

In order to make a comparison of trends and values, inclinometer data close to the screen wall should be taken into account. In the study project the inclinometers were installed inside the screen walls, which suggest a restricted reading of the real behavior of the soil.

During the construction process, it is necessary to use elements that control the verticality when the concrete is melted even on the screen walls, such as the guadua mat. During the construction phase, when choosing to construct screens excavated and melted in situ, the builder did not have precautions, which favored the creation of

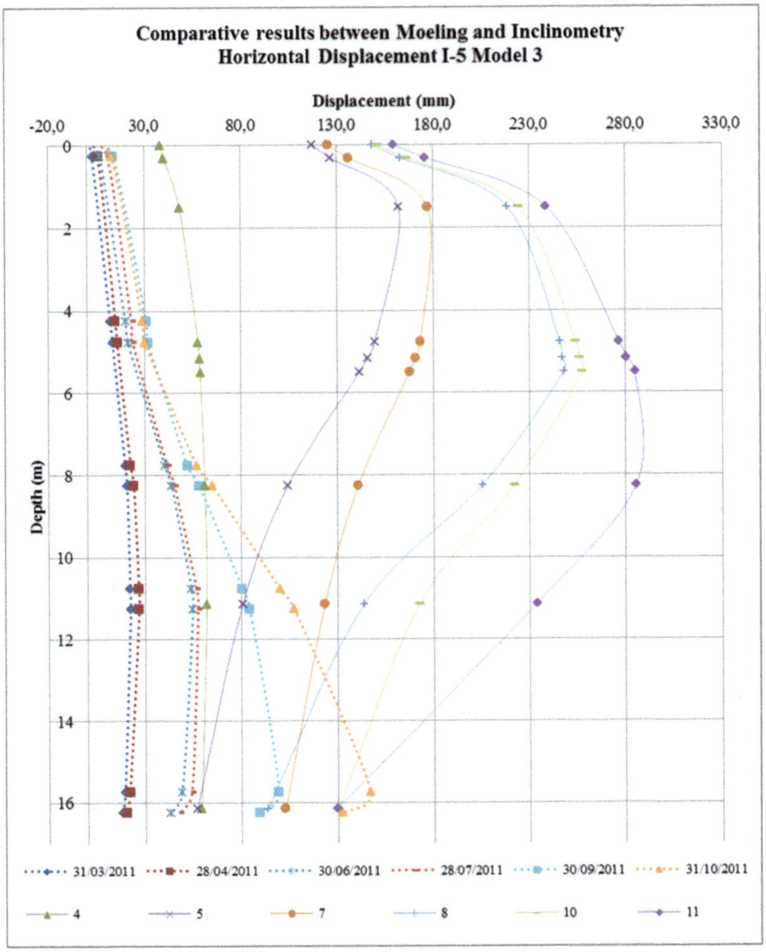

Fig. 13. Comparative graphic between modeling and inclinometry for Inclinometer 5.

deformations. Therefore, as the inclinometers were embedded in the screens, they also lost verticality. Despite all the constructive drawbacks, the comparison tendencies are different from the expected values, but they are not illogical behaviors.

5.2.2 Finite Element Model and Soil-Structure Interaction

As for the finite element model in two dimensions, the excavation processes, the construction of structural elements such as shoring and propping beams, obligates to consider the response of the soil by independent stages and not as a result of processes that depend on excavation and construction, since a two-dimensional model does not assimilate a complete excavation sequence, but as a random process, a factor that a three-dimensional model considers generates more precise results.

When using a finite element modeling method to simulate the study situation, the whole soil structure is considered simultaneously. For the case study, a soil - structure interaction analysis was included under the Mohr - Coulomb constitutive model, which usually yields good results by working with granular soils. Because the practice of geotechnical design in Colombia laboratory tests are conducted in a conventional manner for all types of soil (both for small and large construction projects), more specialized laboratory tests are required, such as triaxial rigidities in load, discharge and modulus of deformation by means of secant modulus of deformation E_{50} and consolidometer respectively. In this way, it is possible to use more specialized constitutive models for the behavior of soft soils such as the elastic - plastic “Hardening Soil” or “Cam Clay” model, where other methods of soil - structure interaction for complex soils are integrated to the model by finite elements.

The soil - structure interaction is easily adapted to the study in a soil where the modulus of elasticity varies linearly with depth and the Poisson ratio is greater than 0.5. For the case of the study project, the modulus of elasticity decreases with depth, and the Poisson ratio was adopted as a smaller value.

5.2.3 Constitutive Model

When modeling with parameters from in situ tests such as the Down Hole test, the finite element model calculates relatively large displacements in all stages of construction. The computed maximum horizontal movements are about twice to those that were measured, and the displacements profiles result as significant and unrealistic movements in the deeper layers of clay. Also, the rigidities of clays based on laboratory data are smaller than those of field.

6 Conclusions

The calculation of substantive failure through classical theories, allows to understand that the soft soil is vulnerable to the rebound if it undergoes deep excavations. That is why the construction process of a project must be planned in a correct way, to minimize the deformations of the walls that support the removal of the material.

The data obtained from the geotechnical information and its obtaining must be carefully considered, since there are shortcomings from the lack of knowledge of the field equipment that takes the samples, which will be reflected in the results of a finite element modeling. The programming of the laboratory tests must be done from a previous study of the characteristic geology at the project location, in order to select the appropriate constitutive model and obtain satisfactory results.

The finite element modeling considers variables based on theoretical calculations generating complete results, but also, predictions with failures, because it is necessary to use specialized models to study each type of soil, as well as the use of methods of observation of back analysis, in order to adjust these with the reality shown by the inclinometry, and with this, generate a more accurate model to reality.

The method of obtaining the traditional modulus of elasticity and used for the design of projects, such as the Consolidation test, generates results that are not coherently applicable, since it shows properties of an already worn out soil. A Down Hole test is more consistent in showing results, as it does not intervene in new soil efforts, but rather evaluates velocities from its normal condition.

References

Ruiz Vásquez, M., et al.: Geology Applied to Civil Engineering. Limusa, Mexico City (2015). (Spanish version)

Ulitsky, V.M., Lisyuk, M.B., Shashkin, A.G.: Soil-Structure Interaction, Underground Structures and Retaining Walls. In: Proceedings of the International Conference on Geotechnical Engineering. "Georeconstruction" Institut, St. Petersburg, Russia (2014)

Bogotá City Hall: By which the seismic microzoning of Bogotá, D.C. is adopted, Decree 523 of December 16, 2010. Department of District Planning, Bogotá, Colombia (2010)

Sanhueza Plaza, C.: Criteria and design parameters for continuous screens in Madrid. Doctoral thesis (Civil Constructor) Universidad Politécnica de Madrid, Madrid, Spain (2008). (Spanish version)

Delgado Vargas, M.: Soil - structure interaction (introduction to static soil interaction - foundations structure). Escuela Colombiana de Ingeniería, Bogotá, Colombia (1998). (Spanish version)

Moya Barrios, J.E., et al.: Excavations in complex situations. Compilation of written material for the course 'Excavations in complex situations'. Escuela Colombiana de Ingeniería, Bogotá, Colombia (1997). (Spanish version)

Cambefort, H.: Geotechnical Engineering: Soil Recognition. Translation by Juan José Sanz Llano. Technical Associates Editorial, Barcelona, Spain (1975). (Spanish version)

Assessment of Granular Soil Failure at the Water Borehole Depth in South Eastern Nigeria by Discrete and Finite Element Methods

Kennedy C. Onyelowe[1(✉)], O.A. Ubachukwu[1], O.C. Ikpemo[1], and F.O. Okafor[2]

[1] Department of Civil Engineering, Michael Okpara University of Agriculture, Umudike, P. M. B. 7267, 440109 Umuahia, Abia State, Nigeria
konyelowe@mouau.edu.ng, konyelowe@gmail.com, konyelowe@yahoo.com

[2] Department of Civil Engineering, Faculty of Engineering, University of Nigeria, Nsukka, Nigeria

Abstract. Failures of water boreholes that have impeded the performance and operation of boreholes have been recorded in recent times across the world and mostly in developing countries and within the south eastern Nigeria. To help solve some of the problems, the role of contact and seepage forces on the failure of water boreholes was explored. This study is important as the scouring of the particles collected at the wall of the transport pipe could damage pumps and result in huge financial loss to owners of boreholes. The combined finite-discrete element method was used to generate model expression from contact and seepage forces considered to be the major forces contributing to the flow of fluid through soil mass and boiling or quick sand effect. Mathematical model was developed for calculating the critical hydraulic head causing critical seepage given as $h_{(x)} = 0.000524r^2[X_5(5 - (2 * SF_{x_2}))]$ while an expression for the safe hydraulic head during well pumping was developed and given as $H_{(x)} = 0.0065814.r^4 X_5 \aleph_{x_2}$. The ability of the model to predict results was verified using the results of the test conducted in the laboratory. Correlation coefficient result has shown that there is strong agreement between model result and the laboratory result which has shown a perfect correlation of 1.00 and 0.99 for the critical state condition and equilibrium state condition respectively. For safe pumping and corresponding yield in the borehole system, inter-granular force between granular particles should equal the seepage force and this is achieved by ensuring that the deduced model expression is used to determine the safe hydraulic head. Finally, irrespective of the fact that an increase in hydraulic head increases discharge, the system should be operated at a head safe for the performance of the well and as long as the model hydraulic head expression deduced is used under the above conditions, safe pumping can be achieved.

Keywords: Assessment · Granular soil failure · Water borehole depth · South eastern Nigeria · Discrete and finite element methods

H. Shehata and Y. Rashed (eds.), *Numerical Analysis of Nonlinear Coupled Problems*, Sustainable Civil Infrastructures, DOI 10.1007/978-3-319-61905-7_17

1 Introduction

Ground water explored through water borehole is the water stored in an aquifer in pore spaces or fractures in rocks or sediments (Baars 1996; Amorin and Broni-Bediako 2010; Chukwurah, 1992; Zhang et al. 2010; Shun-cai et al. 2013). This is a readily available source of water by seepage force throughout populated Africa but the construction costs for sustainable supplies are high. The reasons why groundwater is preferred to surface water include its relative low costs compared to surface water, availability in most areas, potable without treatment, employs low cost technologies, and the frequent drought problems enforce the use of groundwater source as many small intermittent rivers and streams dry out during the dry seasons (Rasoul and Amaneh 2013; Pal 2011; Pan et al. 2014a, b; Eduvie 2006). Groundwater which can be extracted by boreholes and hand-dug wells occurs in permeable geological formations known as aquifers which have properties that allow storage and movement of water through them (Eduvie 2006; Fujiawa et al. 2013; Huihui and Cheng 2014; Hui et al. 2013; Kharagpur 2013; Konikow 2010; Sivakugan 2005). The geological structure of Nigeria gave rise to two types of groundwater; pore-type water in sedimentary cover and fissure-type water found in crystalline rocks (Chukwurah 1992; Eduvie 2006). A water borehole is drilled as a shaft usually vertical, in the earth for bringing ground water to the surface for use. An estimated 91% of total fresh water available to humanity is found as ground water which occurs in aquifers (Eduvie 2006; Soon et al. 2015; Ezomo and Akujieze 2010; Fei et al. 2014; Fox et al. 2010). During water exploitation, few or several years after the drilling of a borehole, sand particles and small sandstone particles start to break away from the borehole surface (Baars 1996; Kumar 2011; Liu and Wang 2014; Pal 2012; Serdar and Recep 2013; Sun et al. 2014; Sutham and Warakorn 2013; Tianshou and Ping 2015). The contact force existing between particles and seepage force causing dislodgment of these particles are studied (Gopal and Rao 2011). The number of particles which are transported by the fluid (water) can reach an amount that the transport pipes, well casings and other equipment will be damaged in a short period of time by the scouring of these particles (Rui and Tangdai 2013; Baars 1996; Zhang et al. 2010; Hailong et al. 2014). Consequently, the only solution up till now has been to change the electro pump (submersible pump) or make a new borehole which is especially a huge financial loss. The main aim of this work is to achieve a state of equilibrium between the opposing forces under investigation. Specifically, the objectives of this study were to: (i) develop hydraulic head equations from physical laws governing contact and seepage forces. (ii) investigate the role played by contact and seepage forces through the simulation of the borehole behavior with the Combined Finite-Discrete Element Method, CFDEM of numerical study, (iii) strike equilibrium condition between the two opposing forces, (iv) validate the programme through experimental and theoretical analysis and (v) compare the hydraulic heads at equilibrium and critical states for both simulated and experimental results.

2 Methodology

Mathematical and experimental modelling was adopted in this study. Firstly, hydraulic head equations were developed from physical laws governing contact and seepage forces using the CFDEM (Sergio et al. 2015; Shun-cai et al. 2015; Ferdowski and Shafipour 2009; Fernando et al. 2008; Onah 2012). The contact zone between three discrete particles; contactor, target and supporter were discretized into finite elements of eight triangular elements, nine nodes and bandwidth of eight, element equations formed and assembled to form the global matrix equation representing the system under study (Federic et al. 2008; Lu et al. 2007; Munjiza 2004; Manne and Neelima 2015; Mahabadi et al. 2012; Chengchun and Mengxi 2014; David and Lidija 2001; Cheng et al. 2003 and 2004; Nwaogazie 2008). By applying relevant boundary conditions, the global matrix equations were solved for the hydraulic head causing dislodgement of the granular particles within the flow region under critical condition and the hydraulic head restoring equilibrium in the system (Nwaogazie 2008; Limin et al. 2015; Jiangping et al. 2015; Fox et al. 2010; Bui and Sako 2008; Bell et al. 2005. Secondly, experimental study was conducted at the soil mechanics laboratory, Department of Civil Engineering Michael Okpara University of Agriculture, Umudike (ASME 2014), Niger Pet Laboratory, Geo-drill Nigeria Limited Laboratory to validate the developed equations and to study the effect of contact force, seepage force (ASTM 2007; Datta and Gulhati 2005; Durlofsky and Aziz 2004; BS 13771995), and power on the hydraulic heads of the flow region. Direct Shear force test was conducted, a 10 lever voltage power control was fabricated and used to conduct a prototype well failure test with a one horsepower submersible pump which discharged water at varying discharge rates until particles started dislodging which was monitored with a downhole sensor (Tianshou and Ping 2015; Soon et al. 2015; Pal 2011). Simulation results and relationships were plotted in graphs with MATLAB 2013 (Sergei 2013) and ten element displacement terms; 0.6, 1.2, 1.8, 2.4, 3.0, 3.6, 4.2, 4.8, 5.4 and 6.0 m were used to simulate the hydraulic heads with respect to equilibrium and seepage force factors of 0.1, 0.2, 0.3, 0.4, 0.5, 0.6, 0.7, 0.8, 0.9 and 1.0. Regression analysis was used to compare experimental and simulated results.

3 Results

In the mathematical model an expression for calculating the critical hydraulic head causing critical seepage was deduced as

$$h_{(x)} = 0.000524r^2[X_5(5 - 2SF_{x_2})] \tag{1}$$

And the safe hydraulic head during well pumping as

$$H_{(x)} = 0.0065814.r^4X_5\aleph_{x_2} \tag{2}$$

where $h_{(x)}$, $H_{(x)}$, r, X_5, SF_{x2}, and $\aleph_{x_2}$ are critical hydraulic head, restoring hydraulic head, radius, fluid displacement, seepage force and equilibrium stress respectively and this has been verified and validated using a laboratory check; prototype well failure test. There is strong agreement between simulation model result and the laboratory check at the element displacement term of 1.8 from the two conditions with maximum critical hydraulic heads of 3.70e^{-8} and 3.27e^{-8} and minimum of 2.31e^{-8} and 2.04e^{-8} respectively and maximum restoring head of 7.90e^{-13} and 7.20e^{-13} and minimum of 7.9e^{-13} and 1.50e^{-13} shown in Tables 1 and 2 and Figs. 1 and 2 and the correlation analysis conducted has shown a perfect and near perfect correlation of 1.00 and 0.99 for the critical state condition and equilibrium state condition respectively.

Table 1. Critical hydraulic head and seepage force model

SF_{x_2}	Critical hydraulic head $h_{(x)} = 0.000524r^2[X_5(5 - 2SF_{x_2})]$ @ X_5 equals										Lab. critical head, h_c
	0.6	1.2	1.8	2.4	3.0	3.6	4.2	4.8	5.4	6.0	
0.1	.123E−7	.247E−7	.370E−7	.493E−7	.616E−7	.739E−7	.862E−7	.985E−7	1.11E−7	1.23E−7	.327E−7
0.2	.118E−7	.236E−7	.354E−7	.472E−7	.590E−7	.709E−7	.826E−7	.944E−7	1.06E−7	1.18E−7	.313E−7
0.3	.113E−7	.226E−7	.339E−7	.451E−7	.565E−7	.677E−7	.790E−7	.904E−7	1.02E−7	1.13E−7	.299E−7
0.4	.108E−7	.215E−7	.323E−7	.431E−7	.539E−7	.646E−7	.754E−7	.862E−7	.970E−7	1.08E−7	.286E−7
0.5	.103E−7	.206E−7	.308E−7	.411E−7	.513E−7	.616E−7	.718E−7	.821E−7	.924E−7	1.03E−7	.272E−7
0.6	.097E−7	.195E−7	.292E−7	.390E−7	.487E−7	.585E−7	.682E−7	.780E−7	.877E−7	.975E−7	.259E−7
0.7	.092E−7	.184E−7	.277E−7	.370E−7	.462E−7	.554E−7	.646E−7	.739E−7	.832E−7	.924E−7	.245E−7
0.8	.087E−7	.175E−7	.262E−7	.349E−7	.436E−7	.523E−7	.611E−7	.698E−7	.785E−7	.873E−7	.232E−7
0.9	.082E−7	.164E−7	.247E−7	.328E−7	.411E−7	.493E−7	.575E−7	.657E−7	.739E−7	.821E−7	.218E−7
1.0	.077E−7	.154E−7	.231E−7	.308E−7	.385E−7	.462E−7	.539E−7	.616E−7	.693E−7	.770E−7	.204E−7

Table 2. Restoring hydraulic head and equilibrium stress model

$\aleph_{x_2}$	Restoring hydraulic head $H_{(x)} = 0.0065814.r^4 X_5 \aleph_{x_2}$ @ X_5 equals										Lab. restoring head, h_R
	0.6	1.2	1.8	2.4	3.0	3.6	4.2	4.8	5.4	6.0	
0.1	.27E−13	.53E−13	.79E−13	1.1E−13	1.3E−13	1.6E−13	1.8E−13	2.1E−13	2.4E−13	2.6E−13	1.5E−13
0.2	.53E−13	1.1E−13	1.6E−13	2.1E−13	2.6E−13	3.2E−13	3.7E−13	4.2E−13	4.8E−13	5.3E−13	1.9E−13
0.3	.79E−13	1.6E−13	2.4E−13	3.2E−13	4.0E−13	4.8E−13	5.6E−13	6.4E−13	7.1E−13	7.9E−13	2.4E−13
0.4	1.1E−13	2.1E−13	3.2E−13	3.6E−13	5.3E−13	6.4E−13	7.4E−13	8.5E−13	9.5E−13	10.6E−13	3.1E−13
0.5	1.3E−13	2.6E−13	4.0E−13	5.3E−13	6.6E−13	7.9E−13	9.3E−13	10.6E−13	11.9E−13	13.2E−13	3.9E−13
0.6	1.6E−13	3.2E−13	4.8E−13	6.4E−13	7.9E−13	9.5E−13	11.1E−13	12.7E−13	14.3E−13	15.9E−13	4.6E−13
0.7	1.8E−13	3.7E−13	5.6E−13	7.4E−13	9.3E−13	11.1E−13	13.0E−13	14.8E−13	16.7E−13	18.5E−13	4.8E−13
0.8	2.1E−13	4.2E−13	6.4E−13	8.5E−13	10.6E−13	12.7E−13	14.8E−13	16.9E−13	19.0E−13	21.2E−13	5.5E−13
0.9	2.4E−13	4.8E−13	7.1E−13	9.5E−13	11.9E−13	14.3E−13	16.7E−13	19.0E−13	21.4E−13	23.8E−13	6.4E−13
1.0	2.7E−13	5.3E−13	7.9E−13	10.6E−13	13.2E−13	15.9E−13	18.5E−13	21.2E−13	23.8E−13	26.5E−13	7.2E−13

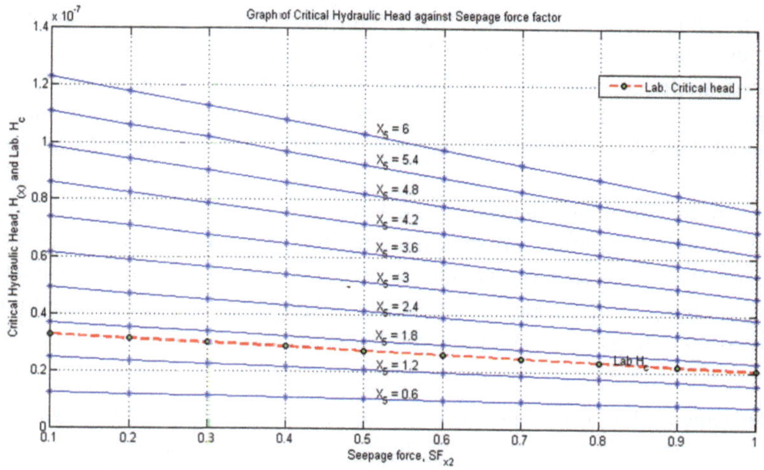

Fig. 1. Model and laboratory critical hydraulic head of system

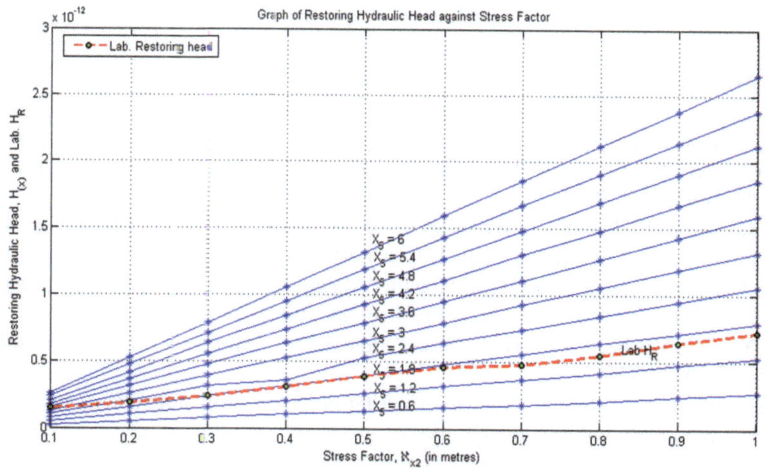

Fig. 2. Model and laboratory restoring hydraulic heads of system

4 Conclusion

Interestingly, the following conclusions can be drawn from the study;

1. The modelling of the effect of contact and seepage forces on the failure of water boreholes has yielded validated results as shown by graphs and equations. An expression has been deduced from the model as shown in Eqs. (1) and (2) to compute critical hydraulic head causing quicksand effect and the hydraulic head at which safe pumping is restored in boreholes across Umuahia and other south eastern urban and suburban dwellings respectively. For safe pumping and

corresponding yield in the borehole system, inter-granular force between granular particles should equal the seepage force and this is achieved by ensuring that the deduced model expression is used to determine the safe hydraulic head.
2. Irrespective of the fact that an increase in hydraulic head resulted in increases in discharge, the system should be operated at a head safe for the performance of the well.
3. And from the correlation analysis carried out on the model and the observed values, the critical state condition has shown a perfect correlation of 1.00 while that of the equilibrium state correlation has also shown a near perfect correlation of 0.99.
4. Finally, so long as the model hydraulic head expression generated is used under the specified conditions, safe pumping can be achieved.

Acknowledgement. I wish to thank my research supervisor and mentor, Prof Fidelis O. Okafor for his immense contribution and help during this research work and especially to my friend Prof Emeka S. Obe for his assistance. I am also grateful to Geo-Drill Nigeria Limited, Umuahia, Civil Engineering Laboratory, Michael Okpara University of Agriculture, Umudike, Niger Pet Laboratory, Uyo and Okey Electrical, Aba for their expertise.

References

Abia State Ministry of Environment (2014), Geological Data of Umuahia, Abia State

American Society for Testing Materilas: Annual Book of ASTM Standards, Part 19. Soil and Rock, Philadelphia (2007)

Amorin, R., Broni-Bediako, K.: Application of Minimum Curvature Method to Wellpath calculations. Int. J. Eng. Math. Intell. **1**(3), 69–80 (2010)

Baars, S.: Discrete Element Analysis of Granular Materials. TROISS, Netherland (1996)

Bardet, J.: Experimental Soil Mechanics. Prentice Hall, New Jersey (1997)

Bell, N., Yu, Y., Mucha, P.J.: Particle Based Simulation of Granualr Materials. Eurographics Association, USA (2005)

BS 1377.: Methods of Testing Soils for Engineering Purposes: British Standards Institution. Her Majesty's Stationery Office, London (1995)

Bui, H.H., Sako, K.: SPH-based numerical simulations for large deformations of geomaterial considering soil-structure interation. In: 12th International conference at International Association of Computer Methods and Advances in Geomechanics, pp. 570–578, 1–6 October, India (2008)

Carbon Capture & Storage Class (UU & NMT), (2010), Autumn semester2010 supplimentary notes about GW flow equation-October 18, 2010, NY

Cheng, Y.P., Bolton, M.D., Nakata, Y.: Crushing and plastic deformation of soils simulation using DEM. Geotechnique **54**(2), 131–141 (2004)

Cheng, Y.P., Nakata, Y., Bolten, M.D.: Discrete element simulation of crushable soil. Geotechnique **54**(7), 633–641 (2003)

Chengchun, Q., Mengxi, Z.: Discrete element modeling of dynamic triaxial test on reinforced saturated sand. Electron. J. Geotech. Eng. **19**(Z3), 10661–10672 (2014)

Chukwurah, C.C.: Basic Borehole Technology and Practice. Olulu Publishers, Ekpoma (1992)

Datta, M., Gulhati, S.: Geotechnical Engineering. Mcgraw-Hill Companies, New Delhi (2005)

David, M.P., Lidija, Z.: Finite Element Analysis in Geotechnical Engineering Application. Thomasstelfold, USA (2001)

Durlofsky, L.J., Aziz, K.: Advanced Technique for Reservior Simulation and Modelling of Non-conventional Wells. Final report to Standford University, USA (2004)

Eduvie, M.O.: Borehole failures and ground water development in Nigeria. A paper presented at a national Seminar held on the occasion of water Africa exhibition lagos, 15th November (2006)

Ezomo, F.O., Akujieze, C.N.: Geophysical investigation of shallow aquifer in idoa village of Edo State, Nigeria. J. Emerg. Trends Eng. Appl. Sci. **2**(5), 759–762 (2010)

Fei, X., Nong, Z., Xiaowei, F.: The estimation method of the borehole stability based on unified strength criterion. Electron. J. Geotech. Eng. **19**(4), 19065–19078 (2014)

Ferdowsi, A.S., Shafipour, R.: Numerical Modeling of Undrained Cyclic Behaviour of Granular Media Using DEM, pp. 485–493. Society for Rock Mechanics and Engineering Geology, Singapore (2009)

Fernando, A., Muhlhaus, H.B., Herrmann, H.J.: Micromechanical investigation of soil plasticity using a discrete model of polygon particels. Theor. Appl. Mech. **35**(1–3), 11–28 (2008)

Fox, G.A., Heaven, D.M., Wilson, G.V., Eddy, J.L., Fox, A.K., Maria, L.C.: Numerically predicting seepage gradient forces and erosion sensitivity to soil hydraulic properties. J. Hydrol. **389**, 354–362 (2010). Elsevier

Frederic, V.D., Vincent, R., Magnier, S.: Advances in DEM applied to soil, rock and concrete mehcanics. Electron. J. Geotech. Eng. Bouquet **108**, 1–34 (2008)

Fujisawa, K., Muvakami, A., Nishimura, S., Shuku, T.: Relations between seepage force and velocity of sand particles during sand boiling. Geotech. Eng. J. **44**(2), 9–17 (2013)

Gopal, R., Rao, A.S.R.: Basic and Applied Soil Mechanics R, 2nd edn. NAI Publishers, New Delhi (2011)

Hailong, L., Haibo, B., Cheng, Y.: Macroscopic rules and application on bingham fluid flowing in the random fracture network. Electron. J. Geotech. Eng. **19**(Z), 2931–2939 (2014)

Huihui, C., Cheng, J.: Numerical simulation of hydraulic fracturing process of rock mass based on F-RFPA. Electron. J. Geotech. Eng. **19**(W), 6879–6886 (2014)

Hui, Z., Deli, G., Wensheng, L.: Risk assessment of relief well based on analytic hierarchy process. Electron. J Geotech. Eng. **18**(U), 5269–5277 (2013)

Jianping, L., Yumong, C., Xinxia, C.: Vibration behaviour of drilling mechanism based on FEM and multibody dynamics. Electron. J. Geotech. Eng. **20**(3), 2334–2355 (2015)

Kharagpur, T.: The science of surface and groundwater; principles of groundwater flow, module 6, lesson 6, version 2, CE IIT, India (2013)

Konikow, L.F.: Representing pump-capacity relations on ground water simulation models. Nat. Ground Water Assoc. **48**(1), 106–110 (2010)

Kumar, S.: Mud Engineering. Sivasagar Assets Publishing, India (2011)

Limin, W., Yancheng, Y., Fei, W.: Fluid coupled-DEM modeling of pile- soil system under seismic loading. Electron. J. Geotech. Eng. **20**(6), 1377–1391 (2015)

Liu, Y., Wang, Z.: Particle flow numerical analysis of cohesive soil in direct shear test. Electron. J. Geotech. Eng. **19**(Z), 9465–9478 (2014)

Lu, H., Anderson, M.T., Likos, W.J., Mustoe, G.W.: A discrete element model for Kaolinite aggregate formation during sedimentation. Int. J. Numer. Anal. Meth. Geomech. **32**, 965–980 (2007). John Wiley and Sons Ltd, USA

Mahabadi, O.K., Lisjak, A., Munjiza, A., Grasselli, G.: New combined finite-discrete element numerical code for geomechanical applications. Int. J. Geomech. ASCE **12**, 676–681 (2012)

Manne, A., Neelima, S.D.: A review on the discrete element modelling of dynamic laboratory tests for Liquefaction assessment. Electron. J. Geotech. Eng. **20**(1), 21–46 (2015)

Munjiza, A.: The Combined Finite Discrete Element Method. Wiley, England (2004)

Nwaogazie, I.L.: Finite element modelling of engineering systems with emphasis in water resources. Chapters 1, 2, 3, 5 and 9. Special Publishers, Enugu (2008)

Onah, H.N.: Different element methods in engineering practice. Niger. J. Technol. **31**(3), 288–292 (2012)

Pal, S.: Drilling Fluid Engineering. Pale Skalle and Ventus Publishing, London (2011)

Pal, S.: Exercises within Drilling Fluid Engineering. Pale Skalle and Ventus Publishing, London (2012)

Pan, H., Chen, M., Qin, R.: Considering the effect of reservoir water level lifting on the slope displacement analysis of seepage field-stress field coupling. Electron. J. Geotech. Eng. **19**(3), 1447–1456 (2014a)

Pan, H., Qin, R., Chen, M., Han, L.: The analysis of the seepage pore water pressure as reservoir water level is being lifted. Electron. J. Geotech. Eng. **19**(R), 6749–6759 (2014b)

Rasoul, A., Ameneh, F.D.: Evaluation of fluid flow from a dam foundation using numerical modeling technique by UDEC software. Electron. J. Geotech. Eng. **18**(I), 1787–1800 (2013)

Rui, H., Tangdai, X.: Pseudo-dynamic analysis for seismic active earth resistance with steady seepage. Electron. J. Geotech. Eng. **18**(M), 2653–2666 (2013)

Serdar, K., Recep, I.: Numerical analysis of groundwater flow through a rectangular cofferdam. Electron. J. Geotech. Eng. **18**(K), 2041–2052 (2013)

Sergei, A.: Introduction to MATLAB (Basics). Technion, Texas (2013)

Sergio, A.B.J., Fabio, A.P., Sergio, T., Fernando, S.A.J.: The influence of frictional coefficient of particles on numerical simulations by the coupled discrete element-fluid dynamics on morphology of density flows. Electron. J. Geotech. Eng. **20**(3), 1003–1010 (2015)

Shun-cai, L., Dan, M., Zhan-qing, C., Yu, L.: Experiment study of water seepage effect on the time-dependent deformation of granular coal gangue. Electron. J. Geotech. Eng. **18**(M), 2605–2616 (2013)

Shun-cai, L., Ming-ming, W., Yu, L.: Grain size effect on the seepage properties of broken coal gangue. Electron. J. Geotech. Eng. **20**(3), 851–860 (2015)

Sivakugan, N.: Permeability and seepage. Int. J. Environ. Eng. Am. Soc. Civ. Eng. **289**, 1165–1181 (2005)

Soon, M., Mohd, A.M.I., Chee, G.T., Rais, Y.: Subsurface characterization using seismic refraction and borehole drilling for slope stability study. Electron. J. Geotech. Eng. **20**(6), 1641–1650 (2015)

Sun, J., Wang, H., Guo, S.: Numerical simulation on coal permeability change after pumping confined water from adjacent seams. Electron. J. Geotech. Eng. **19**(R), 4703–4714 (2014)

Sutham, R., Warakorn, M.: The characteristic of aquifer drawdown and settlement of Bangkok and vicinity area. Electron. J. Geotech. Eng. **18**(W), 5573–5584 (2013)

Tianshou, M., Ping, C.: Wellbore Stability Analysis of Inclined Wells in the AY Field. Electron. J. Geotech. Eng. **20**(4), 1313–1329 (2015)

Zhang, J., Jiang, S., Wang, Q., Hou, Y., Chen, Z.: Critical hydraulic gradient of piping in sand. In: Proceedings of the 20th International Offshore and Polar Engineering Conference, 20–25 June, Beijing, China (2010)

Effect of NBR-Waste on Rheological Properties of Modified Bitumen and Mechanical Characteristics of the Asphalt Mix

Khedoudja Soudani[1(✉)], Véronique Cerezo[2], and Smail Haddadi[1]

[1] Laboratory of Environment Water, Works and Geomechanics (LEEGO), USTHB, Algiers, Algeria
soudani.khedoudja@yahoo.com

[2] Environmental Assessment, Safety and Eco-design Laboratory, IFSTTAR, Nantes, France

Abstract. Under the effect of severe climatic conditions and heavy vehicles traffic, premature and irreversible damages occur on road pavements carried out with conventional bituminous materials.

To deal with this situation, one of the most efficient way is the modification of bitumen by adding polymers. Moreover, use of waste polymers to modify bitumen is a promising solution in view of protecting environment.

The properties of modified bitumen are determined through rheological tests. This led to a better knowledge of bitumen behavior under various thermal and mechanical conditions.

In the present work, properties of modified bitumen by NBR-waste (waste of sole of shoes) are assessed and compared to the properties of 40/50 penetration grade bitumen. Then, mechanical characteristics of asphalt mixes are determined by using Marshall test.

Rheological tests are performed with a Haake Rheostress 1 rheometer. Frequency and temperature sweep tests in oscillatory shear are performed within the linear viscoelastic range of the materials. Results show that NBR-waste influences bitumen rheology, by increasing viscoelastic function and decreasing phase angle. The NBR-waste bitumen is characterized by higher stiffness and elasticity, due to an increase of the elastic (storage) modulus G′. Also, the NBR-waste improves rheological behavior of bitumen. We have also achieved a marked increase in the Marshall quotient, which reflects an improvement of the mechanical performances of waste NBR modified asphalt.

1 Introduction

The increasing demand for transportation generates higher and heavier level of traffic. When combined with severe weather conditions, it leads to the rapid deterioration of pavement. The main approach used for preventing this deterioration is an improvement of the materials properties used for roads construction (Köfteci 2014). In Algeria, flexible pavements are the most commonly used in road construction. The sustainability and the performances of these pavements are strongly related to the characteristics of their components (bitumen and aggregates).

H. Shehata and Y. Rashed (eds.), *Numerical Analysis of Nonlinear Coupled Problems*,
Sustainable Civil Infrastructures, DOI 10.1007/978-3-319-61905-7_18

However, under severe conditions, limits of use for conventional bitumen have been reached (AIPCR 1999). The modification of bitumen with polymer is considered as one solution to overcome this problem (Gonzalez 2004; Airey 2002). Nevertheless, the cost of modified bitumen represents an increase from 60 to 150% by comparison with pure bitumen prices. Thus, the use of recycled materials can be a significant economical alternative (Aguiar-Moya 2013).

During the last years, a huge volume of waste estimated at 13.5 millions of tons per year were produced in Algeria. However, recycling activity remains limited to 5–6% of this volume and mainly concerns paper and plastic materials. The rest is landfilled which raises a lot of issues regarding land use and urban development. Thus, the use of non-biodegradable polymer waste in road construction can effectively contribute to the development of a sustainable national strategy according to Djemaci (2011). On one hand, it would allow diminishing the area dedicated to landfilling and on the other hand it could improve the performances of mixes using local raw materials. Several research projects were recently launched to address this issue.

The purpose of the present work is to study effect of waste industrial polymer (NBR) on bitumen and asphalt mix performances. Three modified blends are tested and the effect on their rheological behavior and on their mechanical properties are investigated, especially at high temperature considering the extreme weather conditions in Algeria.

2 Experiments

2.1 Raw Materials

2.1.1 Bitumen

A 40/50 penetration grade bitumen (pure bitumen) is used. It is supplied by NAFTAL company. The main physical properties of the bituminous material are given in Table 1.

Table 1. Physical properties of bitumen.

Test	Results	Standard	Specification limits (Recommandations Algériennes sur l'Utilisation des Bitumes et Enrobés Bitumineux 2004)
Penetration (1/10 mm)	38	ASTM D5 EN 1426	35–50
Softening point (°C)	52	ASTM D36 EN1427	50–58
Specific gravity	1.029	ASTM D70	1.0–1.1
Ductility (mm)	>1000	ASTM D113	>600

2.1.2 Modifying Agent (Waste Industrial Polymer)

The modifying agent is supplied by S.A.E.L. (Company of Application of elastomers). It is the acrylonitrile butadiene rubber (NBR) obtained from the milling shoe soles. The polymer NBR is an industrial waste blackish color, used as a crumb, composed of particles smaller than 0.8 mm. The density of the crumb is 1.25.

2.1.3 Aggregates

Crushed aggregates are used in this work. They are obtained from el Hachimia quarries located in the north of Algeria. Table 2 summarizes the main characteristics of the aggregates. The chemical analysis shows that these aggregates are calcareous.

Table 2. Chemical analysis of aggregates.

Chemical analysis	Testing results
Insoluble (SiO2+silicates)	13.4%
Perte au Feu à 1050° (PF)	37.04%
(Fe_2O_3+Al_2O_3)	1.80%
Sulfates ($CaSO_4$, $2H_2O$)	Traces
(Na Cl)	0.17%
($CaCO_3$)	81.90%
CO_2	37.04%

2.2 Preparation of Polymer-Bitumen Blends

The strategy for manufacturing blends must take into account environmental impacts based on energy consumption) in order to propose an efficient building process. According to previous researches (Haddadi 2007; Haddadi et al. 2008; Soudani 2010), the blends are prepared by using a mechanical stirring in a vertical low shear mixer with a speed of 600 rpm for 2 h at 180 ± 5 °C.

Different contents of waste NBR are considered (2, 3 and 4% of bitumen weight). The modified blends are denoted by BW_i, were i refers to the waste NBR content.

2.3 Preparation of Asphalt Mixes

The control mix is prepared using Marshall Design Method ASTM D1559. Figure 1 shows the specification limits for aggregates and the selected gradation for the control mix. The standard dimensions of the samples are 63.5 mm height and 101.5 mm diameter, produced with 50-blow compaction per side with Marshall Compactor. The final combined aggregate gradation is given in Fig. 1. The optimum asphalt content was found at 5.8%. The modified asphalt mixes are then obtained by mixing aggregates and waste-NBR modified bitumen. The samples are denoted by Mi where i refers to the waste NBR content used to modify bitumen.

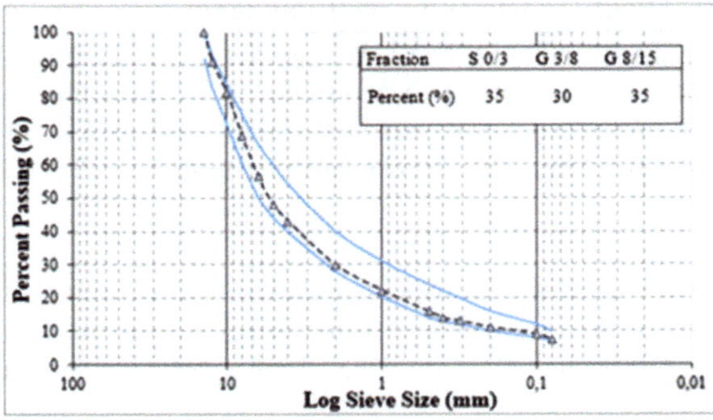

Fig. 1. Grain seize distribution of both control mixtures.

2.4 Testing Program

2.4.1 Conventional Bitumen Tests

Pure and modified bitumens are assessed through standardized penetration test (NF EN 1426) and softening point test (EN 1427).

2.4.2 Dynamic Shear Rheometer Test

The dynamic shear rheometer (DSR) test is carried out in accordance with the procedures described in the AASHTO TP5 standard test method. The rheological properties are measured with a Haake Rheostress 1 using a plate-and-plate geometry (20 mm diameter). Frequency and temperature sweep tests in oscillatory shear measurements are performed. Frequency sweep runs are applied over a range from 0.01 to 10 Hz under isothermal conditions (60 °C). Temperature sweep runs are applied over a range from 50 to 90 °C at 10 rad/s according to AASHTO T315-10 (AASHTO 2010). The DSR is used to measure rheological properties (complex shear modulus G^* and phase angle δ) of neat bitumen and waste (recycled) NBR-modified bitumen. These values are used to determine $G^*/\sin\delta$ value, which is defined as a criterion for high temperature (good viscoelastic) performance of bitumen once in pavement according to SHRP (Strategic Highway Research Program) tests (Gonzalez 2004; Zhang 2009). Moreover, higher $G^*/\sin\delta$ values are well correlated with higher rutting resistance. Thus, $G^*/\sin\delta$ must be superior to 1 kPa for un-aged bitumen to minimize rutting risk (Golzin 2011).

2.4.3 Marshall Test

Mechanical characteristics of the hot Asphalt Mixtures are determined through Marshall test according to ASTM D1559.

Marshall stability value matches to the maximum load stand by the bituminous material at a loading rate of 50 mm/min.

The flow value refers to the vertical deformation when the maximum load is reached.

Marshall stability is related to the resistance of bituminous materials to distortion, displacement, rutting and shearing stresses (Vincent 2010) and the flow value reflects the plasticity and flexibility properties of asphalt mixtures. Flow has a linear inverse relationship with internal friction (Uzun 2012). The ratio of stability to flow is important and not their individual values. It may be used to give an indication of mixture stiffness, and also as a measure of the material's resistance to permanent deformation in service (Malkoc 2012).

3 Results and Discussion

3.1 Conventional Bitumen Tests

Standardized tests are first conducted in order to identify the effect of waste NBR on physical properties of the bituminous binders. The results are summarized in Table 3. They show a slight increase of softening point values and penetration values when increasing waste NBR content, which can be connected to an increase of the viscoelasticity of the modified bitumen.

Table 3. Physical properties of waste NBR modified bitumen

Properties	BW 2	BW 3	BW 4
Penetration (25 °C, 0.1 mm)	27.3	29	31
Softening point (°C)	52	53	54.5

3.2 Dynamic Shear Rheometer Test Results

Some rheological parameters - viscoelastic function (G′ and G″ the elastic and storage modulus) and δ (phase angle) - are determined for both pure and modified bitumens. To analyze the rheological behavior of these different blends, isochronal and isothermal representations are used.

3.2.1 Rheological Properties at High Temperature

Figure 2 shows isochronal curve at 10 rad/s of the viscoelastic function for pure and waste NBR modified bitumens at temperatures ranging from 50 to 90 °C. In all cases, values of G′ and G″ tend to decrease as temperature increases. Also, one can observe that elastic modulus of modified bitumen increases with increasing NBR waste content. Regarding the loss modulus, values are slightly different from those obtained with unmodified bitumen. Moreover, they stay similar for different quantities of NBR.

The phase angle δ values obtained with the different blends are provided in Fig. 3. This parameter is a measure of the viscoelastic behavior of the material (Isacsson 1999). Experimental results clearly show a decrease of δ values when increasing waste NBR content. The highest elastic response (increased G′ and decreased δ) is obtained for the blend containing 4% waste NBR. As NBR waste is an elastomer, its incorporation into bitumen increases the elasticity of the binder at high temperatures. This is recommended to improve resistance to permanent deformation (Navarro 2007).

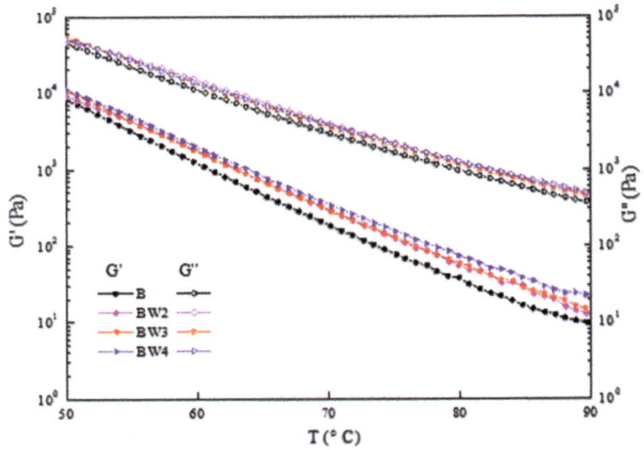

Fig. 2. Temperature dependence of the linear viscoelasticity functions, at 10 rad/s, for the blend tested.

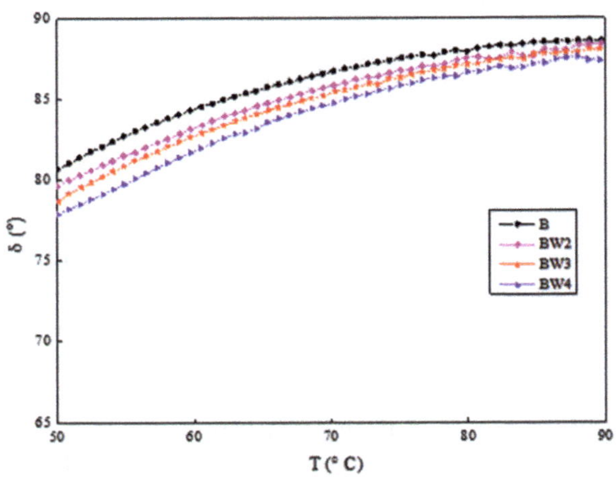

Fig. 3. Temperature dependence of the phase angle, at 10 rad/s, for the blend tested

Plots of G*/sinδ versus temperature are displayed in Fig. 4. According to a SHRP test, the temperature at which G*/sinδ = 1 kPa indicates the maximum temperature for good viscoelastic performances of the binder used in road pavement.

Road pavements in the north of Algeria can undergo very high temperatures (approximately to 60 °C) (Ramond et al. 2000; Merbouh 2010), which can cause permanent deformations or "rutting".

However, waste NBR polymer addition does not lead to any relevant bitumen modification according to this "rutting parameter". This is might be due to the small increase in the loss modulus G″ for these blends which slows the increase of the shear modulus G* values.

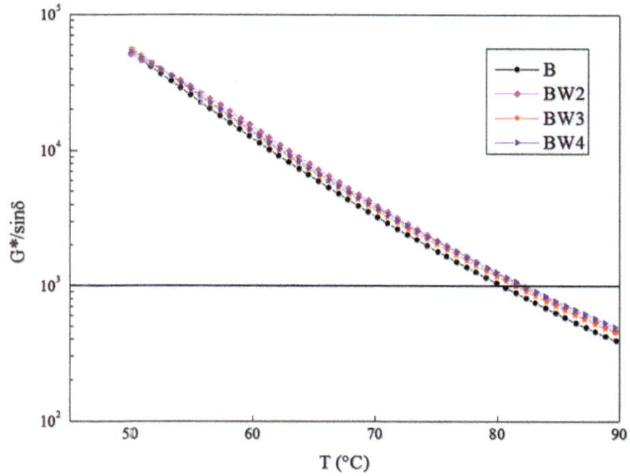

Fig. 4. Curve of G*/sinδ measured at (10 rad/s) as a function of temperature.

Nevertheless, it is important to underline that SHRP rutting criterion was suggested as an indicator of the resistance of the bituminous material to rutting (Bahia 1995) but the rutting phenomenon is influenced by several other factors (Nguyen 2006).

Figure 5 shows the influence of waste NBR on the linear viscoelastic functions with frequency at 60° C for the different blends. At this temperature, incorporation of waste NBR in bitumen increases both storage modulus G′ and loss modulus G″ over the entire range of frequency studied. This increase is higher for the G′ than for G″. This improvement is more visible at low frequency. Moreover, it can be observed that polymer content does not influence the evolution of G″ values.

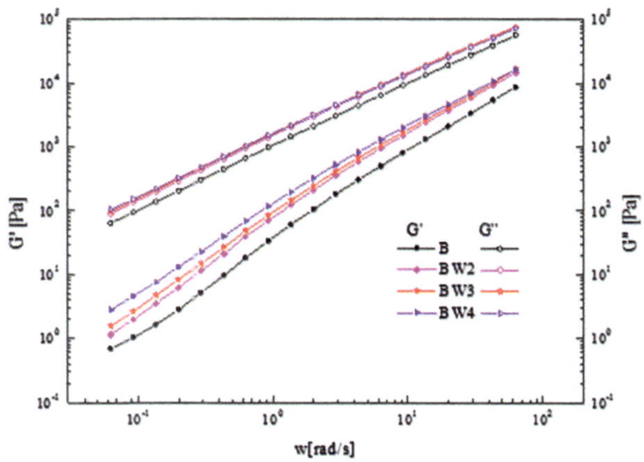

Fig. 5. Evolution of the storage and loss moduli with frequency at 60 °C for neat bitumen and the different blends

Figure 6 shows the evolution of tan δ which represents the inverse of the stored (elastic) energy with frequency. It can be seen that waste NBR tends to reduce tan δ which clearly shows the effect of increasing the elasticity of the binder at high temperatures.

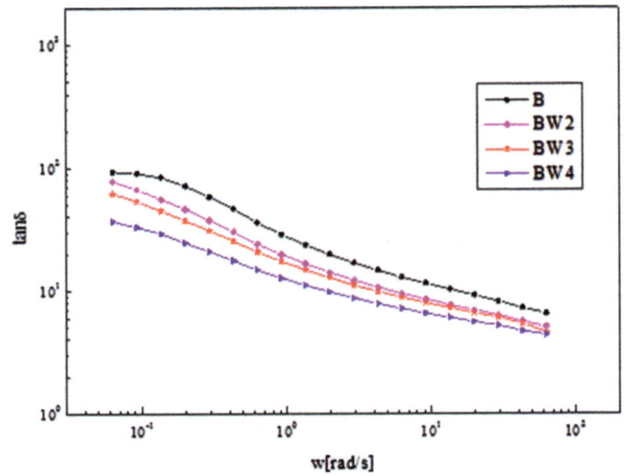

Fig. 6. Isochronal plot of tan δ at 60 °C

3.2.2 Marshall Test

Test was conducted both on mixtures containing waste NBR and mixture prepared with the neat bitumen (control specimens). Each result is obtained from an average of three test specimens (Fig. 7). The Marshall flow values are all within the Algerian specification limits (2004). However, waste NBR contributes to decrease flow values but the values do not significantly change when increasing the quantity of NBR.

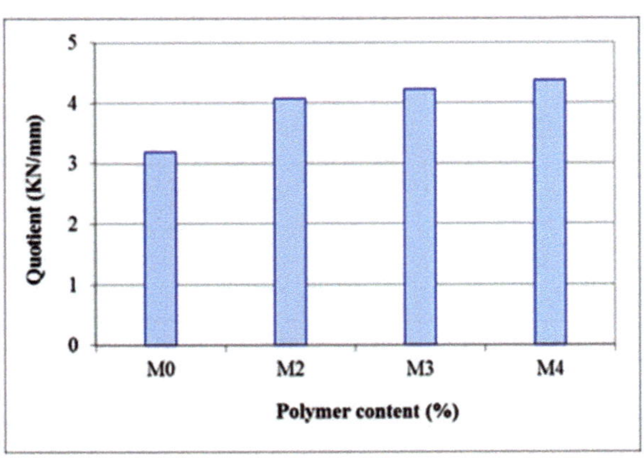

Fig. 7. Marshall Quotient of the different asphalt mixes versus the waste NBR content.

The Marshall quotient values (Fig. 8) indicate higher values for the modified mixtures than for reference mixture. The Marshall quotient tends to increase with the quantity of waste NBR. We can notice that asphalt mix with 4% of waste NBR provides optimal result (i.e. high stiffness and it's likely more resistant to permanent deformation).

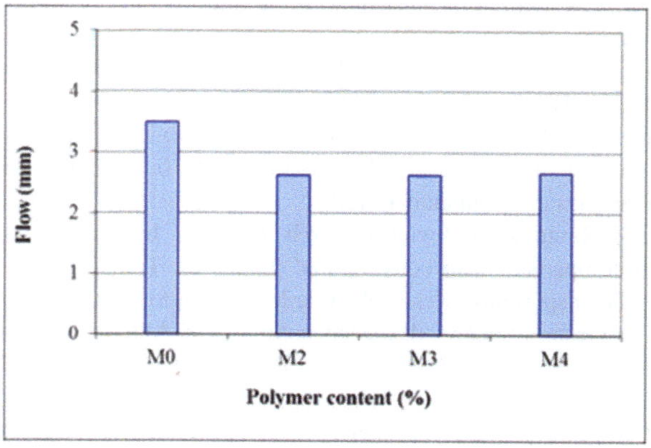

Fig. 8. Marshall flow of the different asphalt mixes versus the waste NBR content.

4 Conclusions

The aim of this study is to assess the impact of waste polymeric compound (NBR) extracted from shoe soles on the rheological properties and mechanical characteristics of road asphalt pavement mixes.

According to manufacturing conditions of modified bituminous binders, rheological properties at high temperature are improved, but the level of improvement is still moderate when compared with the published researches about crumb rubber modified bitumen. This may be due to the low content of modifying agent (<5%) or the low quantity of light fractions of our bitumen (40/50) which are necessary for polymer dispersion.

The mechanical characteristics of modified asphalt mixes show visible improvement, particularly in Marshall quotient. Asphalt mix with 2% of NBR gives the best level of performances. Moreover, relationship can be observed between asphalt mix mechanical characteristics and their rheological properties at high temperature. This indicate that high viscoelastic properties can contribute to improve the stiffness of the asphalt mix and its behavior towards permanent deformations. This conclusion must be confirmed by other experiments such as rutting test. The result would be promising for the recycling operation, due to satisfying results obtained with high content of waste NBR.

References

«Algérie-focus» 28 août 2013. http://www.algerie-focus.com/blog/2013/08/le-non-recyclage-coute-300-millions-deuros-par-an-a-lalgerie/

Aguiar-Moya, J.P., et al. (2013) Use of waste products as bitumen modifiers in Costa Rica. In: 5th EATA Conference, Braunschweig, Germany, 3–5 June 2013

AIPCR-LCPC: Emplois des bitumes spéciaux, des liants bitumineux modifiés et avec additifs en techniques routières. Rapport préparé par un groupe de travail, le Comité technique 8 des routes souples, France, 204 pp. (1999)

Airey, G.D.: Rheological evaluation of ethylene vinyl acetate polymer modified bitumens. Constr. Build. Mater. **16**(18), 473–487 (2002)

AASHTO: Standard method of test for determining the rheological properties of asphalt binder using dynamic shear rheometer (DSR), designation T315-10. American Association of State Highway and Transportation Officials (2010)

Bahia, H.U., et al.: Strategic highway research program binder rheological parameters: Background and comparison with conventional properties. Transportation Research Record 1488, pp. 32–39. Transportation Research Board, Washington, DC (1995)

Djemaci, B., et al.: La Géstion Intégrée des Déchets Solides en Algérie. Contraintes et Limites de sa Mise en Oeuvre (2011). http://www.ciriec.ulg.ac.be/wp-content/uploads/2015/08/WP11-04.pdf

Golzin, Y., et al.: Improving the performance of crumb rubber bitumen by means of Poly Phosphoric Acid (PPA) and Vestenamer additives. Constr. Build. Mater. **25**, 3108–3116 (2011)

Gonzalez, O., et al.: Rheology and stability of bitumen/EVA blends. Eur. Polym. J. **40**, 2365–2372 (2004)

Haddadi, S.: Influence de la poudrette de caoutchouc sur le comportement au fluage des enrobés bitumineux. Ph.D. thesis, FGC/USTHB (2007)

Haddadi, S., et al.: Effects of the manufacturing process on the performances of the bituminous binders modified with EVA. Constr. Build. Mater. **22**, 1212–1219 (2008)

Isacsson, U., et al.: Characterization of bitumens modified with SEBS, EVA and EBA polymers. J. Mater. Sci. **34**, 3737–3745 (1999)

Köfteci, S., et al.: Performance evaluation of bitumen modified by various types of waste plastics. Constr. Build. Mater. **73**, 592–602 (2014)

Malkoc, G., et al.: Improving mechanical properties of hot-mixed asphalt using plastic waste (polyethylene terephthalate). In: 5th Eurasphalt and Eurobitume Congress, Istanbul (2012)

Merbouh, M. : Contribution à la modélisation du comportement rhéologique des enrobés bitumineux: influence des conditions extrêmes de température et de trafic en fatigue. Ph.D. thesis, University of Bordeaux 1 (2010)

Navarro, F.J., et al.: Rheology and microstructure of MDI–PEG reactive prepolymer-modified bitumen. Mech. Time Depend. Mater. **10**(4), 347–359 (2007)

Nguyen, D.-T.: Prédiction des déformations permanentes des Couches de surface des chaussées Bitumineuses. Ph.D. thesis, Ecole Nationale des Ponts et Chaussées, France (2006)

Ramond, G., et al.: Caractéristiques de bitumes utilisés en Algérie. Bulletin de liaison ponts et chaussées, BLPC, N° 225, pp. 3–11 (2000)

Recommandations Algériennes sur l'Utilisation des Bitumes et Enrobés Bitumineux. Organisme de Contrôle Technique des Travaux Publics. Ministère des Travaux Public, Algérie (2004)

Soudani, K., et al.: Influence Process of Modification of Bituminous Concrete by the NBR on the Mechanical and Rheological Characteristics (2010)

Uzun, İ., et al.: Evaluation of andesite waste as mineral filler in asphaltic concrete mixture. Constr. Build. Mater. **31**, 284–288 (2012)

Vincent, T., et al.: A Self-learning Manual Mastering Different Fields of Civil Engineering Works (VC-Q&A Method) U (2010)

Zhang, J., et al.: Evaluation of the improved properties of SBR/weathered coal modified bitumen containing carbon black. Constr. Build. Mater. **23**, 2678–2687 (2009)

Numerical Check of the Meyerhof Bearing Capacity Equation for Shallow Foundations

Stefan Van Baars(✉)

Foundation Engineering and Soil Mechanics, University of Luxembourg, Luxembourg, Grand Duchy of Luxembourg
stefan.vanbaars@uni.lu

Abstract. In 1920 Prandtl published an analytical solution for the bearing capacity of a strip load on a weightless infinite half-space. This solution was extended with a surrounding surcharge by Reissner and with the soil weight by Keverling Buisman. It was Terzaghi who wrote this with three separate bearing capacity factors for the cohesion, surcharge and soil-weight. Meyerhof extended this to the equation which is nowadays used; with shape and inclination factors. He also proposed equations for the inclination factors, based on his own laboratory experiments. Since then, several people proposed updated equations for the soil-weight bearing capacity factor, and also for the shape and inclination factors.

The common idea is that failure of a footing occurs in all cases with a Prandtl-wedge failure mechanism. In order to check the failure mechanisms and the currently used equations for the bearing capacity factors and shape factors, a large number of finite element calculations of strip and circular footings have been made. These calculations proof that for some cases there are also a few other failure mechanisms possible. Also the currently used bearing capacity factors and shape factors are not correct. In fact, for footings on a soil with a higher friction angle, all three bearing capacity factors and all three shape factors can be much lower than the currently used values.

This means that the currently used equations for the soil-weight bearing capacity factors and the shape factors are inaccurate and unsafe. Therefore, based on the finite element calculations, new equations have been presented in this paper.

1 Introduction

In 1920, Ludwig Prandtl published an analytical solution for the bearing capacity of a soil under a load, p, causing kinematic failure of the weightless infinite half-space underneath. The strength of the half-space is given by the angle of internal friction, ϕ, and the cohesion, c. The original drawing of the failure mechanism proposed by Prandtl can be seen in Fig. 1. The lines in the sliding soil part on the left indicate the directions of the maximum and minimum principle stresses, while the lines in the sliding soil part on the right, indicate the sliding lines with a direction of $\alpha = 45° - \frac{1}{2}\phi$ in comparison to the maximum principle stress. This failure mechanism has been validated by laboratory tests, for example those performed by Jumikis (1956), Selig and McKee (1961) and Muhs and Weiß (1972).

H. Shehata and Y. Rashed (eds.), *Numerical Analysis of Nonlinear Coupled Problems*, Sustainable Civil Infrastructures, DOI 10.1007/978-3-319-61905-7_19

Prandtl subdivided the sliding soil part into three zones (see Fig. 2):

1. Zone 1: A triangular zone below the strip load. Since there is no friction on the ground surface, the directions of the principal stresses are horizontal and vertical; the largest principal stress is in the vertical direction.
2. Zone 2: A wedge with the shape of a logarithmic spiral, in which the principal stresses rotate through $\frac{1}{2}\pi$ radians, or 90°, from Zone 1 to Zone 3. The pitch of the sliding surface of the logarithmic spiral $\left(r_\theta = r_1 \cdot e^{\theta \cdot \tan\phi}\right)$ equals the angle of internal friction; ϕ, creating a smooth transition between Zone 1 and Zone 3.
3. Zone 3: A triangular zone adjacent to the strip load. Since there is no friction on the surface of the ground, the directions of principal stress are horizontal and vertical; the largest principal stress is in the horizontal direction.

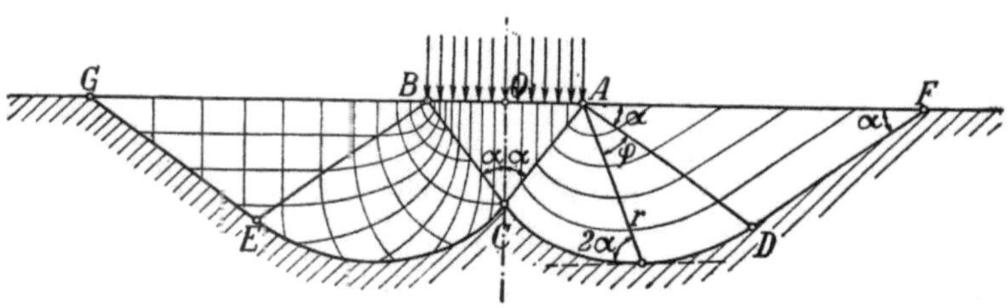

Fig. 1. The Prandtl-wedge failure mechanism (Original drawing by Prandtl).

The failure mechanism does not have to be symmetrical, with sliding in two directions, as proposed by Prandtl in Fig. 1, but can also be unsymmetrical, with sliding in one direction (see Fig. 2).

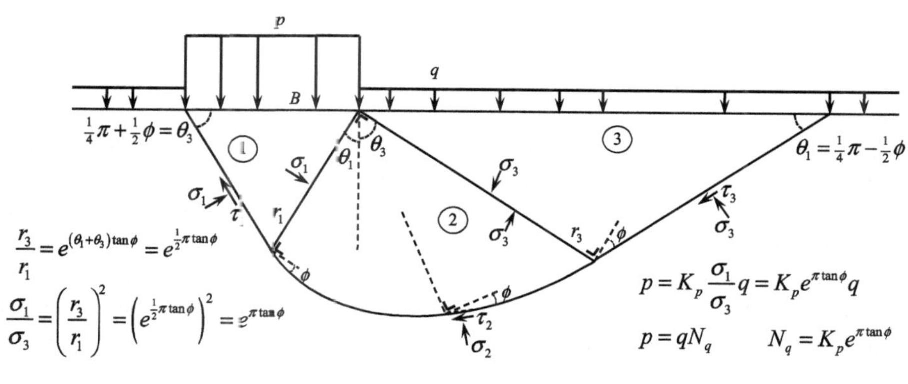

Fig. 2. The Prandtl-wedge with 3 zones.

The solution of Prandtl was extended by Reissner 1924 with a surrounding surcharge, q, and was based on the same failure mechanism. Keverling Buisman (1940) and Terzaghi (1943) extended the Prandtl-Reissner formula for the soil weight, γ. It was Terzaghi (1943) who first wrote the bearing capacity with three separate bearing capacity factors for the cohesion, surcharge and soil-weight. Meyerhof (1953) was the first to propose equations for inclined loads, based on his own laboratory experiments. He was also the first in 1963 to write the formula for the (vertical) bearing capacity p_v with bearing capacity factors (N), inclination factors (i) and shape factors (s), for the three independent bearing components; cohesion (c), surcharge (q) and soil-weight (γ), in a way it was adopted by Brinch Hansen (1970) and it is still used nowadays:

$$p_v = i_c s_c c N_c + i_q s_q q N_q + i_\gamma s_\gamma \frac{1}{2} \gamma B N_\gamma. \tag{1}$$

Prandtl (1920) solved the cohesion bearing capacity factor:

$$N_c = \left(K_p \cdot e^{\pi \tan \phi} - 1\right) \cot \phi, \quad \text{with:} \quad K_p = \frac{1 + \sin \phi}{1 - \sin \phi}. \tag{2}$$

Reissner (1924) solved the surcharge bearing capacity factor with the equilibrium of moments of Zone 2, see Fig. 2:

$$N_q = K_p \cdot \left(\frac{r_3}{r_1}\right)^2 = K_p \cdot e^{\pi \tan \phi}, \quad \text{with:} \quad K_p = \frac{1 + \sin \phi}{1 - \sin \phi}. \tag{3}$$

Keverling Buisman (1940), Terzaghi (1943), Caquot and Kérisel (1953), Meyerhof (1951, 1953, 1963, 1965), Brinch Hansen (1970), Vesic (1973, 1975) and Chen (1975) subsequently proposed different equations for the soil-weight bearing capacity factor N_γ. Therefore the following equations for the soil-weight bearing capacity factor can be found in the literature:

$$\begin{aligned}
N_\gamma &= \left(K_p \cdot e^{\pi \tan \phi} - 1\right) \tan(1.4\phi) && \text{(Meyerhof 1963)}, \\
N_\gamma &= 1.5\left(K_p \cdot e^{\pi \tan \phi} - 1\right) \tan \phi && \text{(Brinch Hansen 1970)}, \\
N_\gamma &= 2\left(K_p \cdot e^{\pi \tan \phi} + 1\right) \tan \phi && \text{(Vesic 1973)}, \\
N_\gamma &= 2\left(K_p \cdot e^{\pi \tan \phi} - 1\right) \tan \phi && \text{(Chen 1975)}
\end{aligned} \tag{4}$$

The equation from Brinch Hansen is, as he writes, "based on calculations first from Lundgren and Mortensen (1953) and later from Odgaard and N.H. Christensen". The equation of Vesic is almost identical to the solution of Caquot and Kérisel (1953) because it is, as he says, based on "the numerical results of an analysis made by them".

All these bearing capacity factors, and also the inclination and shape factors, can be checked nowadays with finite element calculations. There have been several researchers working on this numerical comparison, but their publications often show one or more of the following mistakes:

- Using inaccurate numerical tools, for example the limit-equilibrium analysis method, instead of a more accurate method, like the finite element method.
- Using softening (in Plaxis there is a standard "tension cut off"-procedure which must be switched off).
- Using volume change during failure, for example by selecting an associated flow rule ($\psi = \phi$).

In this study all bearing capacity factors, and also the inclination and shape factors, will be checked for their accurateness, with the software Plaxis 2D for a (bi-linear) Mohr-Coulomb (c, ϕ) soil model without hardening, softening, or volume change during failure ($\psi = 0$).

2 The Bearing Capacity Factors

2.1 The Surcharge Bearing Capacity Factor

A series of finite element calculations have been made, in which the width of the load $B = 2$ m. Plaxis produces incremental displacement plots during failure, indicating the failure mechanism. For low friction angles, the failure mechanism is almost the same as the Prandtl-wedge failure mechanism, see Fig. 3 on the left, which is also the basis of the analytical solution. For high friction angles though, the failure mechanism looks completely different, see Fig. 3 on the right.

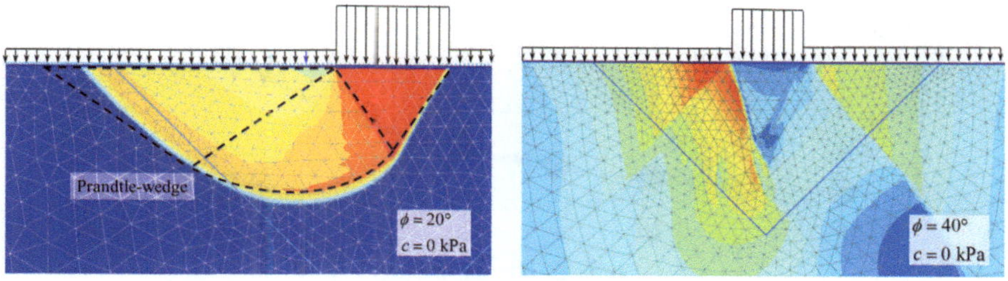

Fig. 3. Failure mechanism; left: low friction angle; right: high friction angle.

For a series of different friction angles, the maximum normalised force, or the surcharge bearing capacity, has been plotted, see Fig. 4.

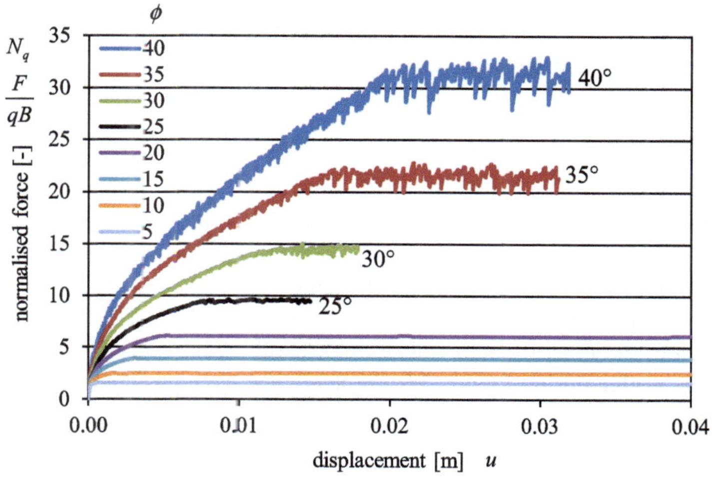

Fig. 4. Normalised force versus displacement for different friction angles.

The change in failure mechanism has consequences; the force versus displacement plot is rather smooth for low friction angles, but becomes very rough for higher friction angles. This is a sign that another and easier failure mechanism exists, in which constantly new shear failure planes are found in the calculation, depending on the internal redistribution of the stresses.

For the Finite Element Modelling (FEM) two different options have been used; both stress controlled and displacement controlled. The two options give as expected (almost) the same results (See Fig. 5).

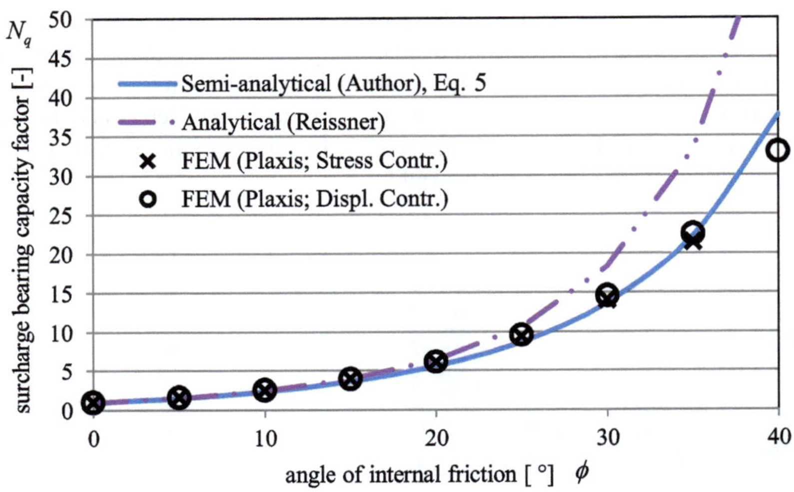

Fig. 5. Surcharge bearing capacity factors: Reissner versus FEM

Remarkable is that the analytical solution of Reissner gives values which are too high, especially for higher friction angles. This must be explained by the existence of another and easier failure mechanism, as was mentioned earlier.

The semi-analytical line in the figure describes much better the surcharge bearing capacity factors, and can be written as:

$$N_q = \cos^2 \phi \cdot K_p \cdot e^{\pi \tan \phi} \quad \text{with:} \quad K_p = \frac{1 + \sin \phi}{1 - \sin \phi} \tag{5}$$

Loukidis et al. (2008) already noticed that non-dilatant (non-associated) soil is 15%–30% weaker than associated soil ($\psi = \phi$), and has a rougher failure pattern.

The difference between the analytical method (Eq. 3) and the numerical results (Eq. 5) has been explained by Knudsen and Mortensen (2016): The higher the friction angle, the wider the logarithmic spiral of the Prandtl wedge and the more the stresses reduce in this wedge during failure. So, the analytical formulas are only kinematically admissible for associated flow ($\psi = \phi$). The problem of assuming associated soil is of course, that such a high dilatancy angle is by far unrealistic for natural soils. This means as well that, a calculation of the bearing capacity of a footing, based on the analytical method is, for higher friction angles, also unrealistic.

2.2 The Cohesion Bearing Capacity Factor

For the cohesion bearing capacity factor, the same results are found as for the surcharge bearing capacity factor (see Fig. 6).

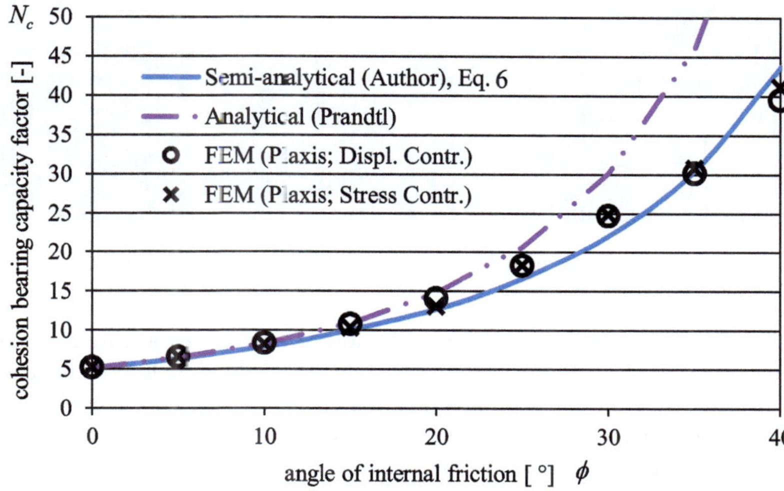

Fig. 6. Cohesion bearing capacity factors: Prandtl versus FEM.

The straight line in the figure can be described by:

$$N_c = \left(N_q - 1\right) \cot \phi \quad \text{with:} \quad N_q = \cos^2 \phi \cdot K_p \cdot e^{\pi \tan \phi} \tag{6}$$

2.3 The Soil-Weight Bearing Capacity Factor

The situation for the soil-weight bearing capacity factor is different from the other two factors because having a constant, or rectangular shaped, load p is impossible, simply because just next to the load, there is no strength for a cohesionless material without a surcharge. So, unlike the effect of the cohesion and the effect of the surcharge, the effect of the soil-weight does not result in a constant maximum load p. So, there will be a maximum load in the middle and a zero load at the edges, where the shear and normal stresses go to zero too.

Caquot and Kerisel (1966) proposed for the soil-weight bearing capacity, a circular failure mechanism, instead of a Prandtl-wedge failure mechanism. This circular failure mechanism was confirmed by laboratory testing of Lambe and Whitman (1969), and also by finite element calculations of Van Baars (2015, 2016a, b), see Fig. 7.

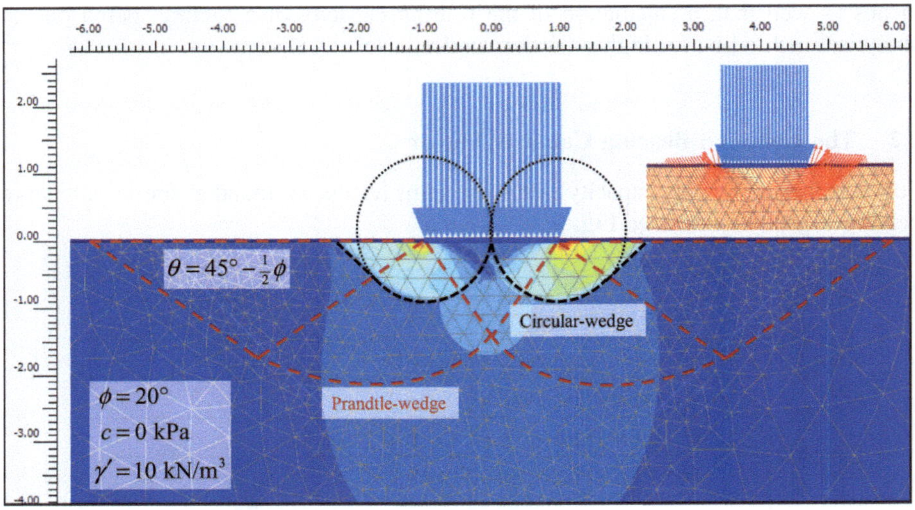

Fig. 7. Numerical modelling: circular failure zones under a footing.

So, the failure mechanism belonging to the soil-weight failure bearing capacity (N_γ) is not the same as the Prandtl-wedge failure mechanism belonging to both the cohesion bearing capacity (N_c) and the surcharge bearing capacity (N_q). This causes a problem for the Meyerhof equation (Eq. 1), because the existence of different failure mechanisms makes it unjustified to apply the used superposition in this bearing capacity equation. The addition of inclination factors and also shape factors makes the risk of superposition errors even larger.

Figure 8 shows the results of the soil-weight bearing capacity factor. For this figure, calculations have been made for both a rough plate (no horizontal displacement of the soil below the plate) and a smooth plate (free displacement).

The highest results are found for the displacement controlled calculations for the rough plate:

$$N_\gamma = 7 \sin\phi \cdot \left(e^{\pi \tan\phi} - 1\right) \quad \text{(Thin line)}. \tag{7}$$

The strange outcome is that the stress controlled calculations with a rough plate give, for the displacement controlled calculations, results which are lower for the rough plate, and are at the same time rather identical for the smooth plate. Nevertheless, the results of the equations in literature (Eq. 4) are, for higher internal friction angles, always too high. Van Baars (2015) proposed therefore to use lower and safer equations, such as for example (see the empirical straight lines in the figure based on FEM):

$$N_\gamma = 4 \tan\phi \cdot \left(e^{\pi \tan\phi} - 1\right) \quad \text{(Thick line)}. \tag{8}$$

Both the stress controlled rough plate calculation results and the displacement controlled smooth plate calculation results can be described by this equation.

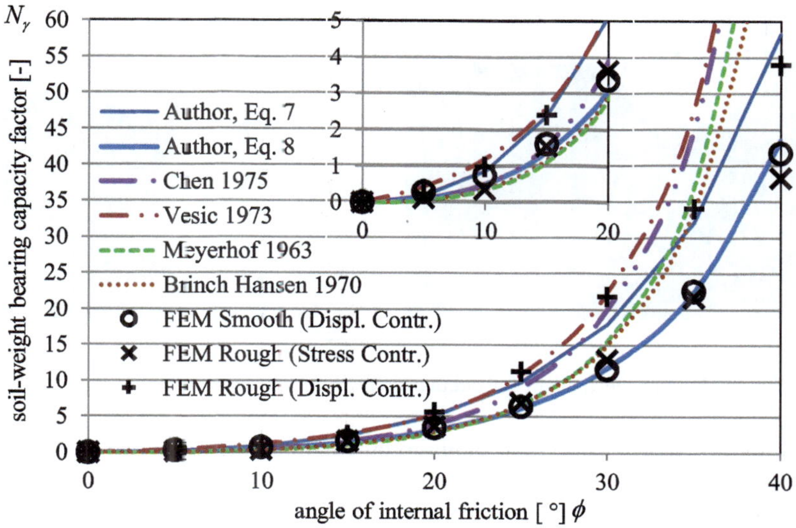

Fig. 8. The soil-weight bearing capacity factor N_γ.

3 The Shape Factors

If the shape of the foundation area is not an infinitely long strip, but a finite rectangular area, of width B and length L (where it is assumed, that the width is the shortest dimension), correction factors for the shape are used. Meyerhof (1963) was the first to publish shape factors (for $L \geq B$):

$$s_q = s_\gamma = 1 + 0.1 K_p \frac{B}{L} \sin\phi, \quad s_c = 1 + 0.2 K_p \frac{B}{L}. \tag{9}$$

A few years later De Beer (1967, 1970) published his shape factors, based on laboratory experiments. Brinch Hansen (1970) based his shape factors on the experimental results from De Beer. So the most commonly used shape factors are (for $L \geq B$):

$$s_c = 1 + 0.2 \frac{B}{L}, \quad s_q = 1 + \frac{B}{L} \sin\phi, \quad s_\gamma = 1 - 0.3 \frac{B}{L}. \tag{10}$$

In this study, circular loaded areas ($B = L$) will be studied with 2D axial-symmetric Plaxis calculations. Figure 9 shows the incremental displacements during failure, for a circular loaded area on two different cohesive soils, without a surcharge, indicating the failure mechanisms. The failure mechanism for high friction angles looks like a punching failure. The failure mechanism for low friction angles starts, like the circular-wedge failure mechanism, in the middle of the loaded area, has 3 zones, like the Prandtl-wedge, and is found for all three bearing capacity cases (cohesion, surcharge and soil-weight). Due to the axial symmetry, the circular wedge is pushed out during failure, causing a declining tangential stress (i.e. circumferential stress or hoop stress).

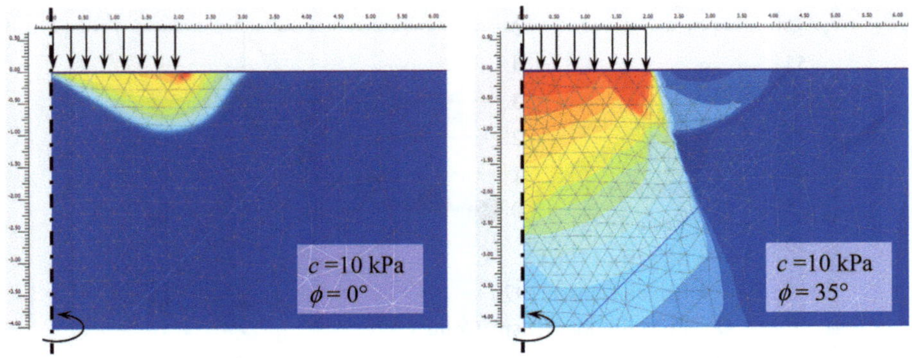

Fig. 9. Failure mechanism for circular areas; left: low friction angle; right: high.

For the strip load (plane strain solution) the minimum stress, in part 3 of the wedge, is the vertical stress, which is zero without a surcharge, and the maximum stress is the horizontal stress (perpendicular to the load).

For the circular load (axial symmetry) however, the minimum stress, in part 3 of the wedge, is not the vertical stress, but the tangential stress, which is zero, or even less in case of a cohesion, which causes even tension. Therefore, the maximum stress, which is still the horizontal stress (perpendicular to the load), will also be far less.

Due to this cleaving failure mechanism, the bearing capacity of a circular load will be far less than of a strip load, resulting in shape factors below "1". In reality, for cohesive materials, there will be radial cracks formed, which can eliminate the tensile tangential stresses and therefore increase the bearing capacity.

3.1 The Surcharge Shape Factor

The shape factor, for a circular loaded area surrounded by a surcharge, has been plotted in Fig. 10, for both Meyerhof and De Beer, together with the outcome of the axial-symmetric FEM calculations.

The results show that the currently used factors from especially Meyerhof, but also De Beer, are too high, so unsafe. Several publications, for example Zhu and Michalowski (2005) and Tapper et al. (2015), show that the shape factors are related to $\sqrt{B/L}$, and not to B/L. This will be adopted here. The shape factor used for the straight line is:

$$s_q = 1 - \left(0.7 - \frac{2}{3}\tan\phi\right) \cdot \sqrt{\frac{B}{L}}. \tag{11}$$

3.2 The Cohesion Shape Factor

The shape factor for a circular loaded area on cohesive soil has been plotted in Fig. 11, for both Meyerhof and De Beer, together with the outcome of the axial-symmetric

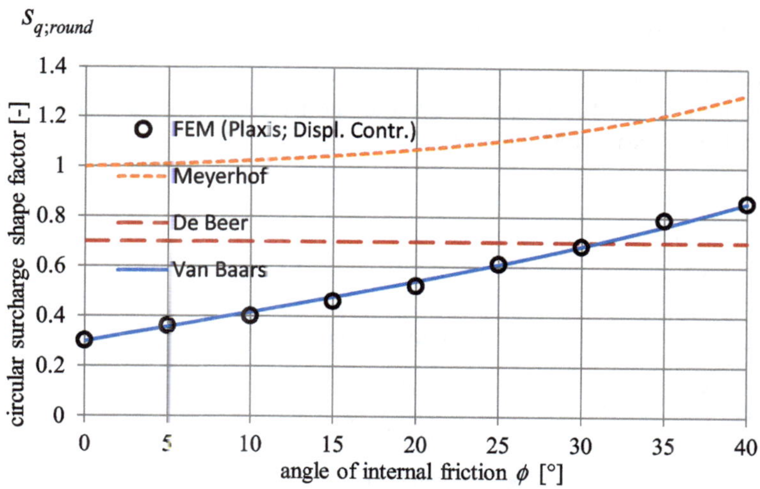

Fig. 10. The surcharge shape factor.

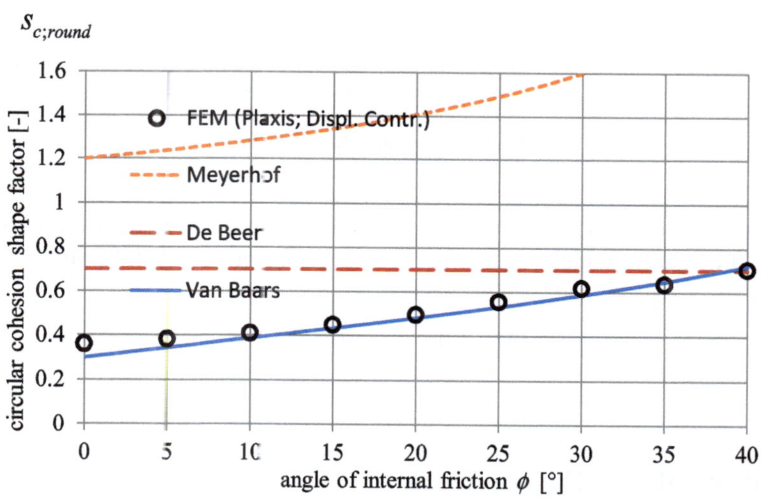

Fig. 11. The cohesion shape factor.

FEM calculations. The FEM results show that the currently used factors from especially Meyerhof, but also De Beer, are too high, so unsafe. The shape factor used for the straight line in the figure is:

$$s_c = 1 - (0.7 - 0.5 \tan \phi) \cdot \sqrt{\frac{B}{L}}. \tag{12}$$

3.3 The Soil-Weight Shape Factor

The shape factor for a smooth circular footing, on a soil with an effective weight, has been plotted in Figure 12, for both Meyerhof and De Beer, together with the outcome of the axial-symmetric FEM calculations.

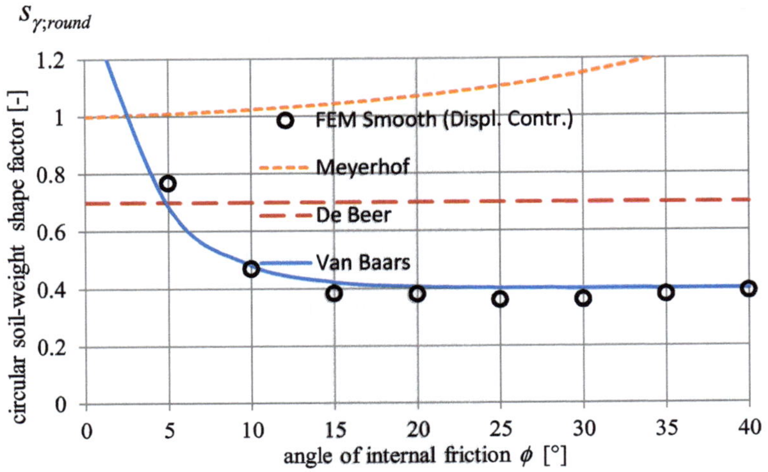

Fig. 12. The soil-weight shape factor.

The results show that the currently used factors from especially Meyerhof, but also De Beer, are again too high, so unsafe. The shape factor used for the straight line is:

$$s_{\gamma} = 1 - \left(0.6 - \exp(-\frac{\phi}{4^{\circ}}\right) \cdot \sqrt{\frac{B}{L}}. \tag{13}$$

4 Additional and Future Work

The set of shape factors is not the only important extension in the bearing capacity equation of Meyerhof (Eq. 1). The other important extension is the set of inclination factors. Therefore additional work about checking and explaining these inclination factors, containing both numerical simulations and analytical derivations, has been published by Van Baars in 2014, and more will be published soon, especially about the analytical solutions.

Another extension is the use of slope factors, which is important for footings in or near a slope. This has been investigated in by Van Baars (2016a, b) and the results will be published soon.

Also the influence of the applied superposition in the bearing capacity equation of Meyerhof, and at the same time the influence of eccentric loading on the bearing capacity of shallow foundations, has been studied and published by Van Baars in 2016.

All this research about the bearing capacity of shallow foundations will be united in a book, dedicated to the first publication of the Prandtl-wedge in 1920. This book entitled "100 Year Prandtl's Wedge" will be released in 2018.

5 Conclusions

A large number of finite element calculations of strip and circular footings have been made in order to check the failure mechanisms and to check the currently used equations for the bearing capacity factors. These calculations proof that in some cases there are a few other failure mechanisms possible as well. And moreover, the currently used bearing capacity factors, and also the shape factors, are not correct. In fact, for footings on a soil with a higher friction angle, all three bearing capacity factors and also all three shape factors can be much lower than the currently used values.

This means that the currently used equations for the soil-weight bearing capacity factors and the shape factors are inaccurate and unsafe. Therefore, based on the finite element calculations, new equations have been presented in this paper.

References

Brinch Hansen, J.A.: Revised and extended formula for bearing capacity. Bulletin No. 28, pp. 5–11. Danish Geotechnical Institute, Copenhagen (1970)

Caquot, A., Kerisel, J.: Sur le terme de surface dans le calcul des fondations en milieu pulverulent. In: Proceedings of the Third International Conference on Soil Mechanics and Foundation Engineering, Zurich, Switzerland, vol. 1, pp. 336–337, 16–27 August 1953

Caquot, A., Kerisel, J.: Traité de Méchanique des sols, pp. 349, 353. Gauthier-Villars, Paris (1966)

Chen, W.F.: Limit Analysis and Soil Plasticity. Elsevier, Amsterdam (1975)

De Beer, E.E.: Proefondervindelijke bijdrage tot de studie van het grensdraagvermogen van zand onder funderingen op staal; Bepaling van de vormfactor sb. Annales des Travaux Publics de Belgique **68**(6), 481–506; **69**(1), 41–88; (4), 321–360; (5), 395–442; (6), 495–522 (1967)

De Beer, E.E.: Experimental determination of the shape factors and the bearing capacity factors of sand. Géotechnique **20**(4), 387–411 (1970)

Jumikis, A.R.: Rupture surfaces in sand under oblique loads. J. Soil Mech. Found. Des. **82**(SM1), 1–26 (1956). ASCE

Keverling Buisman, A.S.: Grondmechanica, p. 243. Waltman, Delft (1940)

Knudsen, B.S., Mortensen, N.: Bearing capacity comparison of results from FEM and DS/EN 1997-1 DK NA 2013. In: Northern Geotechnical Meeting 2016, Reykjavik, pp. 577–586 (2016)

Lambe, T.W., Whitman, R.V.: Soil Mechanics. Wiley, New York (1969)

Loukidis, D., Chakraborty, T., Salgado, R.: Bearing capacity of strip footings on purely frictional soil under eccentric and inclined loads. Can. Geotech. J. **45**, 768–787 (2008)

Lundgren, H., Mortensen, K.: Determination by the theory of plasticity of the bearing capacity of continuous footings on sand. In: Proceedings of the 3rd International Conference on Soil Mechanics, Zürich, vol. 1, p. 409 (1953)

Meyerhof, G.G.: The ultimate bearing capacity of foundations. Géotechnique **2**, 301–332 (1951)

Meyerhof, G.G.: The bearing capacity of foundations under eccentric and inclined loads. In: Proceedings of the III International Conference on Soil Mechanics and Foundation Engineering, Zürich, Switzerland, vol. 1, pp. 440–445 (1953)

Meyerhof, G.G.: Some recent research on the bearing capacity of foundations. Can. Geotech. J. **1** (1), 16–26 (1963)

Meyerhof, G.G.: Shallow foundations. J. Soil Mech. Found. Div. **91**(2), 21–32 (1965). ASCE

Muhs, H., Weiß, K.: Versuche über die Standsicheheit flach gegründeter Einzelfundamente in nichtbindigem Boden, Mitteilungen der Deutschen Forschungsgesellschaft für Bodenmechanik (Degebo) an der Technischen Universität Berlin, Heft 28, p. 122 (1972)

Prandtl, L.: Über die Härte plastischer Körper. Nachr. Ges. Wiss. Goettingen. Math. Phys. Kl. 74–85 (1920)

Selig, E.T., McKee, K.E.: Static behavior of small footings. J. Soil Mech. Found. Des. **87**(SM6), 29–50 (1961). ASCE

Reissner, H.: Zum Erddruck problem. In: Biezeno, C.B., Burgers, J.M. (eds.) Proceedings of the 1st International Congress for Applied Mechanics, Delft, The Netherlands, pp. 295–311 (1924)

Tapper, L., Martin, C.M., Byrne, B.W. Lehane, B.M.: Undrained vertical bearing capacity of perforated shallow foundations. In: Frontiers in Offshore Geotechnics, Fig. 7, p. 816 (2015)

Terzaghi, K.: Theoretical Soil Mechanics. Wiley, New York (1943)

Van Baars, S.: The bearing capacity of footings on cohesionless soils. Electron. J. Geotech. Eng. **20**, 12945–12955 (2015). ISSN 1089-3032

Van Baars, S.: Failure mechanisms and corresponding shape factors of shallow foundations. In: 4th International Conference on New Development in Soil Mechanics and Geotechnical Engineering, Nicosia, pp. 551–558 (2016a)

Van Baars, S.: The influence of superposition and eccentric loading on the bearing capacity of shallow foundations. J. Comput. Mater. Civil Eng. **1**, 121–131 (2016b). ISSN 2371-2325

Vesic, A.S.: Analysis of ultimate loads of shallow foundations. J. Soil Mech. Found. Div. **99**(1), 45–76 (1973)

Vesic, A.S.: Bearing capacity of shallow foundations. In: Winterkorn, H.F., Fang, H.Y. (eds.) Foundation Engineering Handbook, pp. 121–147. Van Nostrand Reinhold, New York (1975)

Zhu, M., Michalowski, R.L.: Shape factors for limit loads on square and rectangular footings. J. Geotech. Geoenviron. Eng. ASCE, Febr., pp. 223–231 (2005)

Utilization of Weathered Rock Mass as the China Three Gorges Dam Foundation

Shirong Xiao(✉), Guodong Zhang, and Qingjun Zuo

College of Civil Engineering and Architecture,
China Three Gorges University, Yichang 443002, Hubei, China

Abstract. The utilization of the weathered rock mass in the Three Gorges Dam Area is one of the key geotechnical issues for the China Three Gorges project (CTGP) construction. Engineering geological investigations and tests had been carried out to study the physical, chemical, mineral and mechanical properties and distribution characteristics in detail for more than 30 years. The classification of the weathered zone has been further divided. The utilization of the lower part of the weakly weathered zone as the Three Gorges dam foundation rock mass had been researched by consolidation grouting test. The research results show that: (1) The weathered rock mass in Three Gorges dam area has a great thickness of 40–70 m, with inhomogeneous behavior, physical and mechanical properties in different weathering zones. (2) It is feasible that the lower part of the weakly weathered zone could be used as the dam foundation after consolidation grouting treatment. This approach can save a large amount of construction investment. (3) The weathered rock mass can be used as different civil engineering foundation, and different weathered zone can satisfy different foundation bearing layer requirements. (4) The weathered rock mass can be utilized as construction material. The completely-weathered rock mass can be used as cofferdam filling sand material. The weakly and slightly weathered granite can be used as filling rock materials of cofferdams.

1 Introduction

The engineering utilization of weathered rock mass at crystalline rock engineering area is related to quality, duration and cost of a project. Therefore, it has always been a key geotechnical issue focused by academic and engineering.

The bedrock of the Three Gorges dam site is granite with crystalline texture, massive structure and the thickness of weathered crust more than 40 m. Influenced by the topography, geomorphology and geological structure, the properties of weathered crust varies greatly. However, it shows that the weathering intensity gradually weakens from the mountain top to bottom, and it is divided into four zones as completely weathered zone, intensely weathered zone, weakly weathered zone and slightly weathered zone. Thus, the selection of dam foundation rock mass and the utilization of the weathered crust are the important engineering geological issues for the Three Gorges Dam Project, and have great meaning both economically and technically.

After decades of engineering geological investigations and experimental researches, through more than 2000 boreholes (Fig. 1), tens of adits and testing for about 30 years,

H. Shehata and Y. Rashed (eds.), *Numerical Analysis of Nonlinear Coupled Problems*,
Sustainable Civil Infrastructures, DOI 10.1007/978-3-319-61905-7_20

a set of scheme for the utilization of the weathered rock mass had been established for the Three Gorges Project. For instance, the lower part of weakly weathered zone was used as the foundation rock mass of the dam, the weathered rock mass of porphyritic granite in Xia-an Stream was used for the artificial concrete aggregate, the weathered sand of dam bedrock was used as cofferdam filling sand material, and the weakly and slightly weathered granite was used as filling rock materials of cofferdams, etc. The Three Gorges Project construction shows that: it is correct and effective for the scheme to use the weathered rock mass, saving a large amount of construction investment and gaining a certain amount of construction time for the Three Gorges Project.

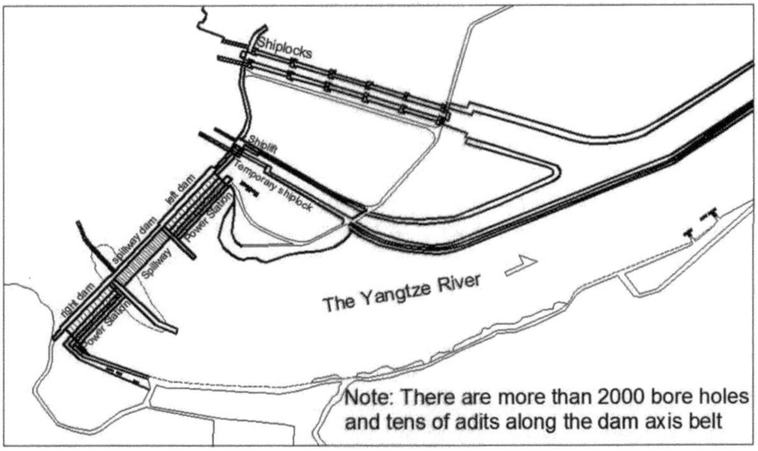

Fig. 1. Sketch map of the layout of CTGP

This paper summarizes the process and results of the survey and experimental study on the scheme of engineering utilization of weathered rock mass at the Three Gorges Dam site.

2 Weathered Crust Classification and the Main Engineering Geological Characteristics of Crystalline Rock

The foundation rock mass of the China Three Gorges Dam is granite with crystalline texture and block structure. In the long geologic time, suffering from the effect of various kinds of weathering agent, it appears inhomogeneous weathered state on the macro and constitutes four weathering zones regularly and orderly, "complete, intense, weak and light" from the outside to the inside on the section (Fig. 2). Due to different weathering intensity in different weathering zones, it shows different comprehensive characteristics in terms of the combination of weathered rock mass, rock mass structure and integrity, physical and mechanical properties and hydro-geological features. Therefore, it forms the basis of weathered crust classification.

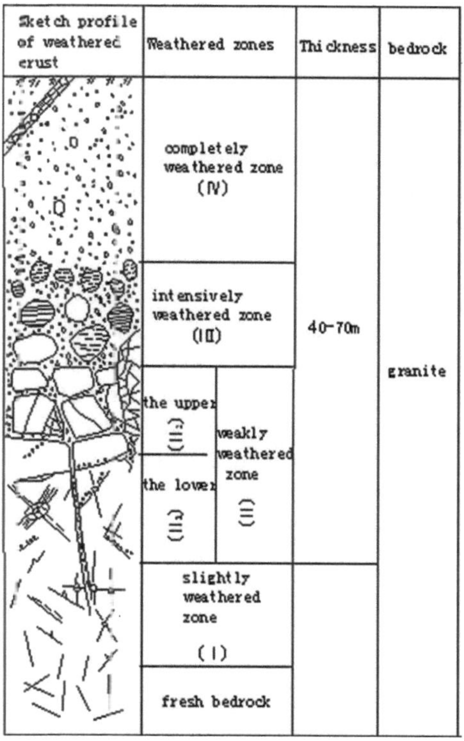

Fig. 2. Sketch weathered profile in the dam site of CTGP

Completely Weathered Zone

The composition is almost loose and half-loose rock, the content of which is more than 95%. It mixes up with a little fragment stone, with some hard weathered block with round shape distributed sporadically on the bottom. The rock has been largely lost crystalline connection.

Generally speaking, the quality of rock is more homogeneous with loose structure, low strength. The rebound index is about 10, and the compressive strength in saturation is less than 10 MPa.

Intensely Weathered Zone

The composition is partly loose to half-loose rock, mixing up with hard and half-hard rock, and the loose to half-loose rock makes up 30%–70%. Among the rock mass, weathered block with round shape of hard and half-hard rock are surrounded by loose and half-loose rock mass. The number of weathered block with round shape is increasing from the up to the down. The whole connection force and strength character is almost equal to the rock mass of completely weathered zone.

The permeability of the rock mass is strong in intensely weathered zone, whose permeability coefficient is 0.1–4 m/d, and some part reaches 11 m/d. Then V_p is generally 2000–3000 m/s, the minimum of which only reaches 500 m/s. And the rebound index is 10–31, average value reaching 19.6; the compressive strength in saturation is less than 20 MPa.

Weakly Weathered Zone
Exposed by a large number of boreholes, heavy-caliber and minor-caliber drilling, and through physical and mechanical parameters tests, it fully shows that the weakly weathered zone can be divided into two subzones, the upper zone and the lower zone, and there appears obviously different engineering geologic features between the two subzones.

Upper Weakly Weathered Zone
It is composed of hard to half-hard rock, mixed up with loose to half-loose clasts, which occupies 9%–20% and reaches 30% at maximum, mainly filling in the structural plane. It results in the junction thickening, and 30%–70% of the structural plane contain clasts, whose thickness is 0.05–0.2 m and the thickest are up to 0.8–1.5 m. The rock mass is dominating weathering along fractures with much inhomogeneous rock mass quality. The structure type belongs to cataclastic-submassive structure, and the integrity of rock mass is medium to poor, appearing very inhomogeneous, and the thickness is 2–4 m.

Lower Weakly Weathered Zone
The main composition is hard rock, mixed up with a little loose to half-loose rock mass. Among these rock mass, the thickness of the loose and half-loose rock mainly are several millimeters and a few reach 1–4 cm, most distributing in structure plane with steep dip angle. The rock mass quality and the engineering geological features of this zone is more homogeneous, with the compressive strength in saturation being 75–90 MPa, modulus of deformation being (20–30) $\times$ 10^3 MPa, and the rebound index reaching 41.

Slightly Weathered Zone
It is made up by hard rock with highly slight weathering. Only along part of fissure surfaces appeared phenomenon of yellow rust or the weathering skin about 1 mm, and a few structural planes were weathered more serious. The hard filling things are occupied 5%, the soften plane occupied 1%–2%.

The behavior of the rock mass in slightly weathered zone is similar to the fresh rock mass, with good quality and weak permeability. The compressive strength in saturation Rc is more than 100 MPa and the value of modulus of deformation is (30–40) $\times 10^3$ MPa. The rebound index value is 46.4 and the Vp is generally 5.39 km/s, the maximum of which can reach 6 km/s. On the whole, the rock mass is homogeneous, complete, with superior engineering geological properties.

The maximum of thickness of the weathered crust composed of completely, intensely, weakly weathered zone in the river bed is 43.12 m and the average is 11.46 m; the maximum of in the floodplain is 68.75 m and the average is 17.45 m; the maximum of in the terrace is 48.92 m and the average is 23.23 m; the maximum of in the valley is 54.95 m and the average is 18.13 m; the maximum of in the hillside is 68.19 m and the average is 26.95 m; the maximum of in the ridge is 83.54 m and the average is 41.34 m. The total thickness of the weathered crust is increasing gradually from the river bed to the mountain ridge.

3 The Use of Weathered Rock Mass as the Dam Foundation

3.1 The Possibility of Using Weakly Weathered Rock Mass as Dam Foundation

Different Opinions for Choosing the Three Gorges Dam Foundation Rock
The several geotechnical investigation reports that had been submitted by Ministry of Geology and Mineral Resources and Changjiang Water Resources Commission (CWRC), put forward the idea of choosing the slightly weathered zone rock mass as the dam foundation. And the specialists in the area of geology, rock mechanics and hydraulic engineering, also agreed with the opinions from the exploration units.

The consulting group of rock mechanics and foundation design, which was composed of the expert Anderson of rock mechanics test from the U.S. Bureau of Reclamation and the Army Corps of Engineers, the experts of engineering geology and the experts of dam foundation design, came to China to consult and took investigation in the construction site. The USA expert thought that it was best for the Three Gorges Dam to set up on the rock mass of slightly weathered zone. The upper rock mass of the weakly weathered zone could not be used. Only with less load condition and after the rock mass consolidated, it has the possibility to make use of the lower zone rock mass of the weakly weathered zone. However, it was better not to use it for the importance of the Three Gorges Project. Because the interface is uneven between the upper part and the lower part, while the lower part was used, the upper part might be left over, which is harmful to the dam foundation stability, and it wastes money to deal with. It is better to dig it up at first.

In the 70s, the Institute of Three Gorges Survey and investigation from CWRC carried out investigation research on the use of weakly weathered zone rock mass at the Tai Pingxi dam site (comparative dam site). According to the information about the geologic description of footrill, small diameter diamond drilling and acoustic wave test, it was divided into two subzones: the upper zone and the lower zone. The use of the lower weakly weathered zone was emphatically studied, taking grouting tests, the large deformation tests and acoustic wave tests at 5# adit.

In the late 70s, the survey and experimental study on the use of the weakly weathered rock mass, was kept on after paying attention to the San Douping dam site (Three Gorges Dam site). The research resulted that the engineering geological properties of the lower rock mass in the weakly weathered zone was similar to the rock mass in the slightly weathered zone, having the possibility for choosing as the dam foundation.

Doctor Campbell, Canadian expert, from the World Bank Group, insisted on the weakly weathered zone as the foundation base, as well as the opinion of the Doctor Muller. They thought that the loose and cracked rock mass was partial, especially the upper part, surrounded by intact rock mass. And it could be treated by excavating like the “dental operation”. The whole rock mass in the weakly weathered zone, could satisfy the requirement of the dam foundation after grouting.

Tackling Key Research on the Possibility of Use of the Weakly Weathered Rock Mass For the possibility of using the rock mass in weakly weathered zone, it had been researched and argued in depth, from the 7 fields: rock mass engineering geological features, rock mass properties, classification and evaluation of rock mass quality, numerical analysis and stability evaluation for different rock plane of dam foundation and the test result of the strengthening. The main conclusions were summarized as follows:

Taking consideration of the overall characteristics and conditions of lower weakly weathered rock mass, it was superior to the upper weakly weathered rock mass as the Three Gorges Dam foundation. It was similar to the slightly weathered rock mass, and it does not have the problem such as foundation sliding along the superficial zone, uneven bearing, deformation and seepage stability. Therefore, the lower weakly weathered rock mass can be used for the project completely, and its characteristic and condition would be better after proper reinforcement. However, the geological flaw, with respect to the lower weakly weathered rock mass on its heterogeneity, was obvious. And the loose debris sandwich, whose thickness was 5–20 cm, is difficult to handle. There exists security and stability problem that can not satisfy the engineering requirement. Therefore, it was needed to carry out experimental study on weathering rock reinforcement.

3.2 The Experimental Study on Weathering Rock Reinforcement

In order to study the effect of the rock consolidation grouting test to the upper and lower weakly weathered rock mass, the Yangtze River Academy and the Institute of Three Gorges Survey and investigation from CWRC chose 3 test areas at 5# adit in the Tai Pingxi dam site in 1978. Thereinto, II, III test area were located in the upper rock mass of the weakly weathered zone (I test area reserved for comparison), and test area was located in lower rock mass. The 3 test areas respectively adopted three different materials, cement, epoxy resin and polymethyl, to carry out grouting tests.

II test area: It is the quartz diorite on the upper weakly weathered zone, with passing through 6 faults, so the rockmass integrity is poor. The area has been arranged 24 grout holes using the ring oxygen resin grouting, starting the experimental grain filling after high-pressure water washing. Because of the lack of the slurry material, only 6 grout holes with location dispersing filled grain. The maximum single-hole grouting pressure reached 0.2 MPa, the minimum less than 0.01 MPa, and the typical value was 0.01–0.1 MPa. The maximum single-hole grouting-absorption capacity reached 116 L, the typical value being about 40 L and the total was 424 L.

III test area: It is the quartz diorite on the upper weakly weathered zone, with many fractures of medium and low-angle dip. The experimental area was layout 22 grout holes using portland cement. The maximum grouting pressure reached 0.8 MPa, the minimum being 0.02 MPa, and the maximum single-hole grouting-absorption capacity reached 955 L/min. The total cement consumption was 1531 kg and unit consumption was 29.46 kg.

IV test area: It was the quartz diorite black diorite rock in the lower part of weakly weathered zone, with 3 small faults whose width was 5–10 cm, fissure ratio about 0.5%.

The test areas were layout 22 grouting holes with polymethyl methacrylate casting material. Since lack of slurry material, 13 grout holes with dispersing location were filled with grain. The maximum single-hole grouting pressure reached 0.6 MPa, and the typical value was 0.3–0.6 MPa. The maximum single-hole grouting-absorption capacity reached 58.9 L, the typical value being about 7–40 L and the total was 260.5 L.

Both took water pressure test before and after grouting, and the changes of their specific water absorption (ω) are shown in Table 1.

Table 1. Comparative list of the specific water absorption before and after grouting

Test area	The specific water absorption before and after grouting $\omega/(\mathrm{L \cdot min^{-1} \cdot m^{-1} \cdot m^{-1}})$			
	Pressure (MPa)	Native state (ω)	After the grouting (ω′)	$\frac{\omega-\omega'}{\omega}$
II	0.3	0.015–0.375	0.008–0.09	0.47–0.76
III	0.3	0.1–0.4	0.011–0.029	0.89–0.93
IV	0.3	0.013–0.078	0.005–0.018	0.59–0.77

It was carried out elastic modulus and deformation modulus tests on the spot before and after grouting, and the results are shown in Table 2.

Table 2. Comparative table of the modulus of deformation and elasticity modulus before and after grouting

Test area	Test method	Modulus of deformation			Elasticity modulus		
		E_1	E_2	$\frac{E_1-E_2}{E_1}$	E_1'	E_2'	$\frac{E_1'-E_2'}{E_1'}$
II	Flexibility	1.15	1.93	0.68	2.63	3.08	0.15
	Stiffness	0.18	0.48	1.67	0.29	0.92	2.17
III	Flexibility	0.99	18.28	17.47	2.11	21.51	2.19
	Stiffness	5.43	11.0	1.03	9.25	17.66	0.89
IV	Flexibility	26.32	47.41	0.8	29.59	57.9	0.96
	Stiffness	1.82			0.78		

All grouting holes were taken ultrasonic test, the wave velocity changes before and after grouting and the dynamic elasticity modulus rock mass are shown in Table 3.

Through consolidation grouting test and comparative analysis, the following elementary opinions were concluded:

Lower part of the weakly weathered rock mass (IV test area), was with better homogeneity and integrity, and each index was enhanced than before. After

Table 3. Comparative table of the wave velocity and the dynamic elastic modulus before and after grouting

Test area	Average wave velocity V_p/(km · s^{-1})			Dynamic elasticity modulus (× 10^3 MPa)		Effect	
	Native state	After washing	After grouting	E_d	E'_d	§	I
II	4.88	4.377	4.927	40.85	51.53	26	0.77
III	5.186	4.432	5.41	47.9	66	38	0.93
IV	5.264	5.04	5.272	57.3	67	17	0.89

Note: $\S = \frac{E_d - E'_d}{E_d} \times 100\%$

consolidation grouting treatment, the deformation modulus and elastic modulus were increased significantly, with obvious low permeability, longitudinal wave velocity increasing, and the integrity of rock mass had been improved. The lower part rock mass quality enhanced from grade B to grade A. Therefore, the lower part rock mass can be as the Three Gorges Dam foundation after consolidation grouting treatment.

Upper part of the weakly weathered rock mass (II, III test area), is weathered intensively along the fractures and fault zones, with more inhomogeneous and less integrity. Through consolidation grouting treatment, the water permeability reduced obviously, with the deformation modulus and elastic modulus enhanced, and the rock mass behavior had been improved. However, due to so much loosen to half-loosen debris distributing along the fractures and fault zones, it was unable to wash clean the loosen to half-loosen debris and the consolidation effect was not obvious. It was difficult to reach the optimizing index of being the high dam foundation rock by consolidation grouting. Thus, it can not satisfy the fundamental demand of the dam.

3.3 The Anti-sliding Stability Evaluation for Lower Weakly Weathered Zone Rock Mass as Foundation Base

The Calculation Condition

(a) Chose the 5# unit on the left dam sections as a typical geological section of the dam, and took lower part of the weakly weathered rock mass as dam foundation and its surface elevation 90 m. Then the elevation of the dam crest was 185 m, designing reservoir level being 175 m in the upstream, and the water level was 66 m in the downstream.

(b) Took two sets of parameters to calculate severally, one with the shear strength of concrete and the other with bedrock at the bottom of the weakly weathered rock mass: $f = 1.2$, $c = 1.5$ MP or $f = 1.0$, $c = 1.5$ MPa.

(c) Calculation load: the upstream water pressure $P_H = 1/2rH^2 = 3621.5$t, the weight of the dam concrete W = 9562.5t, the upstream silt pressure force $P_n = 1/4\,h_n^2 = 25$t, the uplift coefficient a = 0.25, the uplift pressure $U = U_1 + U_2 = 1267$t.

The Calculation Results and Evaluation
Adopting shear-friction formula:

$$K_0 = \frac{f(W - U) + C \cdot L}{P_H + P_n}$$

When selecting parameters:

$$f = 1.2,\ c = 1.5\,\text{MPa},\ K_0 = 5.68$$
$$f = 1.0,\ c = 1.5\,\text{MPa},\ K_0 = 5.22$$

The calculations shows that the safety factor against sliding K_0 was greater than 3. Therefore, the performance can meet the engineering requirements of stability against sliding by using the lower weakly weathered zone rock mass as the dam foundation.

4 The Use of the Rock Mass as Civil Building Foundation

Within the scope of 16 km^2 in the Three Gorges Dam Area, it is necessary to build a large number of engineering facilities such as wharf, highway, house, etc. The load of these buildings on the ground is much less than the dam, and different zones of weathering rock mass can meet different foundation requirements.

Based on the characteristics of civil foundation, the rock mass of the weathered crust is further classified. The original completely, intensely weathered zones are redivided into the upper part of the complete weathered zone, the lower part of the complete weathered zone, intensely weathered zone. And the geologic features, the mechanic indexes of the foundation rock mass and the applicability of each weathered zone are summarized as follow

The upper part of the complete weathered zone: it is brown, tawny, and loose but close-grained. The indoor uniaxial compressive strength test value is 300–600 kPa, and standard value of the bearing capacity is 280 kPa, the bearing capacity of pile endpoint soil standard values being 900–1000 kPa, the friction of pile soil standard being 40–50 kPa, modulus of deformation being 0.014–0.025 GPa. It is applicable to strip foundation, column foundation of the 6 floors and the below.

The lower part of the complete weathering zone: it is flavescens, gray, and half-loose rock including a few half-hard rock blocks. The indoor uniaxial compressive strength test value is 900–1200 kPa, and standard value of the bearing capacity is 800 kPa, the bearing capacity of pile endpoint soil standard values being 2000–2500 kPa, the friction of pile soil standard being 60–80 kPa, modulus of deformation being 0.03–0.05 GPa. It is applicable to strip foundation, column foundation of the 8 floors and the below.

Intensely weathered zone: it is sallow, gray, composed by half- loose rock and hard to half-hard rock blocks, the hard rock mass becoming globes. The indoor uniaxial compressive strength test value is 11–24 MPa, and standard value of the bearing capacity is 1500 kPa, the bearing capacity of pile endpoint soil standard values being 3000–5000 kPa, the friction of pile soil standard being 100–120 kPa, modulus of deformation being 0.5–2.0 GPa. It is applicable to low load pile foundation, but the strip foundation should embedded in stable bearing layer.

The upper part of the weakly weathered zone: it is gray, composed mostly of hard rock, appearing weathering phenomenon of half-loose sandwich whose thickness is 3–30 cm along part of the fissure surface. The indoor uniaxial compressive strength test value is 50–80 MPa, and standard value of the bearing capacity is 2000–3000 kPa, the bearing capacity of pile endpoint soil standard values being 8000–10000 kPa, the friction of pile soil standard being 150–200 kPa, modulus of deformation being 1.0–5.0 GPa. It is applicable to general pile foundation.

The lower part of the weakly weathered zone: it is gray, mostly composed of hard rock, appearing weathering phenomenon of half-loose sandwich whose thickness was 1–10 cm along part of the fissure surface. The indoor uniaxial compressive strength test value is 70–85 MPa, and standard value of the bearing capacity is 4000–5000 kPa, the bearing capacity of pile endpoint soil standard values being 10000–13000 kPa, the friction of pile soil standard being 300–400 kPa, modulus of deformation being 15–20.0 GPa. It is applicable to larger pile foundation. Because of the complexity of the weathered rock mass, it must require a pilot hole for each foundation pile during the construction process, and the determine for the depth of embedding bearing layer should take the pile end rock mass stability into consideration.

5 The Use of the Weathered Rock Mass as Building Materials

According to the design requirements, the Three Gorges Project need natural aggregate concrete reaching 15.09 million m^3, and 26.62 million m^3 artificial sand aggregate, 16.59 million m^3 cofferdam filling sand stone, 1.13 million m^3 soil material, 3 million m^3 monolith. Facing such a huge demand of natural building materials, it is the most economic and reasonable way to use local materials. Thus, the research on the use of the weathered rock mass of dam bedrock as building materials had received great attention.

The Use of the Weathered Porphyritic Granite Rock Mass for Artificial Aggregate
Due to the high content of the mica in granite of damsite, it does not meet standard requirement after made for sand. So the porphyritic granite in Xia-an Stream downstream, 12 km away from the dam site, had been carried on a research for artificial aggregate. Through the identification of binoculars for artificial sample made from complete, intense weathering porphyritic granite rock, the free mica in the sand content is 1.28%–1.41% (weight percentage), basically conforming to the requirements of the quality of fine aggregate.

The Complete Weathered Grit Used for Cofferdam Filling Sand Material
According to the analysis and cast filling test underwater, the indexes such as grain composition, unit weight, silt content of completely weathered sand material can meet with the requirements of the relevant criterions, so it can be used for cofferdam filling sand material.

The Weak and Light Weathered Granite Used as Cofferdam Filling Rock Material
The granite rock in weakly, slightly weathered zones at the dam site is hard, and its

saturated uniaxial compressive strength reaches 40–100 MPa, and the structures are mostly the blocky and sub-massive, so, the quality and technical index can meet the requirements of the relevant criterions as cofferdam filling rock material.

6 Conclusions

The weathered rock mass in China Three Gorges Dam Area has a great thickness of 40–70 m, with different character and greater different physical and mechanical properties in different weathered zones.

It has feasibility to use the lower part of weakly weathered zone rock mass as dam foundation after consolidation grouting treatment, saving a large amount of construction funds for the China Three Gorges Project.

The weathered rock mass can be used as civil building foundation in the dam site area, and the rock mass of different weathered zone can meet different requirements of foundation bearing layers.

The weathered rock mass of porphyritic granite downstream can be used as artificial concrete aggregate; the completely weathered sand of the dam bedrock can be used as a cofferdam filling sand material, and granite in the weakly and slightly weathered zones can be used as cofferdam filling rock material.

References

Zuo, S., Ren, G.: Discussion of using quantitative indexes to study rock mass weathering. Mount. Res. **20**(3), 365–369 (2002). (in Chinese)

Nie, D., Han, A., Ju, G.: Study on integrated zoning of weathering degree of rockmass. J. Eng. Geol. **10**(1), 20–25 (2002). (in Chinese)

Feng, Q., Chen, L., Nie, D.: Research on quantifiable zoning of weathering degree of rockmass. J. Geol. Hazards Environ. Prot. **12**(2), 76–79 (2001). (in Chinese)

Gokay, M.K.: Influence of natural weathering on rock mass. Union Chamb. Eng. Archit. Turk. **42**(1), 3–41 (2003)

Wang, H., Zuo, C.: Utilization of the dam foundation rock-mass about the small and medium hydropower stations in Hunan. Hunan Geol. (Z1), 84–88 (1999). (in Chinese)

Shen, G., Yang, G.: The application of using the weakly weathered rock as a gravity dam foundation plane for transshipment in reservoir engineering. Zhejiang Hydrotech. (01), 29–32 (1997). (in Chinese)

Suo, S., Xu, R., Ye, Y.: Weathered rock and selection of foundation surface of concrete dams. Adv. Sci. Technol. Water Resour. **26**(05), 45–47 (2006). (in Chinese)

Xiao, R., Cai, J.: The issue on utilizing the weathered rock of volcanic in Fu Jian. Water Power (04), 28–32 (1988). (in Chinese)

Ren, Z., Ma, D., Shen, T., et al.: Research on Rock Mass of The Three Gorges Project Dam Foundation. China University of Geosciences Press, Wu Han (1998). (in Chinese)

Huang, Y., Wang, Z.: Understanding and practice of the use for the weakly weathered rock masses. Yangtze River (06), 33–36 (1995). (in Chinese)

Li, Z.: Elastic wave testing technique on the three gorges dam foundation plane. China Three Gorges Constr. (03), 13–14 (1999). (in Chinese)

Displacement Assessment of Rock Socketed Shafts: A Numerical Approach

Asmaa M.H. Mahmoud[1(✉)] and Ahmed M. Samieh[2]

[1] ABCAN Engineering Consultancy, El-Demerdash, Cairo, Egypt
abcams98@gmail.com
[2] Department of Civil Engineering, Helwan University, EL-Naam, Cairo, Egypt
abcan98@yahoo.com

Abstract. Settlement assessment of shallow/deep foundations, under the working load range, has to be assessed to complete the design of any foundation system. The settlement of shafts resting on or socketed into sound rock is generally negligible. Settlement may be significant for shafts socketed into weak rock. Numerical approaches are among the promising tools that can be adopted to assess the settlement of piles socketed into rock. This study presents the results of a proposed numerical modeling scheme to predict the vertical displacement of socketed shafts. A comparison between the predicted and the monitored responses is presented to evaluate the viability of the proposed numerical scheme to assess the socketed shafts vertical displacement. A good comparison is exhibited between the results of the proposed model and those of the monitored test pile. In addition, the results of a sensitivity study are presented to highlight the influence of some numerical analysis parameters such as the rock-shaft interface strength, the shear strength characteristics and the deformation modulus of the rock formation on the predicted numerical response.

1 Introduction

Predicting the load-displacement response of drilled shafts is in some cases as important as, or possibly more critical than, predicting the ultimate bearing capacity. Many theoretical methods have been proposed for calculating the displacement of drilled shafts socketed into rock. Mattes and Poulos (1969) are among the first to investigate the load-displacement behavior of rock-socketed shafts by integrating the Mindlin's equations. Carter and Kulhawy (1988) provided a set of approximate analytical solutions to predict the load-displacement response of rock sockets by modifying the solution of Randolph and Wroth (1978) for piles in soil strata. The majority of the theoretical solutions for predicting the displacement of drilled shafts socketed into rock, however, have been developed using finite element analysis (e.g., Osterberg and Gill 1973; Pells and Turner 1979; Donald et al. 1980; Rowe and Armitage 1987a). Most of the techniques proposed for calculation of the vertical displacements of drilled shafts socketed into rock are based on the theory of elasticity. It has been usual to assume that the drilled shaft is essentially an elastic inclusion within the surrounding rock mass and that no slip occurs at the interface between the shaft and the rock mass, although the solutions of Rowe and Armitage (1987a) and Carter and Kulhawy (1988)

H. Shehata and Y. Rashed (eds.), *Numerical Analysis of Nonlinear Coupled Problems*,
Sustainable Civil Infrastructures, DOI 10.1007/978-3-319-61905-7_21

can consider the possibility of slip. Rowe and Armitage (1987b) performed an elasto-plastic finite element analysis that accounted for slip along the interface according to the technique developed by Rowe and Pells (1980). Leong and Randolph (1994) included dilation, hardening, and softening effects of the rock formation.

This study presents the results of a proposed numerical modeling scheme to predict the vertical displacement of socketed shaft. A comparison between the predicted and the monitored responses is presented to evaluate the viability of the proposed numerical scheme to assess the socketed shaft displacement. To investigate the effect of the load direction on the socket side shear resistance, the developed numerical model results of an O-cell test are, also, compared with the numerically predicted response of the conventional top-down trend of piles loading. In addition, a sensitivity study was carried out to highlight the influence of some numerical analysis controlling parameters such as the rock-shaft interface strength, the shear strength characteristics and the deformation modulus of the rock formation.

2 Numerical Modeling of a Socketed Shaft into a Limestone Formation

The pile load test described by Moon and Cumaraswamy (2008) was considered in the numerical analysis of this study. The shaft was constructed at the Rowlett Creek site, Collin County, Tex., USA. The stratification at the location of the shaft load test consisted of a surface layer of stiff clay, 1.8 m thick, that was underlain by a deposit of weathered limestone extending to a depth of 3.1 m below the ground surface (bgs). The weathered limestone was followed by a solid limestone formation that extended to the end of the carried out borehole, 10 m bgs. The geotechnical properties of the solid limestone formation at the location of the shaft load test were described in details by Moon and Cumaraswamy (2008). The limestone formation had an average rock quality designation (RQD) value of 97% and average unconfined strength of 10 MPa.

The test shaft considered in the current study, Fig. 1, was 0.76 m in diameter and was drilled by an auger tool to a depth of 7.2 m bgs. After completion of the shaft drilling, concrete was poured to a height of 1.5 m above the pile tip level. The Osterberg load cell and the shaft reinforcement were lowered into the shaft that was subsequently concreted to a depth of 2.7 m bgs, 0.9 m below the top of the limestone formation. The remaining length from ground surface to depth of 2.7 m was left void. The O-cell had a maximum load capacity of 16 MN. The socketed shaft was instrumented by using vibrating wire sister bars strain gauges. The load test was carried out after 50 days of the shaft concreting.

Numerical modeling of the considered shaft was carried out adopting the conventional Mohr-Coulomb constitutive model for the stiff clay and limestone formations. The concrete of the shaft was modeled by a non-porous elastic model. Table 1 shows a summary of the parameters that were adopted in the current analysis. Interface elements were used to model the interaction between the shaft and the surrounding soil/rock formations. The interface properties were taken as a fraction of the surrounding rock properties through a strength reduction factor, R_i, taken equal to 0.75 in the current study. Undrained condition was considered in the current study to model the

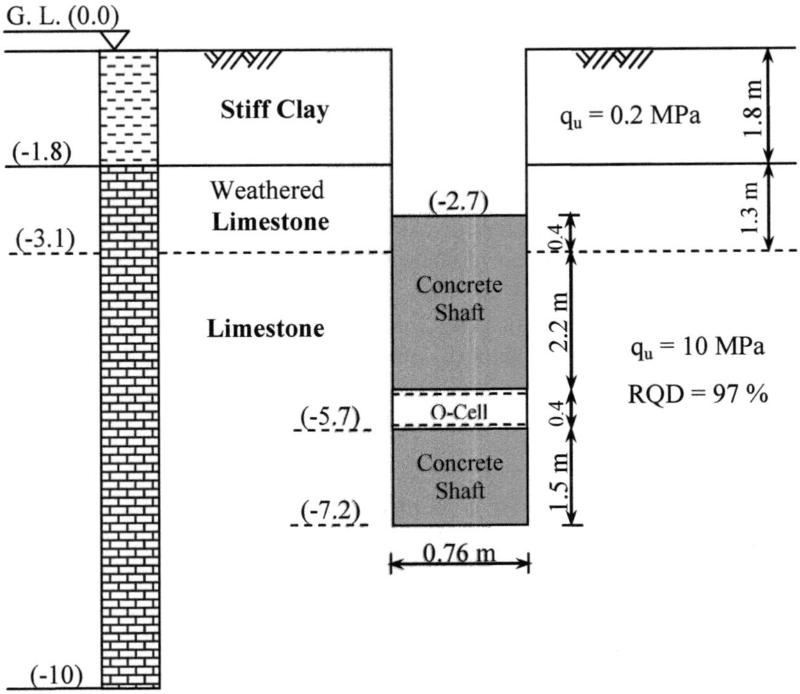

Fig. 1. Configuration of the tested socketed shaft in a limestone formation (Moon and Cumaraswamy 2008).

Table 1. Material properties adopted in the numerical analysis of the shaft socketed into the limestone formation.

Parameter/Layer	Stiff clay layer	Weathered limestone layer	Limestone layer
Model	Mohr-Coulomb model		
Cohesion (MPa)	0.1	0.25	1.0
Friction angle (deg.)	0.0	20	28
Dilatancy angle (deg.)	0.0	0.0	5.0
Deformation modulus (MPa)	10	30	550
Poisson's ratio	0.4	0.2	0.2
Socketed concrete shaft: *Deformation modulus = 20 GPa, Poisson's ratio = 0.2*			

surface clay formation. The O-cell load test device was considered as a void space at the location of the O-cell with thickness equal to the O-cell height and applying pressure load with the same magnitude in two opposite directions.

2.1 Numerical Analysis Results and Comparison with Field Measurements

The established numerical scheme was adopted to predict the load versus displacement response of the shaft under consideration. Figure 2 shows a comparison between the numerically predicted and monitored responses of the shaft. A good comparison is delineated between the numerically predicted load-displacement response and the monitored shaft response until slightly before the shaft failure. The figure shows that the considered shaft was failed at a skin friction load of about 10 MN, an average skin resistance of 1.6 MPa and at a displacement exceeding around 2% of the pile diameter. The curvature of the load-displacement curve was minimal up to about three-quarter of the failure load, displacement level less than 6 mm ($\approx$ 0.8 %) at that load level for the considered shaft, where the displacement started to significantly increase with increase of the loading until failure/full slip occurred. This implies that the slippage/breakdown of the bond between the shaft and the adjacent rock was minimal during the early loading stages and significantly increased near failure. The disagreement between the numerically predicted response and monitored response in the zone near failure is referred to the point that the considered numerical scheme assumes full contact between the shaft and the surrounding rock, i.e. it did not consider the bond breakage development of the pile-rock interface, especially in the loading zone near failure. Taking into consideration that a factor of safety greater than 2 is typically considered on designing rock sockets and applying this to the data shown in Fig. 2, it can be concluded that the proposed numerical approach can be adequately adopted to assess the

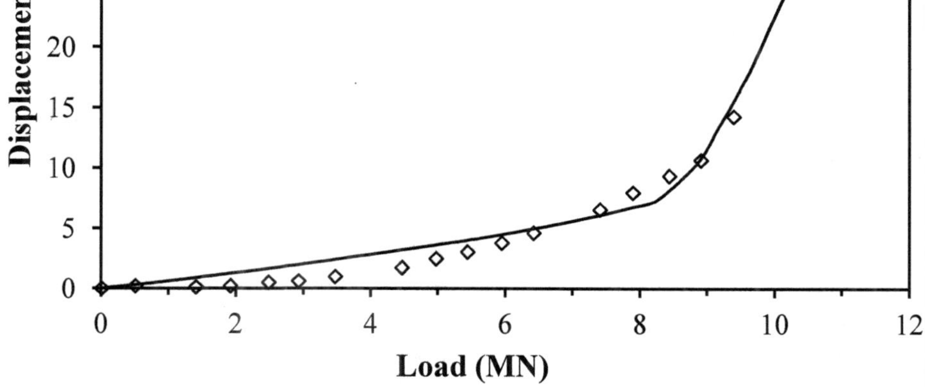

Fig. 2. Monitored versus numerically predicted displacement of the shaft socketed into limestone formation.

displacement of piles under working load conditions, load less than 5 MN for the shaft under consideration.

To predict the conventional top-down deformational response of the considered shaft, the shaft was gradually loaded across its top, at 2.7 m below ground surface. To force the system to resist the applied load by side friction only, a void was considered at the shaft base. Figure 3 shows the model discretization of the top-down loading for the shaft under consideration.

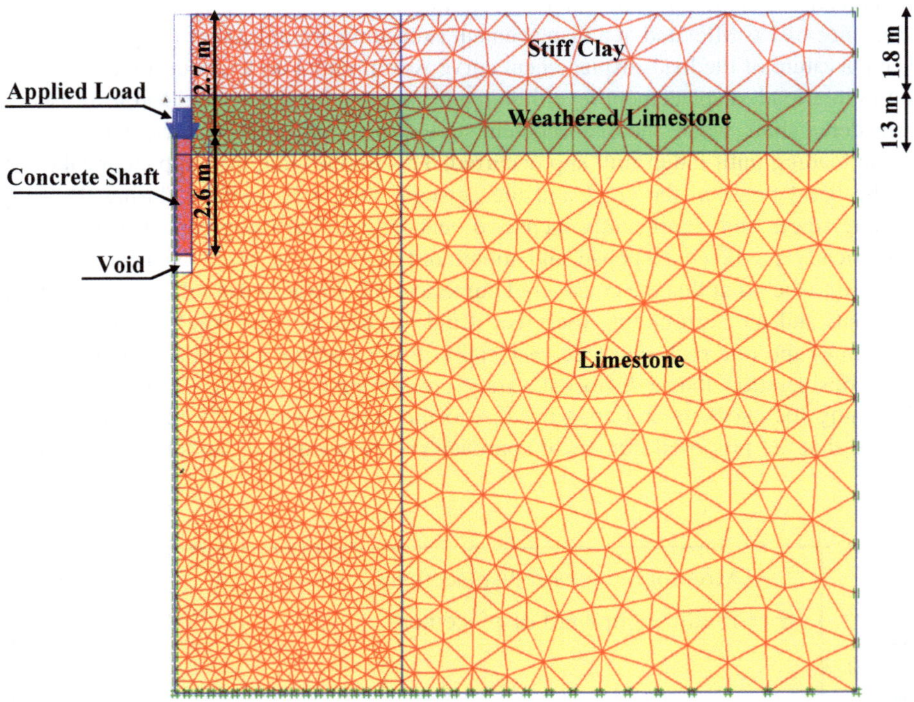

Fig. 3. Two-dimensional discretization of the top-down loading system of the shaft socketed in limestone formation.

Figure 4 presents the numerically predicted load-displacement curves of the bottom-up and top-down loading systems together with the monitored O-cell response, for comparison purposes. As shown in this figure, the numerically predicted load-displacement responses using both the bottom-up and top-down load tests exhibit no significant differences within the working load range, load less than around 5 MN for the considered shaft. For the numerically predicted response, if the ultimate load is considered as that corresponds to the intersection of the initial linear segment of the curve and the last linear segment of the numerically predicted load-displacement curve, the numerically predicted ultimate loads would be around 8.6 MN and 7.5 MN for the bottom-up and top-down loading conditions of the considered case, respectively. These predicted ultimate-values are on the conservative side with respect to the monitored ultimate load value, 10 MN - i.e. a variation of about 14 to 25%.

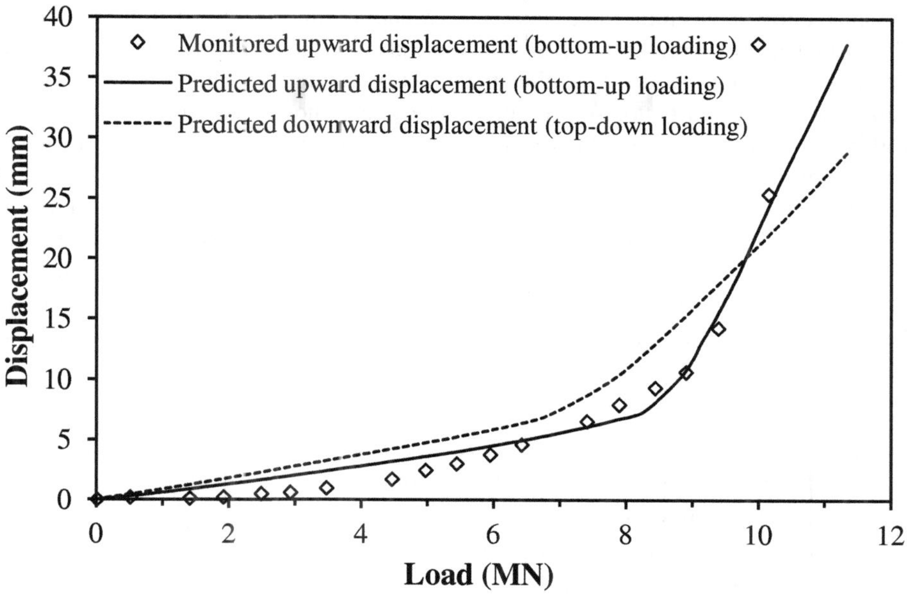

Fig. 4. Monitored versus numerically predicted displacement responses of both of bottom-up and top-down loading tests.

2.2 Sensitivity Analysis

A sensitivity study was carried out to evaluate the influence of the major parameters that are of concern on designing/carrying out a numerical analysis of shafts socketed into rock. The parameters considered in the current study were the shaft-rock interface strength, the shear strength characteristics and the deformation modulus of the limestone formation. In the following figures and discussions, the predicted numerical response that was described previously, considering the O-cell/bottom-up loading condition, is considered as the reference response. The monitored response is also shown in the following figures for comparison purposes. In the following discussions, the empirically predicted ultimate load of the shaft is taken as that corresponding to the intersection of the initial and final nearly linear parts of the numerically predicted load-displacement curve.

2.2.1 Effect of the Limestone-Shaft Interface Strength

The drilled shafts side wall conditions are of significant influence on the shaft response under vertical loading. The value of the side shear mobilized at the concrete-rock interface is affected by many factors. These factors include the socket roughness/construction conditions, strength of the concrete and the rock formation, the interface characteristics, geometry of the socket, in addition to deformational characteristics of the concrete and the rock mass formation (Rowe and Armitage 1987a). In this study, the effect of the rock-shaft interface strength on the socketed shaft response was assessed through varying the interface factor, R_i, from a value of 0.6, which simulates a low interaction between the shaft and the surrounding limestone formation, to a value of 0.9, that nearly represents full interaction between the shaft and the surrounding limestone formation. Figure 5 exhibits the effect of varying the interface factor on the numerically predicted bottom-up response. The figure shows that within the expected working load range, load less than around 5 MN for the considered shaft, the R_i-value has no significant effect on the load-displacement response. In the zone slightly beyond the expected working load range, decrease of the R_i-value to 0.6 significantly increased the shaft displacement, especially in the zone near failure. Increase of the R_i-value to 0.9, decreased the shaft displacement especially in the zone near the monitored failure. In other words, lack of the socket-rock interaction makes to significantly increase the shaft displacement especially in the zone beyond the expected working load range and near failure.

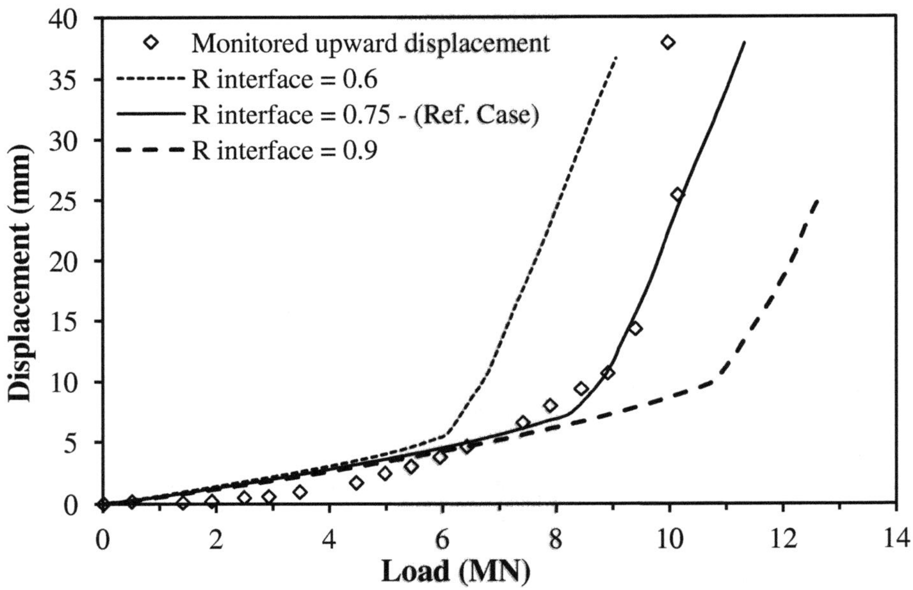

Fig. 5. Effect of the shaft-rock interface strength on the predicted displacement of the shaft socketed into the limestone formation.

Figure 5 reveals that for R_i-values of 0.75 and 0.9, the predicted and monitored responses are not significantly affected up to nearly a load level of around 80% of the monitored failure load. In other words, for good construction practice, a R_i-value in excess of around 0.75 might be adopted on numerically assessing the responses of the socketed shafts. In addition, it is observed that for R_i-values of 0.75 and 0.9, the difference of the empirically predicted ultimate loads of the predicted responses, 8.6 and 10.9 MN – respectively, did not exceed about 20%.

2.2.2 Effect of the Shear Strength Characteristics of the Limestone Formation

The shear strength characteristics of the rock formation surrounding a shaft are expected to influence the load-displacement response of the socketed shaft. For the shaft under consideration, the effect of the variation of the formation internal friction angle and cohesion are discussed in the following.

2.3 Frictional Characteristics Effect

To assess the influence of the rock formation friction angle on the numerically predicted load-displacement response of the considered socketed shaft, the friction angle of the limestone formation was varied from 25° to 35°. Figure 6 shows the effect of the friction angle variation on the predicted load-displacement response. The figure shows that within the expected working load range, the numerically predicted load-displacement responses are not significantly affected by the considered friction angle variation range. However, at high loading levels, beyond the empirically estimated ultimate load conditions, the friction angle of the formation has a significant effect on the shaft response where the socket displacement is reduced with increase of the formation friction angle.

Figure 6 reveals that for the friction angle range considered in this study, from 25° to 35°, the predicted responses and the monitored response are not significantly affected up to nearly a load level of around 75% of the monitored failure load. In addition, it is

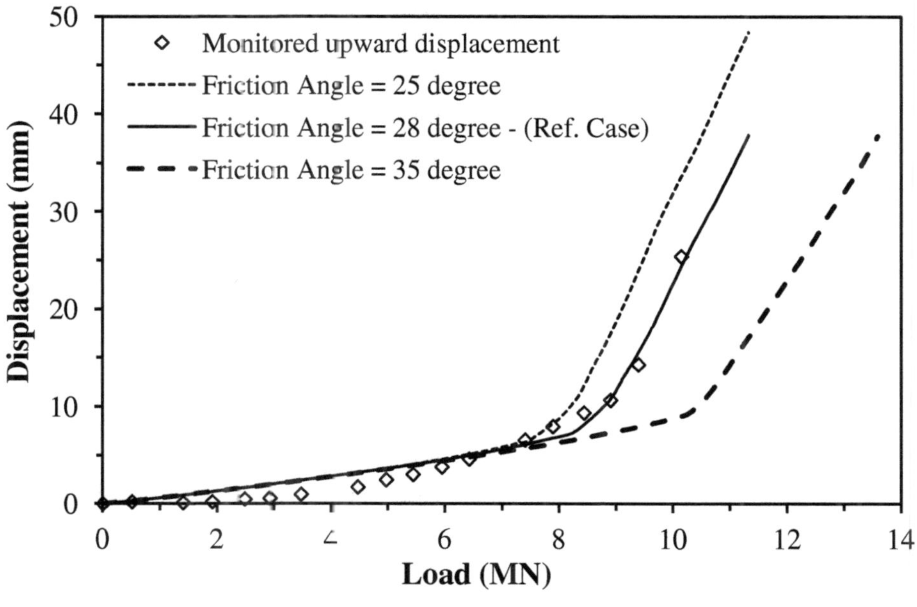

Fig. 6. Effect of the variation of the limestone formation friction angle on the numerically predicted shaft displacement.

noticed that for friction angle values of 25° and 28°, the difference of the empirically predicted ultimate loads of the predicted responses, 8 and 8.6 MN-respectively, did not exceed 7%, while as this difference increases to 20% for the friction angle values of 28° and 35°, empirical ultimate load values of 8.6 and 10.3 MN-respectively.

2.3.1 Cohesive Characteristics Effect

To evaluate the influence of the rock formation cohesion on the numerically predicted response of the considered socketed shaft into the limestone formation, the formation cohesion parameter was varied from 0.75 MPa to 1.25 MPa, ± 25%. Figure 7 shows the variation of the socket load-displacement response with change of the formation cohesive characteristics. The figure exhibits that at the early stage of loading, nearly within the working load range, the numerically predicted load-displacement responses are not significantly affected by the considered variation range of the formation cohesion. For loading range above the aforementioned load level, the formation cohesion has a significant effect on the shaft response where the socket displacement is increased and the ultimate load is decreased with decrease of the formation cohesion. The figure delineates, also, that for the cohesion values of 1.0 and 1.25 MPa, the numerically predicted and the monitored response are not significantly affected up to nearly a load level of around 80% of the monitored failure load.

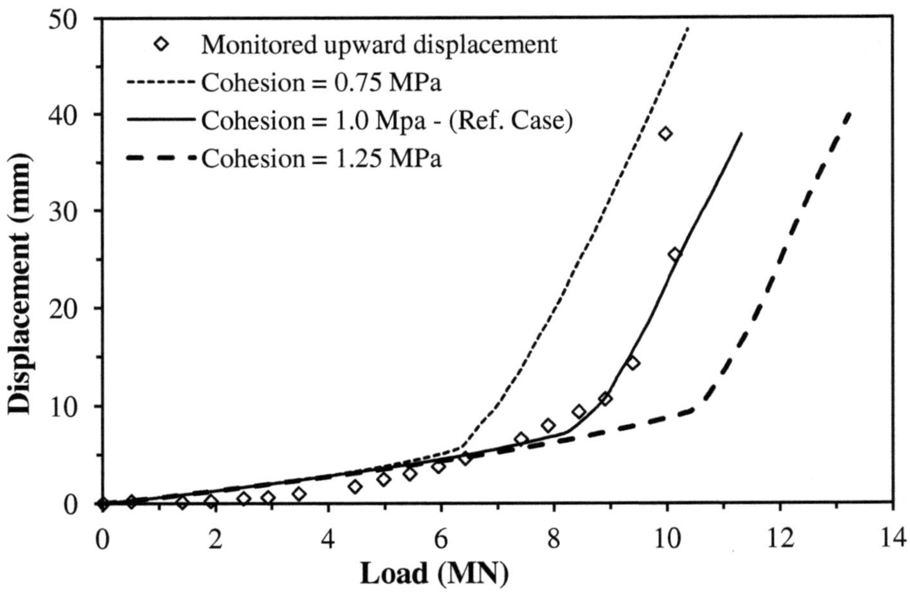

Fig. 7. Effect of the variation of the limestone formation cohesion on the predicted shaft displacement.

2.3.2 Effect of the Limestone Formation Deformation Modulus

A rock formation deformation modulus is one of the parameters that conventionally of a major concern on carrying a deformation analysis of rock sockets. The effect of variation of the considered rock formation deformation modulus on the considered shaft response is shown in Fig. 8. The figure exhibits that the considered modulus variation range, from 400 MPa to 700 MPa, has no significant influence on the predicted response within the expected working load range of the shaft under consideration. However, beyond the expected working load range, the influence of the deformation modulus on the predicted displacement starts to be significant especially in the zone around the estimated ultimate load. The figure shows, also, that for the deformation modulus values considered in this study, from 400 to 700 MPa, the predicted responses and the monitored response are not significantly affected up to nearly a load level of around 80% of the monitored failure load.

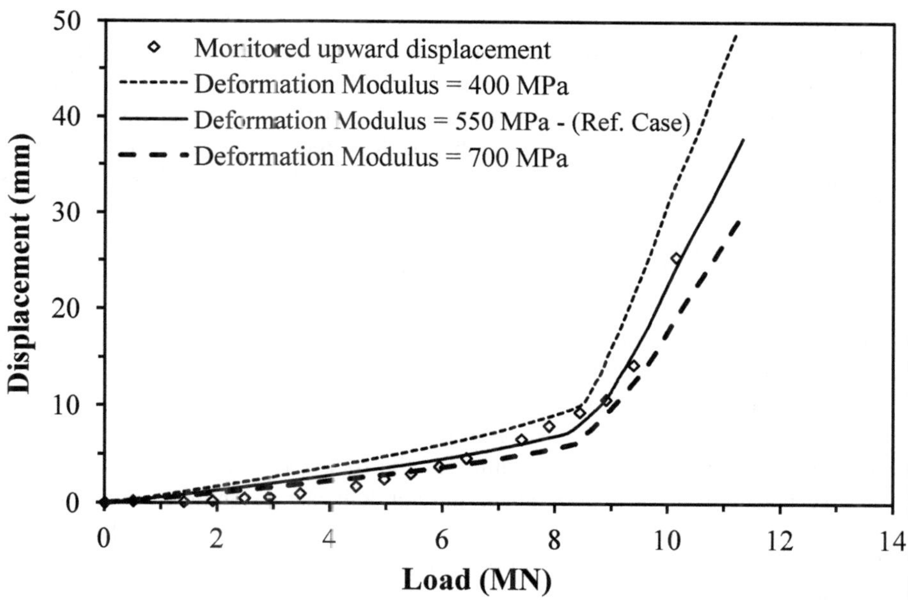

Fig. 8. Effect of the variation of the limestone formation deformation modulus on the numerically predicted shaft displacement.

3 Conclusions

A numerical analysis approach was proposed and utilized in this study to assess the deformational response of socketed shafts, which is a typical requirement of socketed shafts geotechnical design. A numerical analysis was carried out considering two-dimensional numerical idealization of the system. A good comparison was revealed between the predicted and the monitored response of the considered shaft. The developed numerical model results of the O-cell test were, also, compared with the numerically predicted response of the conventional top-down loading trend of sockets. Moreover, a sensitivity study was carried out to highlight the influence of some controlling parameters that are typically of concern on carrying out the numerical analysis for assessing the socket displacement. The parameters considered in this study were the rock-shaft interface strength, the shear strength characteristics of the rock formation and the deformation modulus of the rock formation. This study led to the following conclusions:

- Two-dimensional numerical analysis can be adequately utilized to predict the load-displacement response of rock socketed shafts especially within the working load range.
- On numerically assessing the load-displacement response of socketed shafts, it was noted that the empirically estimated ultimate load value of the predicted top-down loading is always slightly less than that of the numerically predicted bottom-up loading. This difference was generally less than around 14% for the considered this study case.
- Interface strength of rock-socketed shafts has a significant effect on the numerically predicted load-displacement response of the system. For well constructed shafts, a value of the interface factor greater than 0.75 can be adopted in the numerical analysis. In such a case, it is observed that for R_i-values in the range of 0.75 to 0.9, the difference of the empirically predicted ultimate loads of the predicted responses did not exceed about 20% in comparison with the measured value.
- The variation of the rock formations frictional angle, considering a 10% variation range, did not significantly affect the numerically predicted response of socketed shafts, up to nearly a load level of around 80% of the monitored failure load.
- The variation of the rock formations cohesive characteristics, within a range of ±25% of the reference value, did not significantly affect the numerically predicted load-displacement responses. The predicted responses were not significantly affected up to nearly a load level of around 80% of the monitored failure load.
- Variations of the deformation modulus of the rock formation, within a range of around ± 25% of the formation value, did not delineate a significant impact on the numerically predicted load-displacement response within the working load range. However, beyond the expected working load range, the effect of the rock deformation modulus increases with increase of the applied load.
- Considering the expected working load range, where displacement are of major concern for geotechnical designers, the rock shaft interface strength and the rock formation deformation modulus have the most significant influence on the predicted load-displacement response of the socketed shafts. In other words, great care should be given to assessing these values for numerical analysis purposes.

References

Carter J.P. and Kulhawy F.H.: Analysis and design of drilled shaft foundations socketed into rock. Report EL-5918. Palo Alto: Electric Power Research Institute, 190 p (1988)

Donald, I.B., Sloan, S.N., Chiu, H.K.: Theoretical analysis of rock-socketed piles. Proc. Int. Conf. Struct. Found. Rock **1**, 303–316 (1980). Sydney, Australia

Leong, E.C., Randolph, M.F.: Finite element modeling of rock-socketed piles. Int. J. Numer. Anal. Methods Geomech. **18**, 25–47 (1994)

Mattes, N.S., Poulos, H.G : Settlement of single compressible pile. J. Soil Mech. Found. Div. **95** (SM1), 189–207 (1969)

Moon, S.N., Cumaraswamy, V.: Roughness and unit side resistances of drilled shafts socketed in clay shale and limestone. J. Geotech. Geoenviron. Eng. ASCE **134**(9), 1272–1278 (2008)

Osterberg, J.O., Gill, S.A.: Load transfer mechanism for piers socketed in hard soils or rock. In: Proceedings of 9th Canadian Symposium on Rock Mechanics, pp. 235–262, Montreal, ON, Canada (1973)

Pells, P.J.N., Turner, R.M.: Elastic solutions for the design and analysis of rock socketed piles. Canad. Geotech. J. **16**, 481–487 (1979)

Randolph, M.F., Wroth, C.P.: Analysis of deformation of vertically loaded piles. J. Geotech. Eng. Div. **104**(GT12), 1465–1488 (1978)

Rowe, R.K., Armitage, H.H.: Theoretical solutions for axial deformation of drilled shafts in rock. Canad. Geotech. J. **24**, 114–125 (1987a)

Rowe, R.K., Armitage, H.H.: A design method for drilled piers in soft rock. Canad. Geotech. J. **24**, 126–142 (1987b)

Rowe, R.K., Pells, P.J.N.: A theoretical study of pile-rock socket behavior. Proc. Int. Conf. Struct. Found. Rock **1**, 253–264 (1980). Sydney, Australia

Analysis and Modelling of Stiffened Slab Foundation on Expansive Soils

Mohamed A. Shams[1(✉)], Mohamed A. Shahin[1], and Mostafa A. Ismail[2]

[1] Curtin University, Perth, WA, Australia
mohamed.shams@postgrad.curtin.edu.au
[2] Geotechnical Engineering Consultant, Perth, WA, Australia

Abstract. Different methods have been used in the literature for analysis of residential footing slabs on expansive soils; some are adopted by local and international standards. This paper presents a comparison between: (1) Mitchell's method, which is one of the most conventional and common design methods that is currently adopted by the Australian Standard AS2870; and (2) finite element modelling using a coupled flow displacement analysis performed by the commercial software ABAQUS. The comparison is presented for both footing edge and central heave scenarios. The results showed that the finite element method can reliably simulate the foundation behaviour of structures on expansive soils. Consequently, further application of the FE modelling can be used to explore existing limitations of current conventional design methods.

1 Introduction

Expansive soils cause ongoing problems to structures and foundations in many countries around the world, including Australia, USA, Canada, China, India and Middle East. Despite the stringent regulations and design standards enforced by building authorities worldwide, foundations built on expansive soils still suffer severe distress caused by swelling-shrinkage cycles of the foundation soil. For example, the American Society of Civil Engineers (ASCE) reported that about one quarter of all homes in the USA are exposed to some damage caused by reactive soils; the financial losses of property owners are worse than those caused by most natural disasters such as earthquakes, floods, hurricanes and tornadoes combined (Wray 1995). In Australia, expansive soils cover roughly 20% of the continent (Richards *et al.* 1983), causing damaging cracks to nearly 50,000 houses each year, and this forms about 80% of all housing insurance claims (Robert *et al.* 1984). The problem of expansive soils is usually exacerbated for light-weight structures, since the damage is essentially caused by the soil swelling pressure that exceeds in most cases the dead loads of the structures.

Problems associated with expansive soils are usually initiated by soil moisture changes; if the soil water content is kept unchanged, there will be no soil swell or shrink. The soil moisture changes are mainly caused by rainfall, which may not always result in uniform moisture infiltration beneath the foundation. Consequently, the induced soil movement is non-uniform, resulting in differential settlements across the building. The design premise to accommodate this is that foundations on expansive

H. Shehata and Y. Rashed (eds.), *Numerical Analysis of Nonlinear Coupled Problems*,
Sustainable Civil Infrastructures, DOI 10.1007/978-3-319-61905-7_22

soils should be rigid enough to withstand the internal forces initiated by the expected differential settlements; otherwise, cracks will occur to superstructures. In this regard, man-made soil moisture variations are often more severe than water precipitation (Li *et al.* 2014). Ponding due to poor drainage, gardening, vegetation, utility pipes leakage are all direct reasons of soil moisture changes, which in turn cause subsequent foundation differential settlements.

Special foundation systems exist for light-weight structures to control and limit the damages caused by expansive soils. The most common type is the stiffened slab foundation, which is adopted by the Australian and other international codes. This method has demonstrated a documented history of success over the last few decades. However, continuous reporting of structural damages caused by expansive soils reveal that there is still an immense need to improve the current design practice of foundations on expansive soils, by exploring and studying the limitations and effect of the assumptions inherit in the current design methods.

This paper investigates the efficacy of the finite element method (FEM) in simulating the behaviour of stiffened slab foundation for light-weight structures on expansive clays. The paper also compares the FEM with one of the most commonly used conventional design methods that is currently adopted by the Australian Standard AS2870 (2011), known as Mitchell's method. The FEM analysis is carried out through a coupled flow-displacement algorithm considering a realistic water precipitation model using the commercial software ABAQUS. It should be noted that although the FEM technique has been used successfully in myriads for geotechnical applications, the numerical simulation of the behaviour of expansive soils has been always a complicated task involving a large geometric nonlinearity caused by huge deformations. In addition, unsaturated flow solutions are sensitive to accurate and detailed simulation of surface flux conditions, and thus require an extremely tight mesh and time steps which may result in long computational time of several months, even for one dimensional flow assessment (Dye *et al.* 2011).

2 Stiffened Slab Foundation

Many methods are available in the literature for the design of stiffened slab foundations; examples are those developed by the Building Research Advisory Board (1968); Lytton (1971); Walsh (1978a, b); Swinburne Institute of Technology (Holland *et al.* 1980); Post-Tension Institute (1980); and Mitchell (1984). The main premise underlying these methods is to design stiffened slab foundations based on idealized typical patterns of footing movement caused by the soil heave (edge or central), assuming that these two heave scenarios represent the worst loading cases among an infinite number of distortion patterns, with regard to the boundary conditions of each site. In the edge heave scenario, the stiffened slab foundation acts as a simple beam supported by the soil at the edges, assuming that the centre of the footing slab loses its contact with the soil (Fig. 1a). In the central heave scenario, the stiffened slab foundation acts as a double cantilever supported by the soil at the central area while the edges of the slab lose their contact with the soil over a certain edge distance (Fig. 1b). It should be noted that the infiltration of rain water in wet seasons beneath the edges of the footing slab initiates the edge heave, whereas the centre heave is initiated in hot dry seasons by the moisture propagation from the hot area at the edges to the cooler area beneath the centre of footing (Chen 1988). Each of the available design methods simplifies the soil mound shapes using a predicted function involving the boundary conditions and diffusion parameters, based on field observations, with a mathematical best fit to the actual case. Figure 2 summarizes the assumed mound shapes of the soil surface distortion for various design methods.

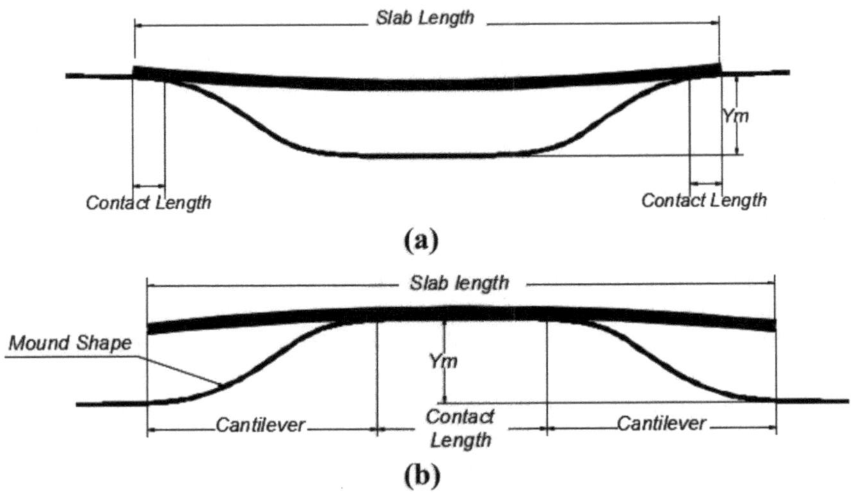

Fig. 1. Footing and soil profile: (a) edge heave scenario; and (b) centre heave scenario.

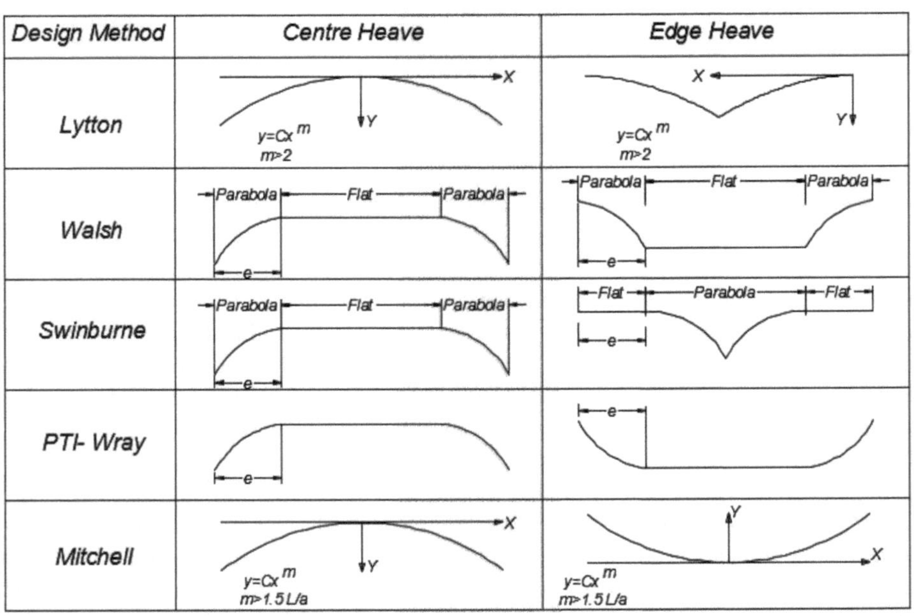

Fig. 2. Initial distorted shapes of soil surface used by different design methods (Mitchell 1984).

By estimating the expected soil surface movement, y_s, the maximum movement of soil mound, y_m (see Fig. 1), can be determined. For example, AS 2870 estimates y_m as $(0.5 - 0.7)\ y_s$; other methods delineate the surface heave with an acceptable accuracy (Adem *et al.* 2015). In AS2870, y_s is defined as the characteristic surface movement and can be determined by estimating the soil movement of all soil layers from 1 to N within the depth of active zone (see Fig. 3). Accordingly, y_s (in millimetres) can be calculated as follows:

$$y_s = \frac{1}{100} \sum_{n=1}^{N} \left(I_{pt} \cdot \Delta u \cdot h \right)_n \tag{1}$$

where, I_{pt} is the instability index (%); Δu is the soil seasonal suction change averaged over the thickness of the layer under consideration (pF) [logarithm to the base 10 of the pressure in centimetres of water]; h is the thickness of the layer under consideration (millimetres); and N is the number of soil layers within the active zone depth.

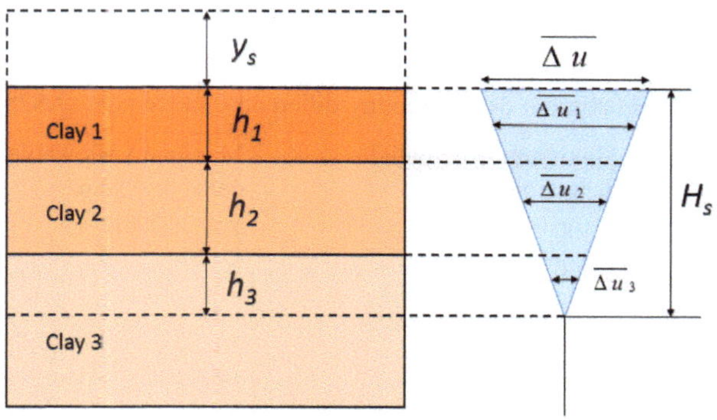

Fig. 3. Estimation of characteristic surface movement.

It should be noted that the depth of the active zone below which no significant change to the soil suction occurs due to seasonal conditions is denoted in the AS2870 as the depth of the design soil suction change, H_s. The depth of the active zone depends on many factors such as the soil permeability, topography, soil suction, and precipitation and evaporation rates. AS2870 provides typical values for both H_s and associated change in soil suction Δu at the soil surface for different regions across Australia, as given in Table 1. By estimating the surface heave, the mound shape can be figured out along with the stiffness needed to maintain the differential movement of the super structure within a relevant acceptable limit. Table 2 shows the allowable differential movement for different construction types listed in AS2870. Implementation of the design adopted by AS2870 usually requires computer programs. Based on the loading data and mound shape profiles, the straining actions and deformations are iteratively calculated to arrive at a satisfactory design.

Table 1. Profile of soil suction for different certain locations in Australia (Australia Standard AS2870).

Location	Ratio of change in soil suction (Δu) pF	Depth of design soil suction change, H_s(m)
Adelaide	1.2	4.0
Albury/Wodonga	1.2	3.0
Brisbane/Ipswich	1.2	1.5 − 2.3*
Gosford	1.2	1.5 − 1.8*
Hobart	1.2	2.3 − 3.0*
Hunter valley	1.2	1.8 − 3.0*
Launceston	1.2	2.3 − 3.0*
Melbourne	1.2	1.8 − 2.3*
Newcastle	1.2	1.5 − 1.8*
Perth	1.2	1.8
Sydney	1.2	1.5 − 1.8*
Toowoomba	1.2	1.8 − 2.3*

*Note: variation in H_s depends largely on climate change.

Table 2. Maximum design footing differential movement (AS2870).

Type of construction	Maximum differential deflection (mm)	Maximum allowable differential deflection (mm)
Clad frame	*L*/300	40
Articulated masonry veneer	*L*/400	30
Masonry veneer	*L*/600	20
Articulated full masonry	*L*/800	15
Full masonry	*L*/2000	10

Typically, the stiffened slab analysis is carried out in both directions of the slab for both the edge and centre heave scenarios, and the result of the worst case scenario is considered for design. This design approach has some limitations, as follows:

(i) The calculated surface heave is assumed to be due to the soil water change caused by the rainfall only. Any other source of soil moisture change, either natural or man-made, is beyond the applicability of this approach. Therefore, the approach imposes harsh restrictions on gardening, tree plantation or removal, and water pipe line fittings.

(ii) The actual three-dimensional distortion problem of slab under swelling is simplified to a symmetrical two-dimensional analysis.

(iii) For irregular foundation plans, the footing is subdivided into overlapped rectangles and each rectangle is analysed separately. The highest internal forces among all rectangles govern the design, resulting in over conservative solutions.

The FE numerical modelling developed in this paper will allow future investigation into the abovementioned limitations.

3 Finite Element Numerical Modelling

As indicated earlier, the design method of stiffened slab foundations on expansive soils drew from both field observations and empiricism. It is accordingly worthy to investigate how realistic the FE numerical modelling will be compared to the current most commonly used conventional design method, i.e. Mitchell's method as adopted by the Australian Standard AS 2870 (2011). The two main parameters that must be considered carefully in ensuing the FE modelling are: (1) the flow of moisture through the soil body, and (2) coupled water-soil-structure interaction. These two aspects are addressed in this paper for a stiffened foundation slab under both the edge and centre heave conditions, and the resulting deformations and internal forces are compared with those predicted by Mitchell's method. An arbitrary footing slab of dimensions (20 m × 20 m) is assumed and the slab thickness required to limit the differential movement to $L/1000 = 20$ mm (where L is the footing length) was first calculated using the conventional method. The same footing slab thickness was then used in the FE modelling. For the edge heave scenario, the calculated thickness using the conventional method was found to be 900 mm, while it was 550 mm for the centre heave. In real design, the maximum thickness will be used; however, in this study, the thickness calculated by the conventional method for each heave scenario was used in the FE analysis for proper comparison. The resulting soil mounds and corresponding footing movement profiles and bending moments were calculated and illustrated for both methods and the results were compared. In this study, the stiffened foundation slab was assumed to rest on a 4 m deep swelling clay layer.

A linear elastic material was used for the footing slab, and was also assumed for the swelling clay layer since the focus is on the volumetric response due to swelling (see Table 3). This assumption is reasonable, because light-weight structures are expected to produce stresses that are low. The permeability of the clay layer and rate of precipitation were assumed to be high at 1.2×10^{-5} m/s and 3×10^{-7} m/s, respectively, so that the saturation process can be accelerated and computational time can be saved. It should be noted that using lower clay permeability and rate of precipitation close to the actual values will not change the final results but will significantly delay the saturation process in the FE modelling and will substantially increase the computational time. The initial void ratio of the swelling clay was taken as 1.5. For clay suction, the initial pore water pressure was taken as −90 kPa and the initial saturation following the sorption profile was taken as 0.1. The flow period was taken as 150 days, which was found long enough to saturate the soil beneath the footing, for both the edge and centre heave scenarios, leading to maximum differential settlement between the footing edge and centre. The boundary conditions were set to restrict the displacement in all three directions at the bottom of the model while no lateral movement was allowed at the vertical edges. Only half of the physical model was simulated in the FE modelling due to symmetry. The water was allowed to precipitate at the edges in the edge heave scenario, while for the centre heave scenario the water infiltration was allowed in the middle 9.5 m of strip of the footing. In general, for proper comparison of footing distortions and bending moments, the soil properties were adjusted in the FE modelling in order to achieve the same amount of heave used in the calculations of the conventional method. The foundation slab was loaded uniformly with a pressure of 6.5 kPa (in addition to its self-weight, comprising the finishing and internal wall loads). At the edge of the footing slab, a line load of 10 kN/m′ was also applied to simulate the edge walls. Interface interaction properties were defined between all surfaces considering the shear friction and normal subgrade reaction.

Table 3. Finite element modelling parameters.

Material	FE element type	Elastic modulous, E (MPa)	Poisson's ratio, ν
Swelling clay	C3D8P: 8-node brick, trilinear displacement, trilinear pore pressure element	80	0.30
Footing slab	S4R: 4-node doubly curved shell element	2.1×10^4	0.16

4 Discussion of Results

Figure 4 shows snapshots of the FE deformed mesh of the analysed footing slab, for both the edge and centre heave scenarios. It can be seen that, in the edge heave scenario, the behaviour is indeed similar to a simple beam supported on the edges, while for the centre heave scenario the behaviour is similar to a double cantilever beam.

Figures 5 and 6 illustrate the footing movements and bending moments for a 1.0 m strip parallel to the flow direction, for both the edge and centre heave scenarios. It can be seen from Fig. 5 that the FE modelling and conventional method produced an overall identical behaviour for both heave scenarios. However, the soil surface movement profile is generally more realistic in the FE modelling than the conventional method, as it is physically more meaningful. Under the edge heave scenario (Fig. 5a), the movement profiles of the footing slab are almost identical for both methods. However, under the central heave condition (Fig. 5b), the differential movement of the footing slab obtained from the FE modelling exceeds that obtained from the conventional method by about 25%. This is due to the difference in the estimated soil mounds between the conventional and FE methods. The conventional method presumes soil mound that limits the movement of the soil surface to a certain edge distance and tends to be flatter in the centre, resulting in a higher contact length than for the FE mound. It has to be noted that the soil infiltration in the central heave scenario was assumed to act only on the middle 9.5 m strip. Increasing this assumed distance can result in a flatter mound similar to the conventional method, whereas reducing this distance leads to a less contact length and higher bending moment. A ratio of $0.6 \times L = 12.0$ m for the middle infiltration was proposed to get the best match with the conventional method. In general, among the estimated mound shapes proposed by the different design methods shown earlier in Fig. 2, Mitchell's method agrees with the soil mound predicted by FE

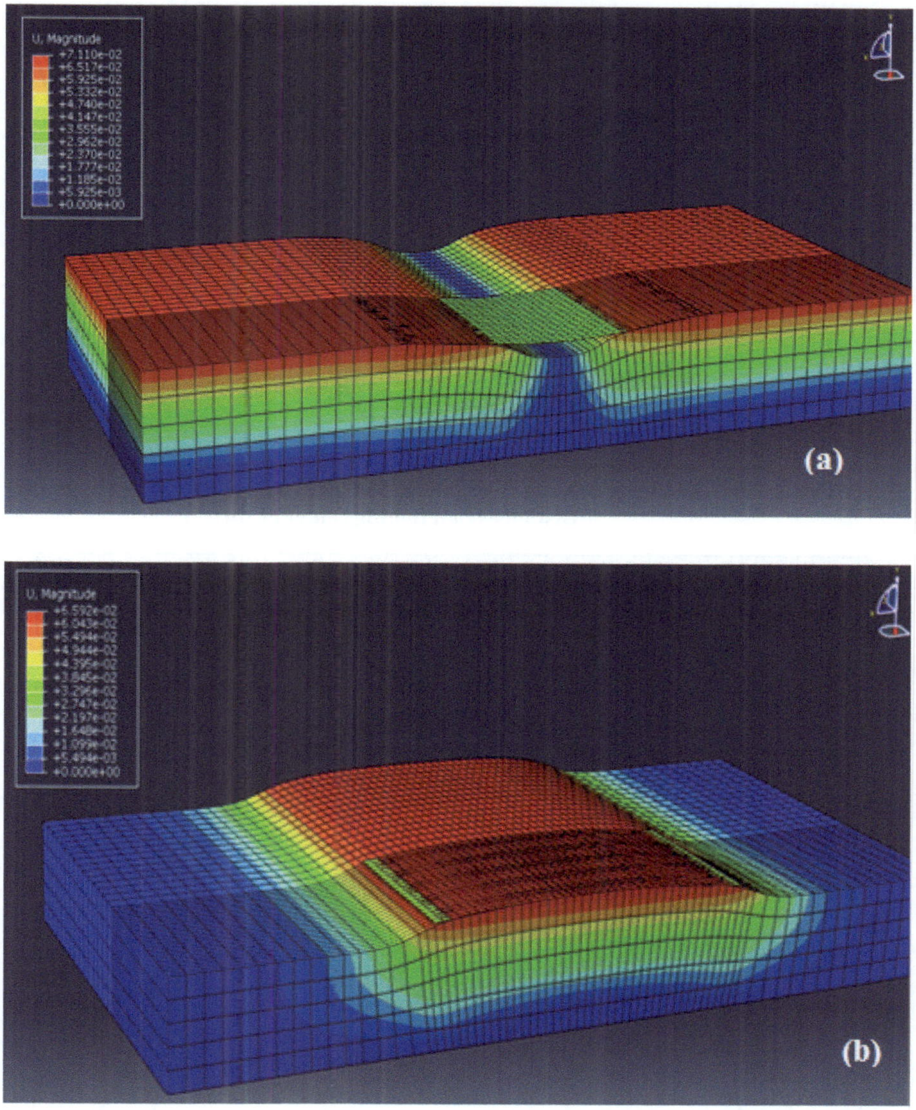

Fig. 4. Deformed mesh of FE modelling: (a) edge heave scenario; and (b) central heave scenario.

modelling for the central heave, while for the edge heave scenario the soil profile of Walsh's method seems to provide a better match.

Figure 6 shows that the maximum bending moment of the conventional method agrees fairly well with that of the FE modelling, for both heave scenarios. Again, the difference in the bending moment values at the middle of the footing in the central heave scenario is due to the difference in the soil mound estimated by the conventional method compared to that calculated using the FE modelling. However, despite this

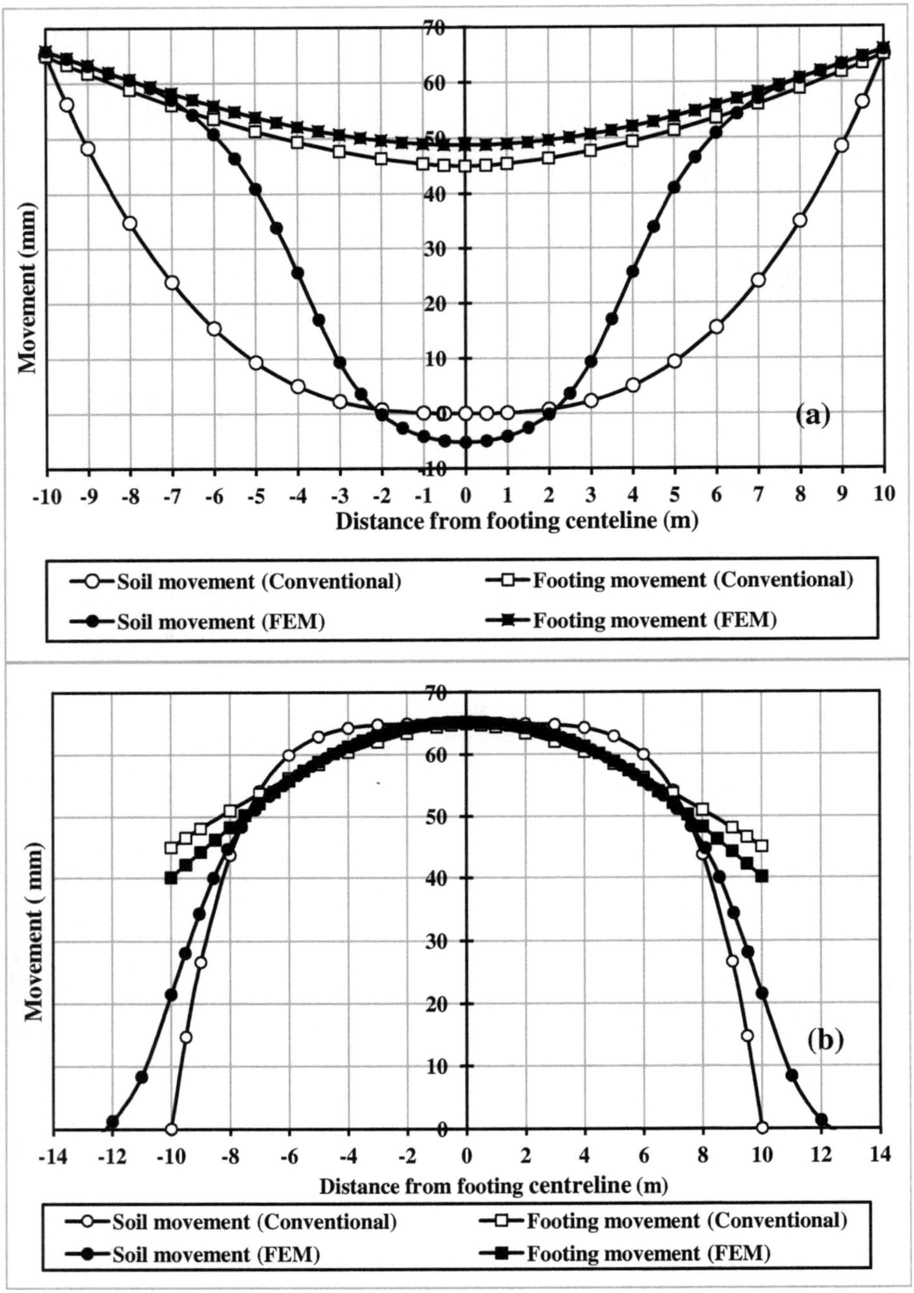

Fig. 5. Comparison between conventional and FE results for soil and footing movements: (a) edge heave scenario; and (b) central heave scenario.

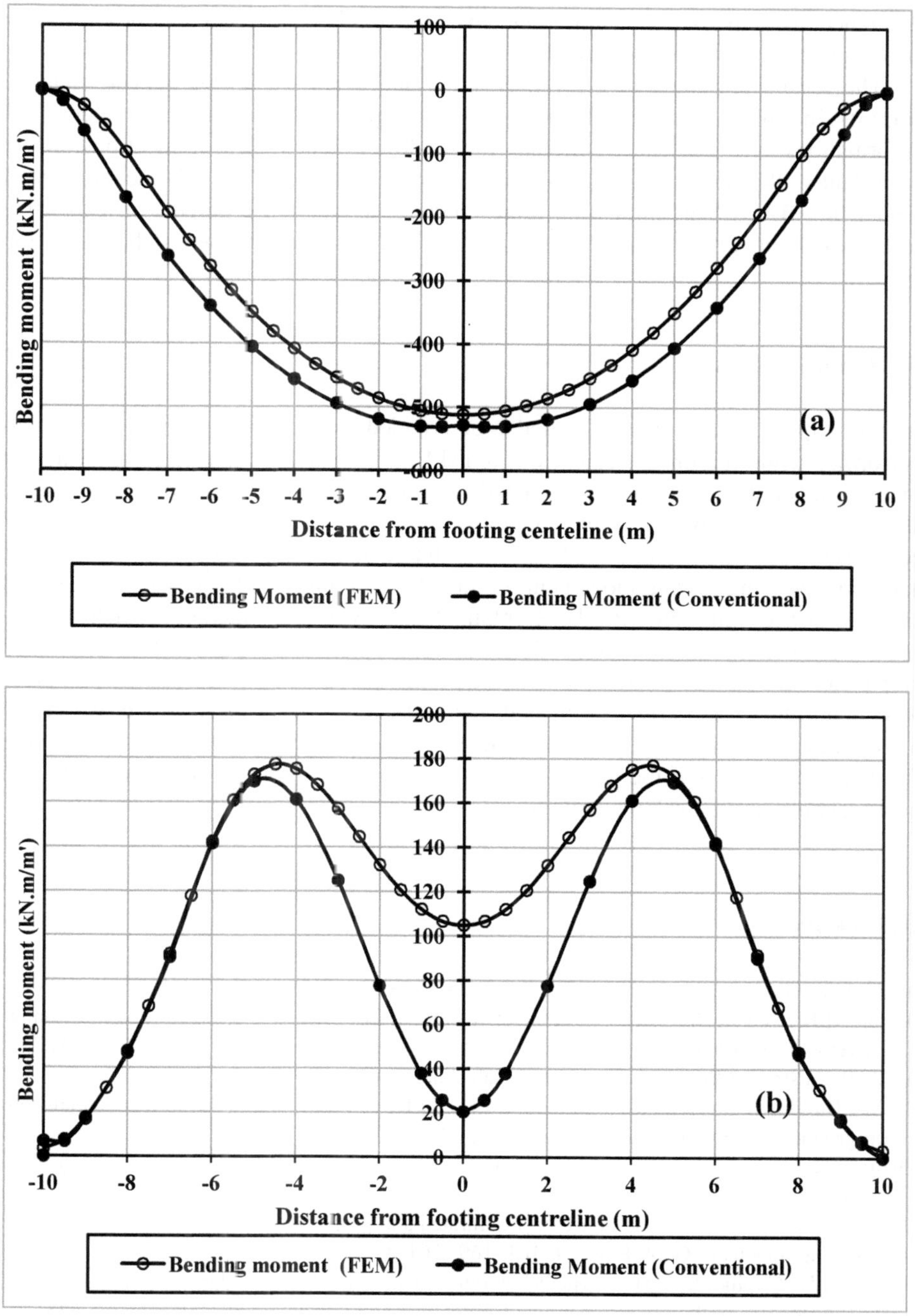

Fig. 6. Comparison between conventional and FEM results for footing bending moment: (a) edge heave scenario; and (b) central heave scenario.

difference, the maximum bending moments and overall bending moment profiles show good agreement between both methods.

The results presented in Figs. 4, 5 and 6 and discussed above indicate that the current simplified conventional design method provides reasonable prediction of deformations and associated internal forces for stiffened slab foundations. More importantly, the results provide insight into the capability of the FE moisture-swell model in ABAQUS in simulating more complex foundation problems on expansive soils, which in turn can involve solutions other than the stiffened slab foundation system.

5 Conclusions

In this paper, a moisture-swell model of coupled flow displacement analysis was performed using the finite element method (FEM) to validate the conventional design method adopted by the Australian Standard AS2870 (2011) for stiffened foundation slabs on expansive soils. Two movement scenarios were considered, namely heave and edge. The study showed that the FEM is a reliable tool for simulating the soil mounds and corresponding footing movement profiles. It was found that the FEM produced soil mounds that agreed well with those obtained from Mitchell's method, which is commonly adopted by the Australian Standard AS2870 (2011), despite some slight discrepancies. On the other hand, the bending moment profiles and their maximum values were found to agree fairly well in both methods, for both the edge heave and centre heave scenarios. The main contribution of this paper is that it paves the way for future FEM investigation into the many limitations of the conventional methods, leading to more reliable future design solutions.

References

ABAQUS (Version 6.14, Moisture Swelling). Analysis User's Manual

Adem, H.H., Vanapalli, S.K.: Review of methods for predicting in situ volume change movement of expansive soil over time. J. Rock Mech. Geotech. Eng. **7**, 73–86 (2015)

AS2870. Australian Standard. Residential slabs and footings (2011)

BRAB. Building Research Advisory Board: Criteria for selection and design of residential slabs on ground. Federal Housing Administration (1968)

Chen, F.H.: Foundations on Expansive Soils. Elseiver Science Publishing Company Inc., New York (1988)

Dye, H., Houston, S., et al.: Influence of unsaturated soil properties uncertainty on moisture flow modeling. Geotech. Geol. Eng. **29**, 161–169 (2011)

Holland, J.E., Cimino, D.J., et al.: The behaviour and design of housing slabs on expansive clays. In: Proceedings of the 4th International Conference on Expansive Soils, Denver, pp. 448–468 (1980)

Li, J., Cameron, D.A., et al.: Case study and back analysis of a residential building damaged by expansive soils. Comput. Geotech. **56**, 89–99 (2014)

Lytton, R.: Analysis for design of foundations on expansive clay. In: Proceedings of the Symposium of Soils & Earth Structures in Arid Climates. Australian Geotechnical Society & Institute of Engineers Australia, Adelaide, pp. 29–37 (1971)

Mitchell, P.W.: The design of shallow footings on expansive soil. Doctor of Philosophy, Department of Civil Engineering-University of Adelaide (1984)

Richards, B.G., Peter, P., et al.: The effects of vegetation on the swelling and shrinking of soils in Australia. Geotechnique **33**, 13 (1983)

Robert, L., Bell, A., et al.: Soils shrink, trees drink, and houses crack. ECOS Mag. **41,** 13 (1984). The Science Communication Unit of CSiRO's Bureau of Scientific Services

Walsh, P.F.: Concrete slabs for houses. CSIRO Division of Building Research. Technical Report No. 25 (1978a)

Walsh, P.F.: The analysis of stiffened rafts on expansive clays. CSIRO Division of Building Research. Tech Report No. 23. (1978b)

Wray, W.K.: So your home is built on expansive soils: a discussion of how expansive soils affect buildings. American Society of Civil Engineers (1995)

Derivation of the Incremental Stress-Strain Relations for Expansive Soils and Implementation into the Boundary Element Method

Jamila El Brahmi[1(✉)] and Mimoun Zoukaghe[2]

[1] Civil Engineering Laboratory (LGC), Mohammadia Engineering School, Rabat, Morocco
elbrahmi@emi.ac.ma
[2] Mohammadia Engineering School, Rabat, Morocco
zoukaghe@emi.ac.ma

Abstract. The Barcelona Expansive Model "BExM" (Gens and Alonso 1992; Alonso et al. 1999) is a constitutive model describing the mechanical behaviour of expansive soils formulated in the framework of hardening elasto-plasticity.

This paper demonstrates the derivation of the incremental stress-strain relations for expansive soils using BExM model, and discusses the implementation of these relations into boundary element method.

1 Introduction

Expansive soils have a particularly complicated structure. Given this complicated structure, the available models are unsuitable for the modelling of their particular behaviour. As a result, Alonso et al. (1999) suggested a new model intended for unsaturated expansive soils, this new model is referred to as the Barcelona Expansive Model (BExM model) and derived from the original Barcelona Basic Model (BBM), which was designed for unsaturated but non-expansive soils (Alonso et al. 1999).

There are two levels to be taken in consideration regarding expansive soil structure: micro and macro. Firstly, the microstructure corresponds to the active clay minerals, which constitute the aggregates. Secondly the macrostructure, which includes the rest of the soil structure and aggregates arrangement (Alonso et al. 1999).

2 Barcelona Expansive Model (BExM Model)

2.1 Assumptions (Alonso et al. 1999)

- The microstructure remains saturated regardless of suction value, the effective stress concept is then applicable to this scale. The region of microstructural swelling is separated from that of microstructural shrinkage by the neutral line "NL" defined in the p*-s plane while being at 45° relative to both "p*" axis and "s" axis (Fig. 1).

H. Shehata and Y. Rashed (eds.), *Numerical Analysis of Nonlinear Coupled Problems*, Sustainable Civil Infrastructures, DOI 10.1007/978-3-319-61905-7_23

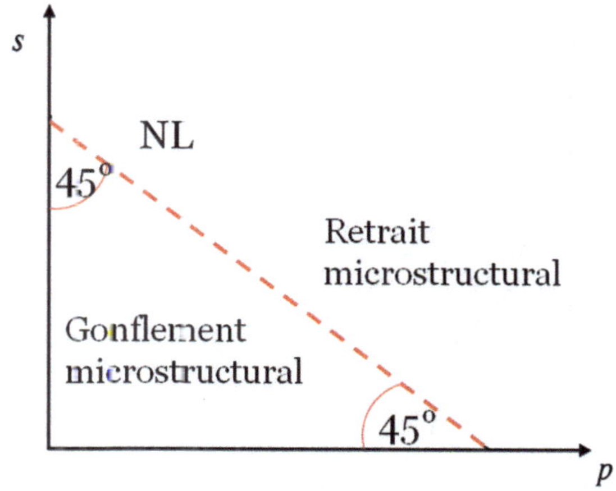

Fig. 1. Neutral line NL

- The microstructural strains are considered elastic and volumetric.
- The microstructure is coupled to the macrostructure: there is a possibility of creating macrostructural plastic deformations from microstructural elastic deformations. Against by, the microstructural behaviour is independent of the macrostructural behaviour.

2.2 Choosing of Stress Variables

The concept of effective stress of saturated soils was initially intended for unsaturated soils in several forms (but the stress formula proposed by Bishop remains the most famous (1959)). Later, this concept has been contested (Bishop and Blight 1963; Aitchison 1967) because it cannot describe correctly the collapse phenomenon. Two independent variables are then required.

Fredlund and Morgenstern triaxial tests (1977) showed that the net stress and matric suction are the most appropriate variables that have to be, constitutively, used to describe unsaturated soil behaviour.

2.3 Isotropic Stress State

2.3.1 Yield Surfaces

The limit between the elastic region and the region of irreversible deformations of the macrostructure is defined by two parallel lines to NL noted: IS (Suction Increase) and SD (Suction Decrease) over and above the LC curve defined at Barcelona basic model (BBM) (Fig. 2).

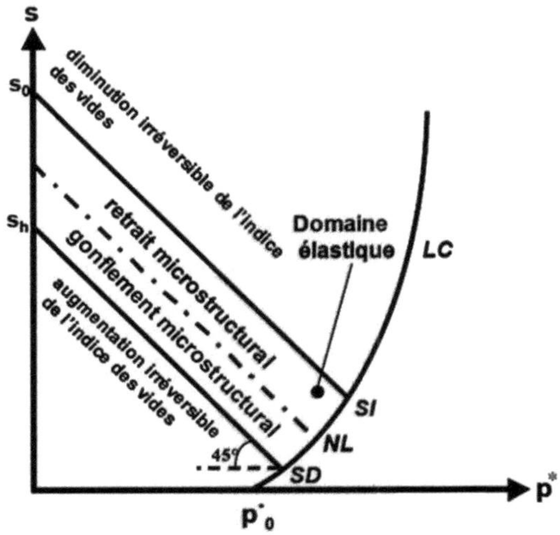

Fig. 2. Yield surfaces in BE × M model

The LC yield surface is expressed by:

$$LC: \quad \frac{p_0}{p_c} = \left[\frac{p_0^*}{p_c}\right]^{\frac{\lambda(0)-k}{\lambda(s)-k}} \tag{1}$$

SD and SI yield surfaces are expressed as follows:

$$SD: \quad p^* + s - s_h = 0 \tag{2}$$

$$SI: \quad p^* + s - s_0 = 0 \tag{3}$$

with:

S_h: suction from which plastic deformations occur during wetting at zero mean net stress.

S_0: suction from which plastic deformations occur during desiccation at zero net mean stress.

2.3.2 Volumetric Strains

In the elastic region, we can distinguish:

- The macrostructural strain due to stress increment (BBM model):

$$d\varepsilon^e_{vMp} = \frac{\kappa}{1+e_M} \cdot \frac{dp^*}{p^*} \tag{4}$$

- The macrostructural strain due to suction increment (BBM model):

$$d\varepsilon^{e}_{vMs} = \frac{\kappa_s}{1+e_M}\cdot\frac{ds}{s+p_{atm}} \tag{5}$$

- The microstructural strain:

$$d\varepsilon^{e}_{vm} = \frac{\kappa_m}{(1+e_m)}\frac{d(p^*+s)}{(p^*+s)} \tag{6}$$

The total volumetric and elastic strain increment is thus:

$$d\varepsilon^{e}_{v} = d\varepsilon^{e}_{vm} + d\varepsilon^{e}_{vM} = d\varepsilon^{e}_{vm} + d\varepsilon^{e}_{vMp} + d\varepsilon^{e}_{vMs} \tag{7}$$

In the plastic region, strains are only macrostructurals.

- If the exceeded yield surface is LC (BBM model):

$$d\varepsilon^{p}_{vM} = \frac{\lambda(0)-\kappa}{1+e_M}\cdot\frac{dp^*_0}{p^*_0} \tag{8}$$

- If SD or SI that is activated, the macrostructural plastic strains depends on the microstructural deformations and also the ratio p^*/p^0 (Fig. 3)

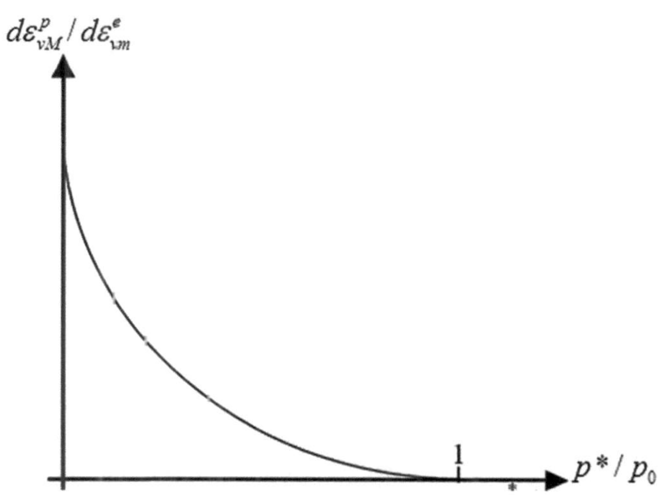

Fig. 3. Representation of the report $\frac{\varepsilon^{p}_{vM}}{\varepsilon^{e}_{vm}}$ depending on $\frac{p^*}{p_0}$

- If SD is activated (Alonso et al. 1999):

$$d\varepsilon_{vM}^{p} = f_D.d\varepsilon_{vm}^{e} \tag{9}$$

where $f_D = f_{D0} + f_{D1}\left(1 - \frac{p^*}{p_0}\right)^{n_D}$

- If SI is activated (Alonso et al. 1999):

$$d\varepsilon_{vM}^{p} = f_I.d\varepsilon_{vm}^{e} \tag{10}$$

where $f_I = f_{I0} + f_{I1}\left(\frac{p^*}{p_0}\right)^{n_I}$ and $F_{DO}, f_{D1}, n_D, F_{IO}, f_{I1}$ and n_I are characteristic parameters.

2.3.3 Hardening Rules

The hardening rules of the three yield surfaces are expressed by two internal variables:

$$d\alpha_1 = d\varepsilon_{vSI}^{p} + d\varepsilon_{vSD}^{p} \tag{11}$$

$$d\alpha_2 = d\varepsilon_{vSI}^{p} + d\varepsilon_{vSD}^{p} + d\varepsilon_{vLC}^{p} \tag{12}$$

where:

$d\varepsilon_{vSI}^{p}$: Plastic volumetric strains due to the activation of SI.
$d\varepsilon_{vSD}^{p}$: Plastic volumetric strains due to the activation of SD.
$d\varepsilon_{vLC}^{p}$: Plastic volumetric strains due to the activation of LC.

- The LC hardening is expressed by the variable $d\alpha_2$:

$$\frac{dp_0^*}{p_0^*} = \frac{(1 + e_M)}{\lambda(0) - \kappa}.d\alpha_2 \tag{13}$$

- The SI and SD hardening are governed by $d\alpha_1$:

$$ds_h = \frac{K_m.d\alpha_1}{f_D} \tag{14}$$

$$ds_0 = \frac{K_m.d\alpha_1}{f_I} \tag{15}$$

2.4 Case of Triaxial Stress State

When there is deviatoric stresses, BExM model takes into account the effects of the deviator q. A yield surface is added and can be described in the p^*-q-s plane, with a succession of ellipses along suction axis and which are inclined at 45° with respect to the two axes p^* and s (Fig. 4).

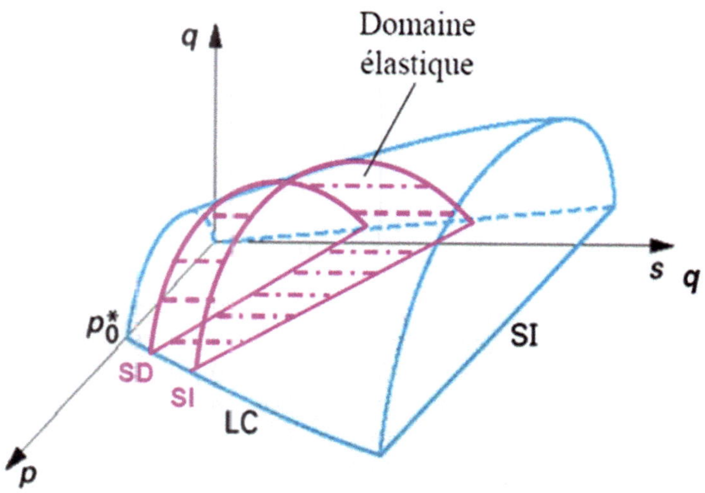

Fig. 4. Three-dimensional representation of the yield surfaces in the p^*-q-s space

The microstructural elastic strains remain volumetric. As for the macrostructural elastic strains, they have an additional component of deformation given by:

$$d\varepsilon^e_{qM} = \frac{1}{3G} dq \tag{16}$$

where $d\varepsilon^e_{qM}$ means the macrostructural deviatoric strain in the elastic region and G is the shear modulus.

In terms of hardening rules (which depend only on the plastic volumetric strains), there is no change from the formulation of isotropic stress state. However, the expression of the distance of the LC curve of the state of stress (contained in the expressions of f_I and f_D) is no longer expressed by the ratio p^*/p_O but by the report p^{**}/p_O where:

$$p^{**} = p^* + \frac{q^2}{M^2(p^* + ks)} \tag{17}$$

3 Constitutive Equations in Elastoplasticity

Using the BExM model, the stress - strain – suction relation in the incremental form can be written, in the elastic range, as follows:

$$\dot{\underline{\underline{\sigma}}} = D^e \dot{\varepsilon}^e - D^e (W^e)^{-1} ds = D^e \dot{\varepsilon}^e - F^e ds \tag{18}$$

with:

$$(W^e)^{-1} = \begin{pmatrix} \frac{K_s}{1+e} \cdot \frac{1}{s+p_{atm}} \\ \frac{K_s}{1+e} \cdot \frac{1}{s+p_{atm}} \\ \frac{K_s}{1+e} \cdot \frac{1}{s+p_{atm}} \\ 0 \\ 0 \\ 0 \end{pmatrix};$$

$$D^e = \begin{pmatrix} 2\mu+\lambda & \lambda & \lambda & 0 & 0 & 0 \\ \lambda & 2\mu+\lambda & \lambda & 0 & 0 & 0 \\ \lambda & \lambda & 2\mu+\lambda & 0 & 0 & 0 \\ 0 & 0 & 0 & \mu & 0 & 0 \\ 0 & 0 & 0 & 0 & \mu & 0 \\ 0 & 0 & 0 & 0 & 0 & \mu \end{pmatrix}$$

and

$$F^e = D^e (W^e)^{-1} = \begin{pmatrix} \left(\lambda + \frac{2}{3}\mu\right) \frac{K_s}{1+e} \cdot \frac{1}{s+p_{atm}} \\ \left(\lambda + \frac{2}{3}\mu\right) \frac{K_s}{1+e} \cdot \frac{1}{s+p_{atm}} \\ \left(\lambda + \frac{2}{3}\mu\right) \frac{K_s}{1+e} \cdot \frac{1}{s+p_{atm}} \\ 0 \\ 0 \\ 0 \end{pmatrix}$$

$\underline{\underline{\dot{\varepsilon}^e}}$:	incremental elastic strain matrix
$d\sigma$:	stress increment
ds:	suction increment
D^e:	a fourth-rank tensor quantity, the incremental elastic modulus
W^e:	the elastic suction - strain vector;
λ, μ:	Lamé constants
G, K, v and E:	respectively: the shear modulus, bulk modulus, the Poisson's ratio and Young's modulus.

Decomposing the total strain increment into elastic and plastic strains:

$$\underline{\underline{\dot{\varepsilon}}} = \underline{\underline{\dot{\varepsilon}^e}} + \underline{\underline{\dot{\varepsilon}^p}} \tag{19}$$

and according to the flow rule:

$$d\underline{\underline{\varepsilon^p}} = \underline{\underline{\dot{\varepsilon}^p}} = \dot{\lambda}.\frac{\partial g}{\partial \underline{\underline{\sigma}}} \tag{20}$$

where $\dot{\lambda}$ is the plastic multiplier ($\dot{\lambda} \succ 0$) and g is the plastic potential.

Since the flow rule is considered associative, the plastic potential g coincides with the yield surface F.

we can write:

$$d\underline{\underline{\sigma}} = D^e\left(d\underline{\underline{\varepsilon}} - \dot{\lambda}\frac{\partial g}{\partial \underline{\underline{\sigma}}}\right) - F^e.ds \tag{21}$$

or with the index notation:

$$d\sigma_{ij} = d\sigma_{ij}^{ep} = \lambda\delta_{ij}\dot{\varepsilon}_{kk} + 2\mu\dot{\varepsilon}_{ij} - \dot{\lambda}\left(\lambda\delta_{ij}\frac{\partial g}{\partial \sigma_{kk}} + 2\mu\frac{\partial g}{\partial \sigma_{ij}}\right) - \Gamma\delta_{ij}ds \tag{22}$$

where:

$$\Gamma = \left(\lambda + \frac{2}{3}\mu\right)\frac{\kappa_s}{1+e}\cdot\frac{1}{s+p_{atm}}$$

and $\delta_{ij} = \begin{cases} 0 & si \quad i \neq j \\ 1 & si \quad i = j \end{cases}$ (the Kronecker delta).

Introducing the hardening modulus $H(\sigma, s, \chi)$ and exploiting the consistency condition $(dF = 0)$ (where F is the yield function and χ is the hardening parameter), we find that:

$$\dot{\lambda} = \frac{\left(\frac{\partial F}{\partial \sigma_{ij}}\right)^T(\lambda\delta_{ij}\dot{\varepsilon}_{kk} + 2\mu\dot{\varepsilon}_{ij})}{H + \left(\frac{\partial F}{\partial \sigma_{ij}}\right)^T\left(\lambda\delta_{ij}\frac{\partial g}{\partial \sigma_{kk}} - 2\mu\frac{\partial g}{\partial \sigma_{ij}}\right)} - \frac{\left(\frac{\partial F}{\partial \sigma_{ij}}\right)^T\Gamma\delta_{ij} + \frac{\partial F}{\partial s}}{H + \left(\frac{\partial F}{\partial \sigma_{ij}}\right)^T\left(\lambda\delta_{ij}\frac{\partial g}{\partial \sigma_{kk}} + 2\mu\frac{\partial g}{\partial \sigma_{ij}}\right)}ds \tag{23}$$

The Eq. (22) can be written us:

$$d\sigma_{ij} = \dot{\sigma}_{ij}^{ep} = \dot{\sigma}'_{ij} - \dot{\sigma}_{ij}^{0} - (\Gamma\delta_{ij} - S_{ij})ds \tag{24}$$

where:

$$\dot{\sigma}'_{ij} = \lambda\delta_{ij}\dot{\varepsilon}_{kk} + 2\mu\dot{\varepsilon}_{ij} \tag{24 – a}$$

$$\dot{\sigma}_{ij}^{0} = \frac{\left(\frac{\partial F}{\partial \sigma_{ij}}\right)^T(\lambda\delta_{ij}\dot{\varepsilon}_{kk} + 2\mu\dot{\varepsilon}_{ij})\left(\lambda\delta_{ij}\frac{\partial g}{\partial \sigma_{kk}} + 2\mu\frac{\partial g}{\partial \sigma_{ij}}\right)}{H + \left(\frac{\partial F}{\partial \sigma_{ij}}\right)^T\left(\lambda\delta_{ij}\frac{\partial g}{\partial \sigma_{kk}} + 2\mu\frac{\partial g}{\partial \sigma_{ij}}\right)} \tag{24 – b}$$

$$S_{ij} = \left(\frac{\left(\left(\frac{\partial F}{\partial \sigma_{ij}}\right)^T\Gamma\delta_{ij} + \frac{\partial F}{\partial s}\right)\left(\lambda\delta_{ij}\frac{\partial g}{\partial \sigma_{kk}} + 2\mu\frac{\partial g}{\partial \sigma_{ij}}\right)}{H + \left(\frac{\partial F}{\partial \sigma_{ij}}\right)^T\left(\lambda\delta_{ij}\frac{\partial g}{\partial \sigma_{kk}} + 2\mu\frac{\partial g}{\partial \sigma_{ij}}\right)}\right) \tag{24 – c}$$

Solving the problem requires the determination of the yield function equation. To do so, it is convenient to approach the three yield surfaces proposed by the BExM model with a single yield surface (Mrad 2005), this is to calculate the equation of an ellipse for each given stress state (point A of Fig. 5).

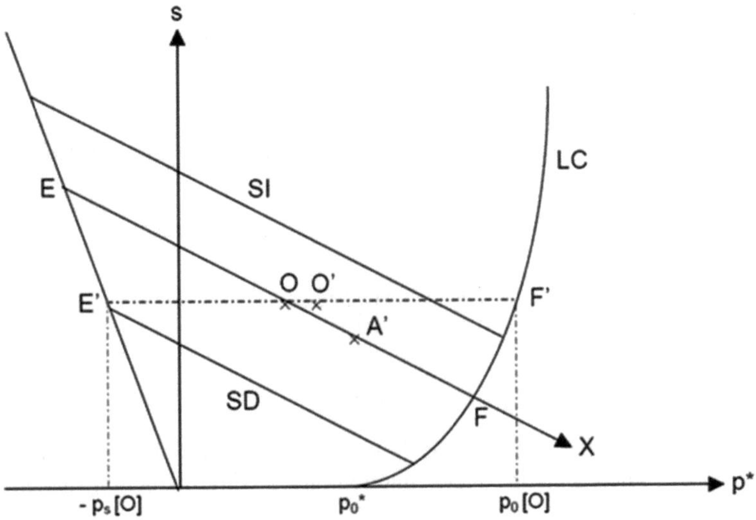

Fig. 5. Projection of yield surfaces in the plan p^*, s (Mrad 2005)

Let A′ be the projection of A in the plane p^*, s. The line passing through the point A′ parallel to SI and SD cut the cohesion line in the point E and the LC curve in the point F. Equation of the ellipse that has as axis the segment EF will be the equation of the yield function in the space p^*, q, s (Fig. 1).

The yield surface has the following expression (Mrad 2005):

$$F = -1 + \frac{2k_1^2}{\left(\frac{I}{2}\right)^2} - \frac{4k_1}{\left(\frac{I}{2}\right)^2} p^* + \frac{2}{\left(\frac{I}{2}\right)^2} p^{*2} + \frac{q^2}{\left(\frac{J}{2}\right)^2} \tag{25}$$

where:

$$I = \|EF\| = \sqrt{\left(p_F^* - p_E^*\right)^2 + \left(s_F - s_E\right)^2}$$

$$J = 2\sqrt{M^2[k_1 + p_s(O)] + [p_0(O) - k_1]}$$

$$p_0(O) = p_c \left[\frac{p_0^*}{p_c}\right]^{\frac{\lambda(0)-k}{\lambda(0)\left[(1-r)e^{-\beta k_2} + r\right] - k}}$$

$$p_s(O) = kk_2$$

$$k_2 = \frac{s_F + s_E}{2}$$

$$k_1 = \frac{p_F^* + p_E^*}{2}$$

The flow rule is considered associative; so the plastic potential g coincides with the yield surface F.

4 Equilibrium Equations

An elastoplastic material must obey the equilibrium equations for the stress increment, which are, in the absence of time-dependent body forces:

$$\frac{\partial \dot{\sigma}_{ij}^{ep}}{\partial x_j} = 0 \tag{26}$$

Substituting Eq. (22) into Eq. (26), we obtain:

$$\frac{\partial \dot{\sigma}'_{ij}}{\partial x_j} + \dot{F}_i = 0 \tag{27}$$

where $\dot{F}_i$ is the modified body force defined by:

$$\dot{F}_i = -\left(\frac{\partial \dot{\sigma}_{ij}^0}{\partial x_j} + \Gamma \delta_{ij} \frac{\partial s}{\partial x_j} - \frac{\partial (S_{ij} ds)}{\partial x_j}\right) \tag{28}$$

By using the strain-displacement relations:

$$\dot{\varepsilon}_{ij} = \frac{1}{2}\left(\frac{\partial \dot{U}_i}{\partial x_j} + \frac{\partial \dot{U}_j}{\partial x_i}\right) \tag{29}$$

where $\dot{U}_i$ and $\dot{U}_j$ are, respectively, the displacements in the i and j directions.

We obtain the governing differential equations for the elastoplastic problem:

$$(\lambda + \mu) \frac{\partial^2 \dot{U}_j}{\partial x_i . \partial x_j} + \mu \frac{\partial^2 \dot{U}_i}{\partial x_j . \partial x_j} + \dot{F}_i = 0 \tag{30}$$

The elastoplastic problem may therefore be posed as the solution of an elastic problem governed by the differential Eq. (30) over the volume V and subject to the boundary condition:

$$\dot{t}'_i = \dot{\sigma}_{ij} n_j - \dot{t}^0_i \tag{31}$$

where:

$$\dot{t}^0_i = \dot{\sigma}^0_{ij} n_j + \Gamma ds \delta_{ij} n_j - S_{ij} ds.n_j \tag{32}$$

and n_j the outward normal to the boundary.

5 Integral Equation Formulation: Boundary Element Method

By applying the direct formulation of the Boundary Element Method, we can obtain the displacements $\dot{U}_j(\xi)$ at an interior point (ξ) of the domain V due to any admissible combinations of $\dot{t}'_i$ and $\dot{U}_i$ over S and a given distribution of $\dot{F}_i$ within the volume:

$$\dot{U}_j(\xi) = \int_S \left[\dot{t}'_i(x) G_{ij}(x,\xi) - F_{ij}(x,\xi)\dot{U}_i(x)\right] dS(x) + \int_V \dot{F}_i(x) G_{ij}(x,\xi) dV(x) \tag{33}$$

Where:

$$G_{ij}(x,\xi) = C_1\left(C_2 \delta_{ij} \ln r - \frac{y_i y_j}{r^2}\right) \tag{34}$$

With:

$$\begin{cases} C_1 = -\frac{1}{8\pi\mu(1-v)} \\ C_2 = 3 - 4v \end{cases}$$

$$y_i = x_i - \xi_i \; : \; \begin{cases} y_1 = x_1 - \xi_1 \\ y_2 = x_2 - \xi_2 \end{cases}$$

$$r^2 = y_i\, y_i = (x_1 - \xi_1)^2 + (x_2 - \xi_2)^2$$

and:

$$F_{ik}(x,\xi) = \frac{C_3}{r^2}\left[C_4(n_k y_i - n_i y_k) + \left(C_4 \delta_{ik} + \frac{2 y_i y_k}{r^2}\right) y_j n_j\right] \tag{35}$$

with:

$$\begin{cases} C_3 = -\frac{1}{4\pi(1-v)} \\ C_4 = 1 - 2v \end{cases}.$$

6 Results

Using the divergence theorem (in addition to some simplifications and calculations), we could write Eq. (33) as follows:

$$
\begin{aligned}
C_{ij}\dot{U}_j(\xi) = & \int_S \left[\dot{t}_i(x)G_{ij}(x,\xi) - F_{ij}(x,\xi)\dot{U}_i(x)\right]dS - \int_S G_{ij}(x,\xi)[\Gamma s.n_i - S_{ik}ds.n_k]dS \\
& + \Gamma\left(\int_V (1+2\ln r)y_j\nabla^2(s)dV + \int_S \left(s\left(\frac{2}{r^2}y_jy_mn_m + (1+2\ln r)n_j\right)(1+2\ln r)y_j\frac{\partial s}{\partial n}\right)dS\right) \\
& + \int_V \frac{C_1}{r}\left(C_2\delta_{ij}\frac{y_k}{r} + \frac{2y_iy_ky_j}{r^3} - \frac{y_i}{r^2}\delta_{kj} - \frac{y_j}{r^2}\delta_{ik}\right)(\dot{\sigma}^0_{ik} - S_{ik}ds)dV
\end{aligned}
\qquad (36)
$$

Equation (36) provides the displacement in the direction "j" at any point, ξ, within the problem domain. The free term C_{ij} is dependent on the location of point ξ. If the point ξ is inside the domain then $C_{ij} = \delta_{ij}$; if ξ lies on a smooth boundary then $C_{ij} = \frac{1}{2}\delta_{ij}$.

The first line integral in (36) contains $\dot{U}_i$ and $\dot{t}_i$ which are prescribed as boundary conditions. That involves, in general, the specification of displacements $\dot{U}$, or tractions $\dot{t}$, on the boundary S of the problem domain. Either conditions may be specified over any portion of the boundary. Note, however, that both cannot be prescribed over the same boundary segment in a well-posed problem. To obtain the missing value of either $\dot{U}_i$ or $\dot{t}_i$, Eq. (36) must first be solved by taking the source point, ξ, to the boundary S. Note, however, when a field point on the boundary coincides with the source point, ξ, the integrals containing terms such as 1/r, ln(1/r), etc. become singular. This singularity is removable, and the integrals exist in the ordinary and Cauchy principle-value senses, respectively (Banerjee and Butterfield 1981).

It is seen in Eq. (36) that the internal suction distribution, s, and internal suction gradient, $\frac{ds}{dn}$, are required. This can be secured by solving the governing diffusion equation for steady state flow moisture.

Finally, For the solution of the problems in plasticity we will use the "initial stress" method. In this method, we compute initial stresses and apply this as loading. For this we have to amend the discretisation of the problem. In addition to surface elements, we require the specification of volume cells in the parts of the domain that are likely to yield, for the integration of initial stresses.

Since all quantities appearing on the right of Eq. (36) are now available, solutions at interior points can then be obtained by moving the source point, ξ, into the domain. the strains, ε_{jk}, for interior points are then given by Eq. (29).

7 Conclusion

The constitutive equations in elasto-plasticity can be written for unsaturated and expansive soils using the BE × M model. However, this is the application of the equilibrium equation that allows us to obtain the governing differential equations for the problem. To solve this last, we use the Boundary Element Method and we can write the integral equation formulation for the displacement at any point within the problem domain if boundary conditions ($\dot{U}_i$ and $\dot{t}_i$) and the internal suction distribution, s, and internal suction gradient, $\frac{ds}{dn}$, are prescribed.

References

Aitchison, G.D.: Separate roles of site investigation, quantification of soil properties and selection of operational environment in the determination of foundation design on expansive soils. In: Proceeding of 3rd Asian regional Conference on Soil Mechanics and Foundation Engineering, vol. 3, pp. 72–77 (1967)

Alonso, E.E., Vaunat, J., Gens, A.: Modelling the mechanical behaviour of expansive clays. Eng. Geol. **54**, 173–183 (1999)

Banerjee, P.K., Butterfield, R.: Boundary Element Methods in Engineering Science. McGraw Hill Book Company, New York (1981)

Bishop, A.W., Blight, G.E.: Some aspects of effective stress in saturated and unsaturated soils. Géotechnique **13**(3), 177–197 (1963)

Fredlund, D.G., Margenstern, N.R.: Stress state variables for unsaturated soils. J. Geotech. Eng. **103**, 447–466 (1977)

Gens, A., Alonso, E.E.: A framework for the behaviour of unsaturated expansive clays. Can. Geotech. J. **29**, 1013–1032 (1992)

Mrad, M.: Modélisation du comportement hydromécanique des sols gonflants non saturés. Thèse de doctorat. Institut national polytechnique de Loraine, Ecole Nationale Supérieure de Géologie de Nancy (2005)

The Behaviour of Shallow Foundation Near Slope Under Inclined Loading

Messaoud Baazouzi(✉), Mekki Mellas, Djamel Benmeddour, and Abdelhak Mabrouki

Laboratory of Research in Civil Engineering LRGC, University of Biskra, BP145, 07000 Biskra, Algeria
messaoud.baazouzi@gmail.com, m_mellas@yahoo.fr, benmeddourdj@yahoo.fr, a.mabroukii@yahoo.fr

Abstract. In civil engineering, major studies had examined the failure conditions of shallow foundations were based on the application of a combined state of loading in a homogeneous soil.

The present study aims to define the inclined loading effects on the undrained bearing capacity of a strip footing located on a homogeneous clay soil. This paper has presented a numerical study using the finite-difference code Fast Lagrangian Analysis of Continua (FLAC), in order to observe the distribution of stresses and the pattern of failure mechanism under an eccentric loading. The results obtained are presenting the effect of geometrical and material properties on the bearing capacity by using the shape of the failure envelope in vertical and moment loading plan. The numerical values have been compared with those available in the literature.

1 Introduction

Generally, the bearing capacity of shallow foundations is determined using Terzaghi equation (1943). This equation is a superposition of three terms; surface term, cohesion term, overloading term. In the case of a purely cohesive soil, the bearing capacity is restricted on the cohesion term and it represented by the following expression:

$$q_{ult} = cN_c \tag{1}$$

In order to take the effect of the load and the ground surface inclination into account, empirical factors introduced on this term. (Hansen 1970) proposed the following semi empirical equation:

$$q_u = \frac{V_{Ult}}{A'} = (\pi + 2)\,c_u\,(1 - i_c - g_c) \tag{2}$$

H. Shehata and Y. Rashed (eds.), *Numerical Analysis of Nonlinear Coupled Problems*, Sustainable Civil Infrastructures, DOI 10.1007/978-3-319-61905-7_24

Where:

$$i_c = 0.5 - 0.5\sqrt{1 - H/Ac_u} \tag{3}$$

$$g_c = \frac{2\beta}{\pi + 2} = \frac{\beta^{\circ}}{147^{\circ}} \tag{4}$$

In the equation of Hansen (1970), Vesic (1975) proposed to include the effect of the weight of soil (Georgiadis 2010).

$$q_u = \frac{V_{Ult}}{A'} = (\pi + 2)\, c_u\, (1 - i_c - g_c) - \gamma B \sin \beta \tag{5}$$

Where:

$$i_c = 1 - \frac{2H}{Bc_u N_c} \tag{6}$$

Georgiadis (2010) proposed an expression in terms of normalised loads v and h, depends the bearing capacity failure.

$$v = \frac{1}{2} + \frac{\cos^{-1}(\xi.h) + \sqrt{1 - (\xi.h)^2} - \beta}{\pi + 2 - 2\beta}. \tag{7}$$

Where:

$$\xi = 1 - \beta(1 - e^{-\lambda}). \tag{8}$$

$$\xi = 0, \quad \lambda \,\text{or}\, \beta = 0 \tag{9}$$

Where: V_{ult}, A′, Nc, i_c, g_c, β, ζ, λ, v, h are the ultimate vertical loads, effective area of the footing, undrained bearing capacity factor, load inclination factor, ground slope inclination factor, slope angle, parameter of load interaction diagram, normalized footing distance, normalized vertical failure load, normalized horizontal failure load, respectively.

In this study, series of numerical computation, using the finite difference code $FLAC^{2d}$, are carried out to investigate the influence of the slope and load inclination angles on the ultimate bearing capacity. The results of the current study are compared with results available the literature.

2 Problem Definition

The geometry of the analyzed problem is show in Fig. 1. In the current study, we consider a rigid strip footing of width B = 2 m located on a homogeneous clay soil which makes a various angle β° with the horizontal and slope height H/B at a several

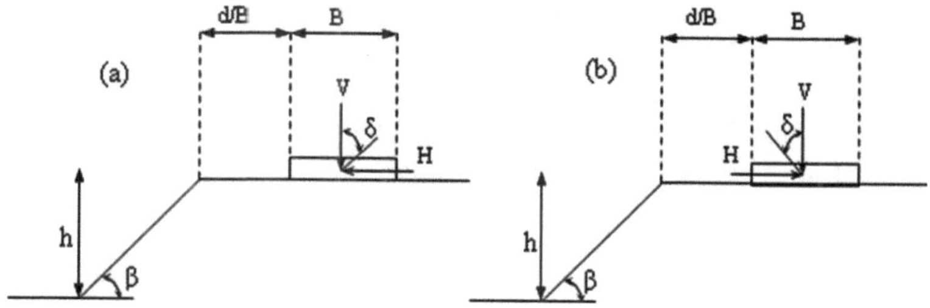

Fig. 1. Probleme geometry

distance λ/B measured from the edge of the foundation to the crest of the slope. The soil is modeled as a Tresca material using the Mohr-Coulomb elastic-perfectly plastic constitutive model with a shear strength C_u ($\varphi = 0$), the undrained Young's modulus $E_u = 30$ MPa, Poisson's ratio $v = 0.49$ and unit weight of soil $\gamma = 20$ KN/m^3 which has affected (original: affects to) the overall stability of the slope. The footing assumed smooth and rigid.

The undrained bearing depends on a wide range of dimensionless ratios, namely the slope angle β, Cu/γB, λ/B, H/B. Furthermore, these parameters are affecting the failure mode of slope, which can to take place according to either of two modes: the first two failure modes Figs. 2(a–b) which governed by either the local foundation failure, where the shear surface does not intersect the slope and can be referred to as "bearing capacity" failure modes. However, the second is overall slope failure where the critical shear surface extends beyond the crest and therefore; involves part of the slope as shown on Fig. 2c.

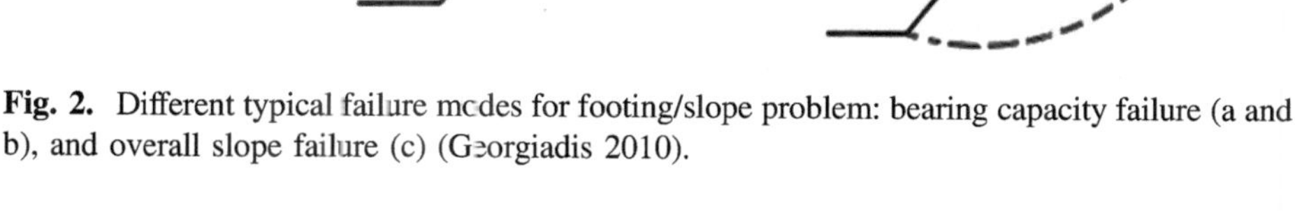

Fig. 2. Different typical failure modes for footing/slope problem: bearing capacity failure (a and b), and overall slope failure (c) (Georgiadis 2010).

3 Numerical Modeling Procedure

The finite-difference code FLAC (2005) was used to estimate the undrained bearing capacity of a strip footing on/or near slopes under conditions of plane strain and subjected to a centered vertical load. FLAC (Fast Lagrangian Analysis of Continua) is a two-dimensional explicit finite-difference program for engineering mechanics computations. Many researchers have used the finite difference code FLAC (2005) to study the bearing capacity of strip and circular isolated footings (e.g., Frydman and Burd 1977; Baazouzi et al. 2016).

Because of the nonsymmetrical problem, the entire model is considered in the computations. Figure 3 shows a typical finite-difference mesh, used for the case of footing located at the crest of slope (λ/B = 0).

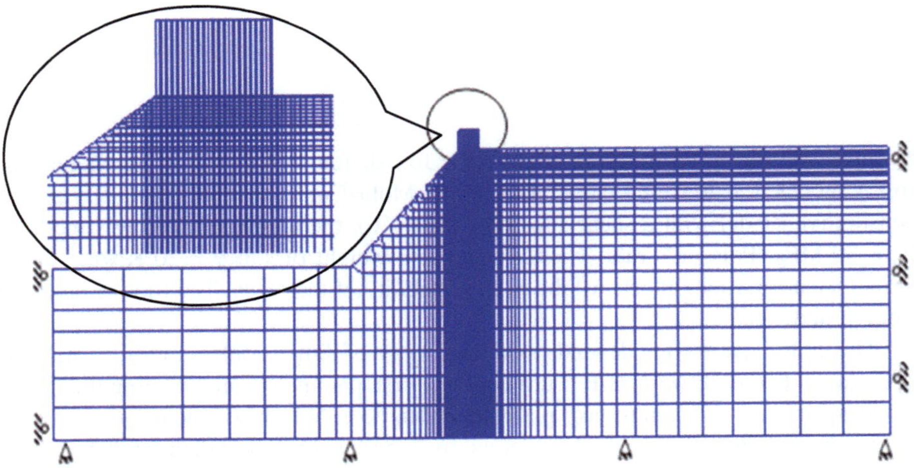

Fig. 3. Finite difference mesh and boundary condition for the case: β = 30°, H/B = 3, and λ/B = 0.

For evaluated bearing, capacity with FLAC is based on dividing the soil into a number of zones, and applying vertical velocities (displacement-controlled method) onto the zone representing the footing. The mesh consists of 177 by 71 zones for width and depth respectively; it has been refined at the region most close to the boundaries of the foundation, under the base and near the crest of the slope. The overall mesh dimensions were selected to ensure that the zones of plastic shearing and the observed displacement fields were contained within the model boundaries at all times.

The boundary condition for this problem the displacement of the left and right vertical sides is constrained in the horizontal direction and full fixities to the base of the mesh.

The soil is modeled as a Tresca material using the Mohr-Coulomb elastic-perfectly plastic constitutive model with a shear strength $C_u(\varphi = 0)$, the undrained Young's modulus E_u = 30 MPa, Poisson's ratio v = 0.49 and unit weight of soil γ = 20 KN/m^3 which has affects to the overall stability of the slope.

The footing was assumed linear elastic material with concrete Young's modulus of E_c = 29 GPA and Poisson's ratio v = 0.21. The properties of the interface elements are related to the properties of the adjacent soil elements, for the smooth case, the cohesion is set as C = 0, these ensures that the shear stress underneath the footing is zero and the displacement of nodes in the horizontal direction is free, while for the rough case C = C_u, it is characterized with normal stiffness K_n = 10^9 Pa/m and shear stiffness K_s = 10^9 Pa/m.

Several tests were done for chosen the optimal vertical velocity, because the time and rapidity of analysis is very important part of numerical modeling, also must be not affect the accuracy of results. Finally, the vertical velocity chosen for all analyses is $2 \times 10-7$ m/step.

4 Results and Discussions

4.1 Bearing Capacity

The loading of the rigid strip footing is simulated by applying equal vertical velocities at the area representing the footing. The progressive displacement of the footing induced by the vertical velocity applied at the footing nodes is accompanied by the increase of the pressure in the soil. Finally, the pressure under the footing stabilizes for a value that indicates a limit load (Fig. 4).

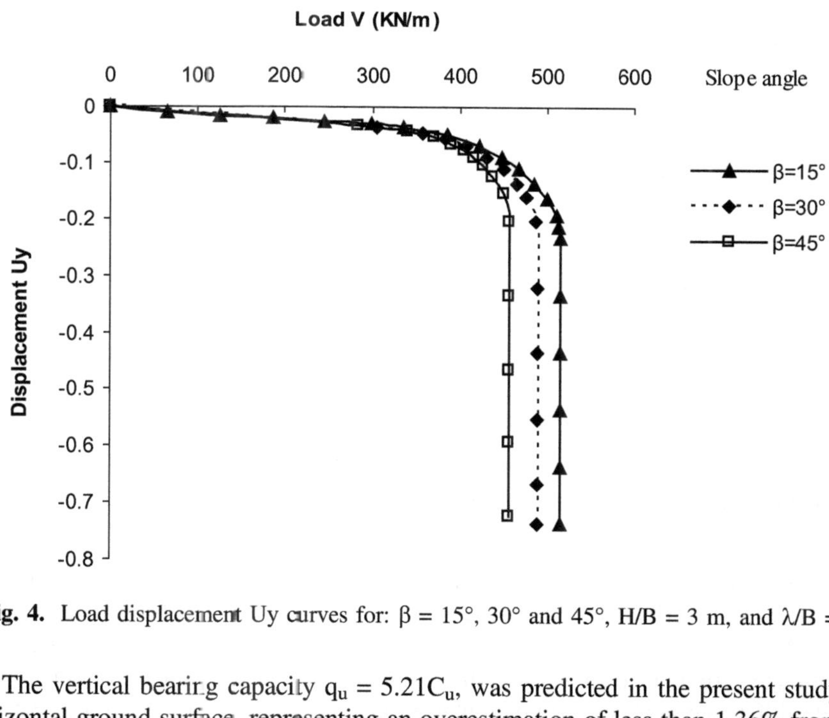

Fig. 4. Load displacement Uy curves for: β = 15°, 30° and 45°, H/B = 3 m, and λ/B = 1.

The vertical bearing capacity $q_u = 5.21C_u$, was predicted in the present study for horizontal ground surface, representing an overestimation of less than 1.36% from the analytical solution of $q_u = 5.14C_u$. It means that numerical prediction is in excellent agreement with a solution of Prandtl.

Figure 4 shows the load-displacement curves for rough footing at distance λ/B = 1 from the slope, (i.e., the curves of the vertical load V versus the vertical displacement). The comparison made for three values of slope inclination, it can be seen that the limit load V decreases substantially with increasing β even to stabilised, this is the vertical failure load.

4.2 Inclined Load Cases

This section study the influence of geometrical and properties of soil on the failure load, subjected to an inclined load. Four parameters were studied which are as follows: $c_u/\gamma B$, height of the slope *H/B*, slope angle β and normalised distance λ, specifically: $c_u/\gamma B = 1, 2.5$ and 5, $H/B = 0.5, 1, 2, 3$ and 4, $\beta = 0, 15, 30$ and 45, $\lambda = 0, 0.5$ and 1. Plurality analyses were realised with a slope height o f $H = 6$ m, finally to enable a fair comparison with other's results.

Typical vertical load displacement curves obtained from the finite difference analyses with different loads inclination θ are illustrated in Fig. 5 for $C/\gamma B = 2.5$ kPa, $\beta = 30°$ and $\lambda = 1$. The case of $\theta = 0$ corresponds to pure vertical loading.

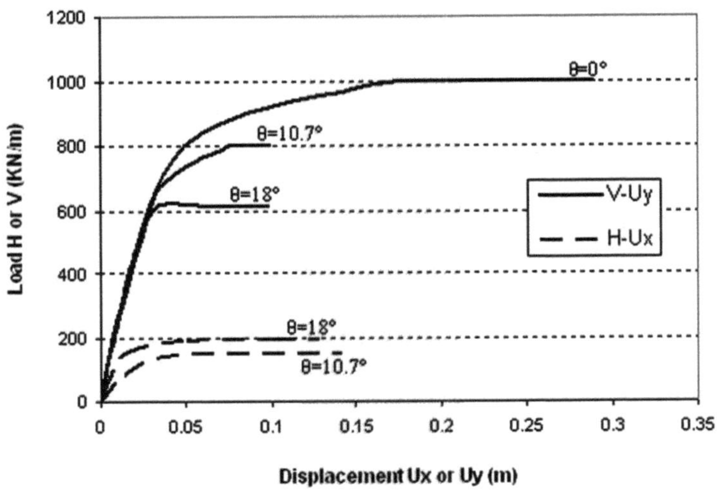

Fig. 5. Load-displacement curves.

4.3 Comparison of Available Solutions

The predicted failure envelope in the *VH* plane for the footing at the horizontal ground and at the top of the slope given by the conventional methods of bearing capacity calculation, as described by Meyerhof (1963) (Eq. 3), Hansen (1969) (Eq. 4), and Vesic (1973) (Eq. 5), is presented in Fig. 7 in order to validate the modeling procedure. The ultimate vertical load capacity of the foundation V_{ult} was obtained from the results of finite difference analysis for horizontal ground surface we have fond $V_{ult} = 5{,}186\ Ac_u$, this value greater than the exact solution of (Prandtl 1921) by just 0.89%. We can see that sliding along the soil-footing interface can occur for vertical load V less than some critical magnitude (approximately $0.5V_0$). These corresponding the ultimate horizontal failure $H = B.c_u$. However, if the vertical load is greater than $0.5V_0$ requires a decrease in horizontal load.

The results of the numerical analyses are a good agreement with Hansen and Meyerhof equation, but Vesic equation underestimates the failure loads. As seen in Fig. 6.

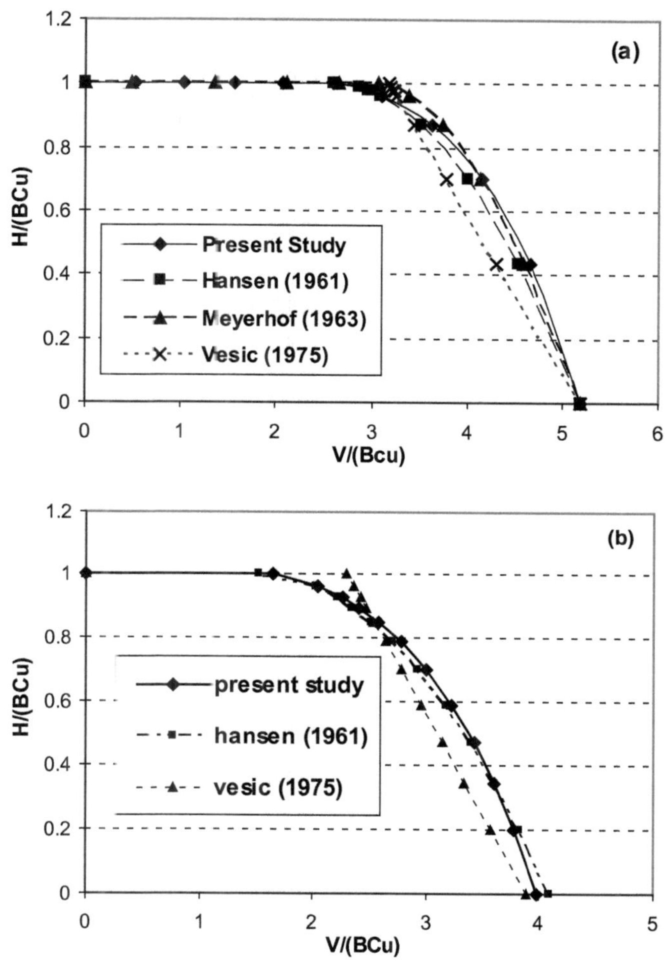

Fig. 6. Comparison of Failure surfaces of footing resting on the horizontal ground surface (a) and at the crest of a slope (b).

4.4 Parametric Analyses

4.4.1 Influence of the Ratio $C_u/\gamma B$

Influence of ratio $c_u/\gamma B$ on the failure surface is considered for the two normalized footing distances $\lambda = 0$ and 1, and four slope angles $\beta = 0$, 15°, 30° and 45°. Three ratios are investigated $cu/\gamma B = 1$, 2.5 and 5. These results are compared to the results obtained by the FE analyses (Georgiadis 2010). Figures 7, 8 and 9 illustrated the normalized failure load surface for the normalised distance $\lambda = 0$ and $\beta = 15°$, 30°, 45° respectively. For positive load (load directed away from the slope), the results of present study are a good agreement to those obtained by FE analyses (Georgiadis 2010), both these analyses reaches the horizontal ground surface $H = c_u.B$. For the load directed from the slope, the expansions of the failure envelope indicate the increase load capacity with increasing of the ration $c_u/\gamma B$. A greater reduction is observed for $c_u/\gamma B = 1$. The horizontal loads govern failure of the foundation and a pure sliding (where H = H0) for the vertical loads less than the ultimate vertical loads about 0.55V0 and 0.66V0 for $\beta = 30°$and 45°, respectively. The result obtained by FE analyses slightly overestimate the failure load than present study, for the case of a slope $\beta = 45°$ and $cu/\gamma B = 1$, the maximum vertical load predicted by the present study FD is

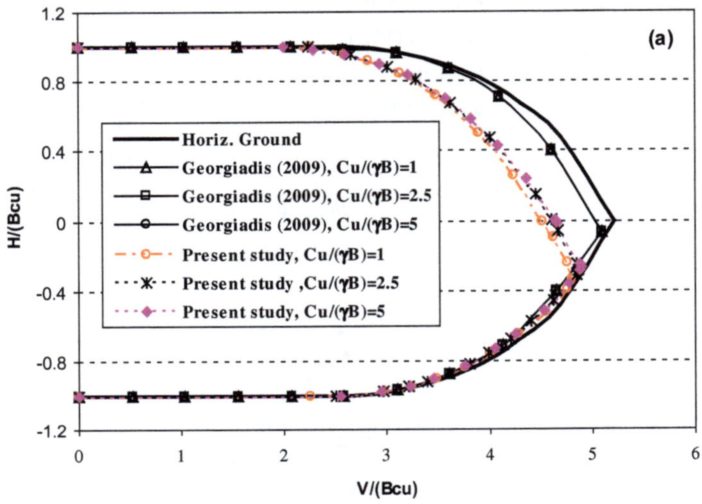

Fig. 7. Normalised failure load surface for $\lambda = 0$; (a) $\beta = 15$ (Figure modified from Georgiadis 2010).

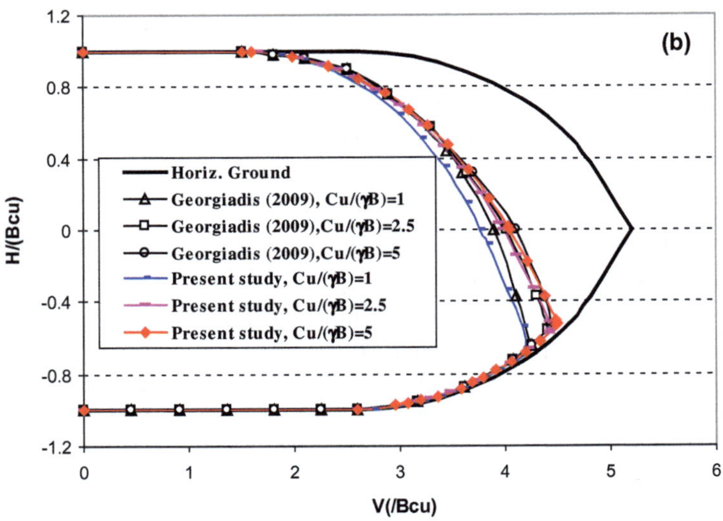

Fig. 8. Normalised failure load surface for $\lambda = 0$; $\beta = 30$ (Figure modified from Georgiadis 2010).

1.7 ACu, about 5% smaller than that given by FE (Georgiadis 2010), which is 1.8 ACu. Failure load interaction curve for the case slope angle $\beta = 45°$ and $\lambda = 0$ are presented in VH load space in Fig. 10, defined the same shape failure envelope, this shows no effect of the ratio $c_u/\gamma B$ in the failure envelope.

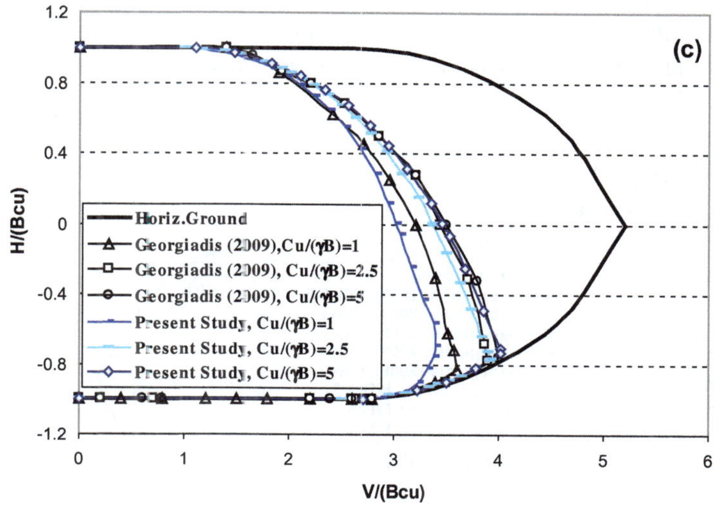

Fig. 9. Normalised failure load surface for $\lambda = 0; \beta = 45°$ (Figure modified from Georgiadis 2010).

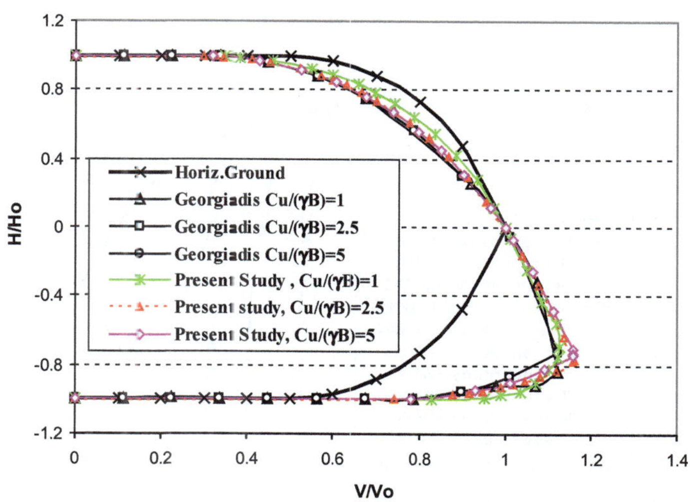

Fig. 10. Load interaction curves for $\lambda = 0$ and $\beta = 45°$ (Figure modified from Georgiadis 2010).

4.4.2 Influence of Slope Height

The capacity of foundation under pure vertical load they show that the low ratio $cu/\gamma B$ and height value of normalised distance λ effect on the shaped failure load surface (overall slope failure). Meyerhof (1957) was proved the undrained bearing capacity of great height, the undrained bearing capacity factor Nc, depend λ, β, and the stability number of the slope N_s, which is defined as, $\gamma H_c/C_u$ where H_c is critical height of the slope. For a "safe" design, we must include a factor of safety $F_s(N_s = \frac{F_s \gamma H_c}{c})$, the critical height $H_c \approx 11.025$ m corresponding $N_s = 5.51$, which is Close than Taylor's stability number of 5.52.

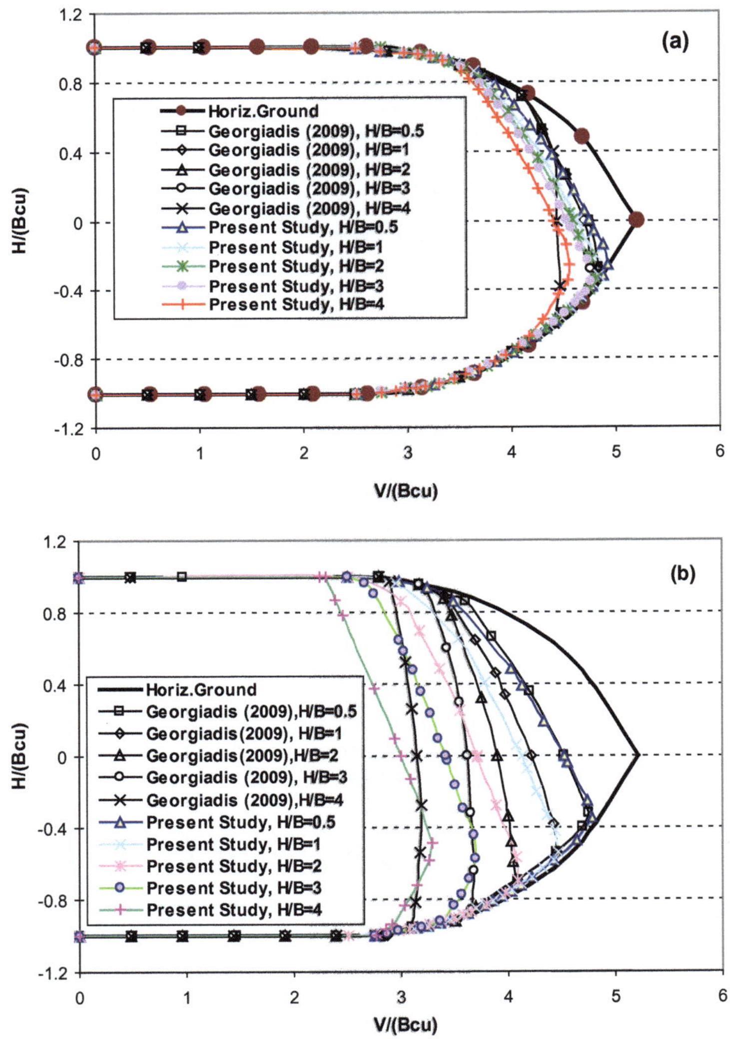

Fig. 11. Influence of slope height: normalised failure load surface for λ = 1 and $C_u/\gamma B = 1$: (a) β = 30°, (b) β = 45° (Figure modified from Georgiadis 2010).

In order, investigate the influence of the slope height on the failure load under inclined loading. Figures 11 and 12 shows the failure envelope plotted in terms of normalised failure load and load interaction for the low $c_u/\gamma B = 1$ ratio, $\lambda = 1$ and two angles $\beta = 30°$, $45°$ and five slopes height $H/B = 0.5$, 1, 2, 3, and 4. For $\beta = 30°$ show little decrease of the failure load with greater height of the slope, but for $\beta = 45°$ decreases considerably with greater height of the slope.

The results of FE analyses for the low horizontal load are in excellent agreement with present study.

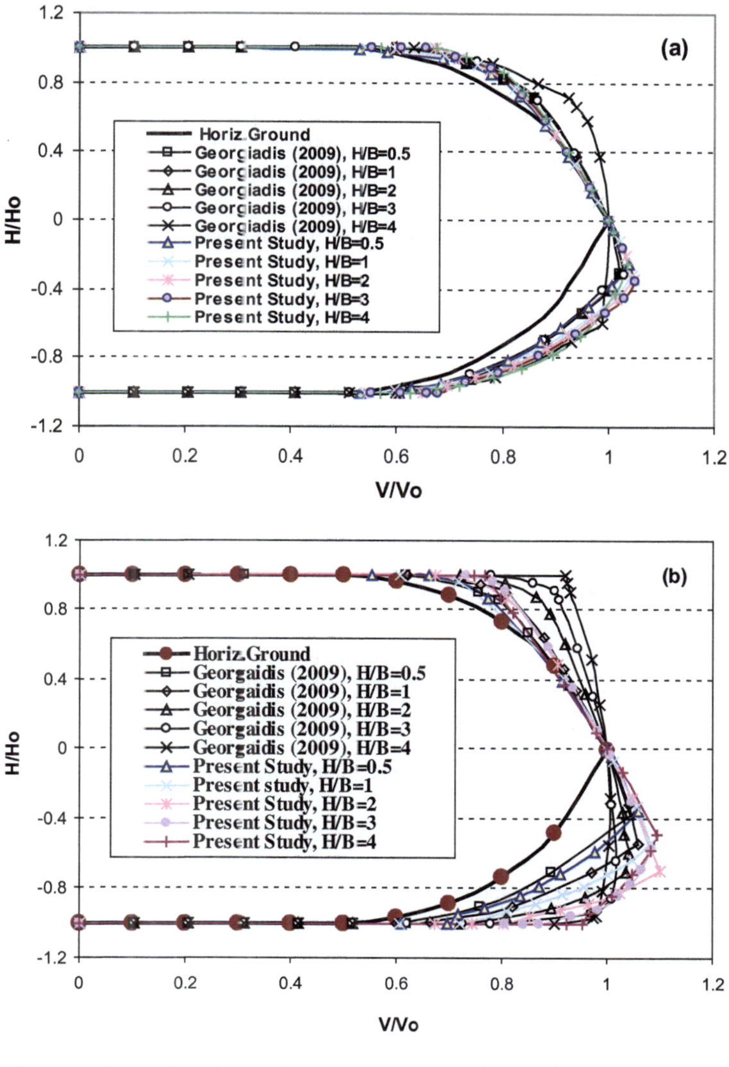

Fig. 12. Influence of slope height: Load interaction curves for $\lambda = 1$ and $C_u/\gamma B = 1$:(a) $\beta = 30°$ and (b) $\beta = 45°$ (Figure modified from Georgiadis 2010).

For β = 30° FE analyses slightly overestimate the failure loads with Present study, but for β = 45° overestimate the failure loads.

The load interaction curve for β = 30°, the same shape of the failure load for all value of the height of the slope and the results presented by FE (Georgiadis 2010) analyses are excellent agreement with the present study for H/B $\leq$ 3 (Fig. 12a).

Figure 12b show load interaction for β = 45°, FE results are excellent for the H/B ratio $\leq$ 1 with present study, but for higher height overestimate the failure loads.

5 Conclusion

A series of finite difference analyses were carried out to investigate the undrained bearing capacity of strip footings on or near slopes under inclined load. Various geometries were considered, varying the slope inclination angle and the normalised footing distance. The numerical results were found to be in good agreement with those obtained from Hansen's (1961), Georgiadis's (2010) equations. For positive load inclination, the results showed increasing failure surface with increasing slope angle and decreasing normalised distance. However, it is the opposite case for negative load inclination.

References

Baazouzi, M., Benmeddour, D., Mabrouki, A., Mellas, M.: 2D numerical analysis of shallow foundation rested near slope under inclined loading. Procedia Eng. **143**, 623 (2016). doi:10.1016/j.proeng.2016.06.086

FLAC: Fast Lagrangian Analysis of Continua, version 5.0. ITASCA Consulting Group Inc., Minneapolis (2005)

Georgiadis, K.: The influence of load inclination on the undrained bearing capacity of strip footings on slopes. Comput. Geotech. **37**, 311 (2010). doi:10.1016/j.compgeo.2009.11.004

Hansen, J.B.: A general formula for bearing capacity. Dan. Geotech. Inst. **11**, 38–46 (1961)

Meyerhof, G.G.: The ultimate bearing capacity of foundations on slopes. In: Proceedings 4th International Conference on soil Mechanics and Foundation Engineering, vol. 1, 384–386 (1957)

Meyerhof, G.G.: Some recent research on the bearing capacity of foundations. Can. Geotech. J. **1** (1), 16–26 (1963)

Terzaghi, K.: Theoretical Soil Mechanics. Wiley, New York (1943)

Vesic, A.S.: Bearing capacity of shallow foundations. In: Winterkorn, H.F., Fang, H.Y. (eds.) Foundation Engineering Handbook. Van Nostrand Reinhold, New York (1975)

Case Study About Erosion in Elmo SerejoAvenue, Taguatinga/Federal District – Brazil

Mariane Rodrigues da Vitória[1(✉)], Rideci Farias[2], Haroldo Paranhos[2], Itamar de Souza Bezerra[3], and Roberto Pimentel de Sousa Júnior[4]

[1] Universidade Católica de Brasília (UCB), Brasília, Brazil
marianervm@gmail.com
[2] UCB Reforsolo Engenharia UniCEUB IesPlan, Brasília, Brazil
rideci.reforsolo@gmail.com, reforsolo@gmail.com
[3] Maccaferri, Goiânia, Brazil
itamar@maccaferri.com.br
[4] UnB, Brasília, Brazil
eng.robertopimentel@gmail.com

Abstract. The land occupation without proper planning leads to many urban problems. One of the problems of cities that are in the process of industrialization, economic growth and population is related to land use and environmental impacts. Poor land use and non-compliance with existing laws cause devastating consequences to the environment, since adverse environmental impacts are often irreversible, leaving the alternative to mitigate them. Take as example the Federal District had a steady increase of the population that could not keep up satisfactorily the migration process, creating many problems for the region, both because of the occupation as the capacity of goods such as energy, water supply, sewage, public health and education with impacts to the environment and people's quality of life. Another effect of urbanization with many problems is the increased production of solid waste that are dumped inappropriately on the surface reaching the urban drainage networks, these issues related to deficiencies in services and low awareness of the population. In general, the water and the winds are the main triggering agents of erosion of the soil, being caused erosion, often by heavy rainfall runoff that generate the leverage. The erosion can occur for years and intensified with human action. Given the above, a field incursions program for verification of the geotechnical conditions in the study area were made, for understanding the region soil and be used as parameters for the slope stability tests.

1 Introduction

Brasilia was a planned city, being built to be the new Federal Capital. The construction of Brasilia held for purely political objectives, economic and social objectives these the president of the time Juscelino Kubitschek, who was developing Brazil, with the slogan fifty years in five years of government.

H. Shehata and Y. Rashed (eds.), *Numerical Analysis of Nonlinear Coupled Problems*,
Sustainable Civil Infrastructures, DOI 10.1007/978-3-319-61905-7_25

The federal capital was seen as a place of hope, getting people from different regions of Brazil, seeking opportunity of a better life in the city planned. With the coming of these people came the first cities satellites and consequently the disorderly occupation of the territory.

According to the Brazilian Institute of Geography and Statistics (IBGE), the population of the Federal District in 2013 was 2,789,761 inhabitants and today, in 2016, is with 2,960,129 inhabitants.

This increase in population density with planning that can not keep up with growth creates several problems as in the case of erosion occurred at Avenida Elmo Serejo which is inserted in the basin of the Cut Stream.

With the disordered population growth, soil and drainage system has not been able to withstand the load generated d'water. With this, the rainwater that were directed to the stream by pipeline, overloaded the system causing the site does not bear the pressure, generating spillovers and the resulting saturation of the soil with the induction and enhancement of the erosion occurred on site. The impact of urbanization in that interfere most urban drainage are relative to runoff, since the networks were undersized and receive a greater flow than expected.

The water in the urban environment has many aspects and this recurring population growth entail several negative factors such as misuse and pollution of soil by several factors, generating inadequate environmental conditions, fostering the development of waterborne diseases (dengue, zika), air pollution and noise, temperature rise, groundwater contamination among others. This aggravating factor shows that urban development without full environmental planning results in significant and often irreversible damage to the environment and society.

The soil saturation can occur in various ways, such as heavy rainfall, changes in water levels in the soil on the banks of lakes, reservoirs, canals and rivers, water leakage infrastructure such as water pipes or sewer. Many cases of landslides are linked to floods, causing erosion on the banks of streams and rivers also caused by saturation and runoff.

The term erosion comes from the Latin (erodere) and means "eat away", and the wear land surface (soil or rock) by the action of water, wind, ice and living organisms, plants and animals, as well as human action. And removing a soil mass to a location and its deposition elsewhere. The erosion can occur for years and intensified with human action. The process of erosion is caused by active forces such as rainfall characteristics, slope and length of the slope or slope and the ability of the soil to absorb water, and passive forces, such as resistance holding the soil to the erosive action water and the density of the vegetation cover (Bertoni and Lombardi Neto Bertoni 1999).

Water and wind are the main agents of soil erosion, water erosion being caused, often by rainwater or runoff. In Brazil, the erosion is very aggravating because of the tropical climate with rainfall frequentes. Segundo authors, among the main causes of triggering eevolution of erosive processes in urban areas are:

(a) the absence or disability of rainwater drainage systems and served in most popular and housing subdivisions;

(b) Implementation of inadequate road system, with perpendicular streets to the contours and lack of paving, curbs and gutters; and,
(c) Implementation of allotments in inappropriate places, under the geotechnical point of view.

In the watershed areas as the study area, it is important to keep vegetação. Um bare soil or with insufficient vegetation cover, will be subjected to erosion. The rain will start falling a laminar erosion process, forming gullies which greatly increases the ability to transport the soil particles. To avoid this environmental degradation, it is necessary to monitor local development and indicate possible faults in the planning and construction management, noting that the soil plays an important role in the activities of society.

2 Location of the Study Area

The study area is located in Serejo Elmo Avenue, close to the Samdu Avenue between satellite cities of Ceilândia and Taguatinga (Administrative Region III), in the passage of the stream Cut. Figures 1, 2 and 3 show aerial views, macro and micro the said area, and Figs. 4, 5 and 6 views are shown with the indication of break of slope process.

Fig. 1. Macro view of the study area. Source Google Earth (access 10/02/2016).

Fig. 2. Macro view of the study area. Source Google Earth (access 10/02/2016).

Fig. 3. Micro view of the erosion area. Source Google Earth (access 10/02/2016).

Fig. 4. Talus in rupture process.

Fig. 5. Area rupture process.

With the breakdown of embankments occurred in November 2013 there was the need to ban the road that is one of the main links between the regions of the satellite cities of Ceilândia and Taguatinga to the center of Brasilia (Pilot Plan), generating major traffic jams and inconvenience to thousands of people there by passing it diariamente. Ressalte in umadas occasions a bus crossed the block and almost.

Fig. 6. Area after rupture process.

The estimated flow on the avenue is close to 50 000 vehicles/day in each direction. Figure 7 shows the continuous congestion due to the erosion that has formed. In a period of execution of the restoration work it was necessary to completely interrupt the route in one of the senses.

Fig. 7. Congestion on the road Elmo Serejo Wansbeck.

3 A Brief History of the Process Erosive

The following presents a brief history of the development of erosivoobjeto process of this work:

11/25/2013 - Erosion caused by rain on the banks of Elmo Avenue Serejo reached the local sewage network of the Company water and sewage Brasilia (Caesb) that broke potentiating the erosion process in the area and the consequent ban on local traffic;

11/26/2013 - With the advancement of the process was necessary to ban two-way Elmo Serejo.

In 2014 Avenida Elmo Serejo, was closed several times due to local. Ainda repair works in 2014 the governor signed the contract to start the construction works of a bridge under erosion in Serejo Elmo Avenue and recovery area;

05/06/2014 - Completion of the 1st stage of the works on Avenida Elmo Serejo;

03/13/2015 - The Civil Defense of the Federal District issued an alert on new risk of collapse in Elmo Avenue Serejo in Wansbeck.

4 Materials and Methods

The study consisted in the execution of a field incursions program for verification of the geotechnical conditions in the study area, but also the execution of laboratory tests for material characterization and determination of strength parameters (cohesion and friction angle), and analysis of stability of slopes.

When considering the issue at hand, there are several ways to identify occurrences of erosion and soil mass slips in a certain area. Amongst them:

(a) verify the possible occurrence of infiltrations - if the soil is moist or saturated;
(b) The drainage of rainwater, damage to water or sewage pipes and other underground structures;
(c) The presence of excessive inclination or cracks on the crest and slope;
(d) Planting of tree species in incompatible slopes. (The manual slips - a guide to understanding the landslides-2008 1st ed.).

In order to assess the erosive process in question and possible causes, it proceeded to the analysis of geotechnical conditions, stability of slopes, but also to check the registration of network interference of water and sewage of the area in question.

These analyzes were necessary for the collection of disturbed and undisturbed samples with a view àcaracterização and determination of strength parameters (cohesion and friction angle) from the ground; analysis maps; field recognition; poll; and computational analysis to verify the stability of slopes.

For the implementation of the characterization tests and resistance, followed by osprocedimentos recommended by standards NBR/ABNT, then, and the adoption of commonplace procedures geotechnical laboratory at the Catholic University of Brasilia.

- NBR 6459/ABNT - Determination of Soil Liquidity Limit;
- NBR 7180/ABNT - Determination of Plasticity of Soils Limit;
- NBR 6457/ABNT - Preparation of soil samples and characterization test;
- NBR 6508/ABNT - Determination of bulk density of the grains;
- NBR 7181/ABNT - Particle size analysis of soils;

Table 1 shows a summary of the results of tests performed, and Fig. 8 shows the granulometry of the ground in the condition with and without dispersant. There it is a soil with silt features sandy clay, when performed without dispersant in sedimentometric test, and little sandy silty clay when run with dispersant.

Point out that are characteristic soils of the Federal District and, in general, are susceptible to erosion process.

Table 1. Summary of results of tests carried out geotechnical and certain parameters.

Summary of results for the soil collected		
Specific weight natural	γ_{nat}(kN/m^3)	16,0
Specific mass of grains	(kN/m^3)	2,66
Natural moisture	(%)	36,87
Hygroscopic moisture	(%)	12,29
Liquidity limit	(%)	57,4
Plasticity limit	(%)	39,2
Plasticity Index	(%)	18,2
Clay fraction (without dispersant)	(%)	19,2
Clay fraction (with deflocculant)	(%)	46,5
Silt fraction (without dispersant) dispersantdisˈpərsənt Definições de dispersant substantivo a liquid or gas used to disperse small particles in a medium. Further, Mr Knowles knew nothing of the idea of powdered ingredients or particles acting as dispersants in the dry state from his long experience in the agrochemical industry.	(%)	65,4
Silt fraction (with dispersant)	(%)	45,2
Sand fraction (without dispersant)	(%)	15,4
Sand fraction (with dispersant)	(%)	8,2
Effective cohesion	C (kPa)	20 (Flooded Condition) Flooded Definições de flood verbo cover or submerge (a place or area) with water. the dam burst, flooding a small town sinônimos: inundate, swamp, deluge, immerse, submerge, drown, engulf arrive in overwhelming amounts or quantities. congratulatory messages flooded in sinônimos: pour, stream, flow, surge, swarm, pile, crowd Veja também flood Traduções de flood verbo

(*continued*)

Table 1. (*continued*)

Summary of results for the soil collected		
		inundar flood, inundate, drown, overflow, submerge, deluge alagar overflow, flood, waterlog, deluge transbordar overflow, boil over, flood, overrun, deluge, overbrim encher fill, stuff, full, flood, charge, pad jorrar gush, spout, flood, flush, outpour irrigar irrigate, water, flood, canal, float afluir flow, flock, flood, run into sofrer hemorragia uterina flood
Effective friction angle	ϕ' (graus)	30 (Flooded Condition) Flooded Definições de flood verbo cover or submerge (a place or area) with water. the dam burst, flooding a small town sinônimos: inundate, swamp, deluge, immerse, submerge, drown, engulf arrive in overwhelming amounts or quantities. congratulatory messages flooded in sinônimos: pour, stream, flow, surge, swarm, pile, crowd Veja também flood Traduções de flood verbo inundar flood, inundate, drown, overflow, submerge, deluge alagar overflow, flood, waterlog, deluge transbordar

(*continued*)

Table 1. *(continued)*

Summary of results for the soil collected		
		overflow, boil over, flood, overrun, deluge, overbrim encher fill, stuff, full, flood, charge, pad jorrar gush, spout, flood, flush, outpour irrigar irrigate, water, flood, canal, float afluir flow, flock, flood, run into sofrer hemorragia uterina flood
Density Solid	γs (g/cm3)	2,66

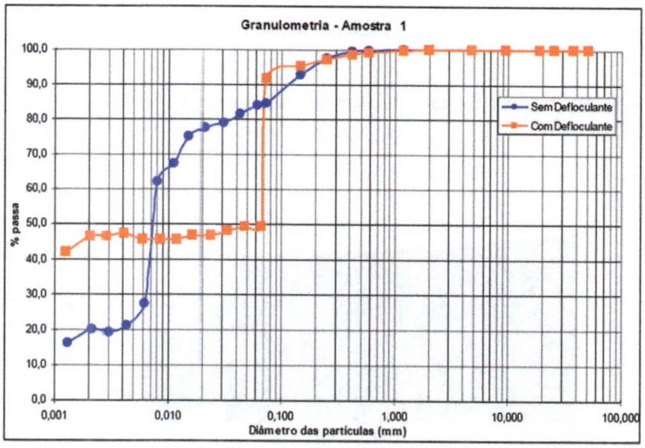

Fig. 8. Composition of granulometric curves (with and without dispersant).

Possession of surveys on documents provided by the Technical Cadastre of Environmental Sanitation Company of the Federal District (CEBS), it was found that in the region there are sewage networks and water supply network does not interfere in the area of erosion. This is in line to meet the existing systems in the region as a form of preventive maintenance in order to avoid possible leaks that could compromise again the works performed to recover the site.

5 Solution Adopted for Recovery of Affected Area

The solution for recovery of erosion considered crimping metal profiles along the road side in the region of the erosive process to enable the lateral confinement of the track, protection and stabilization of slopes with the adoption of the technique in soil nailing, shotcrete eem complementation, insertion gabion (Figs. 9 and 10).

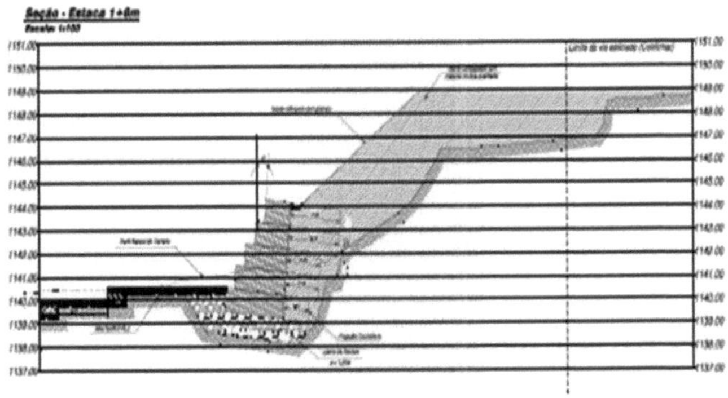

Fig. 9. Slope recovery project in gabion, stapled and shotcrete soil - Cutaway view. (Source: Arp Engineering).

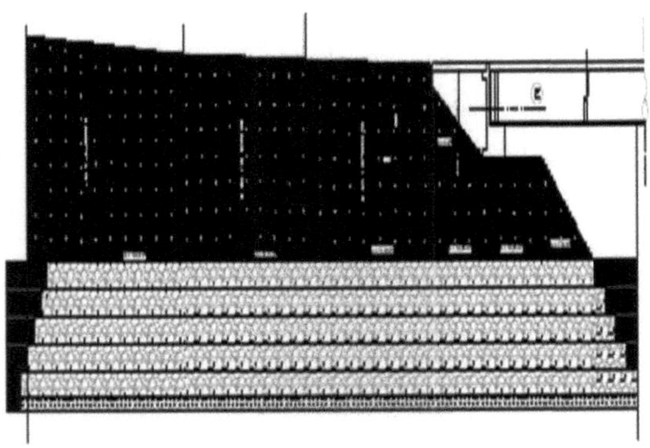

Fig. 10. Slope recovery project in gabion, soil nailing and shotcrete (front view).

In addition to all these aspects, we proceeded to the implementation of a system of drainage works with the necessary refunds flow, plumbing and device to break the energy of the water flow (Figs. 9, 10, 11, 12, 13, 14, 15 and 16).

In accordance with the works if citadaspartiu-to recompose the road system and the consequent normalization of traffic in the region (Fig. 17).

na região (Fig. 17).

Fig. 11. Slope recovery work in gabion, soil nailing and shotcrete.

Fig. 12. Slope recovery work. concrete application designed.

Fig. 13. View of the work to recover the eroded section

Fig. 14. View of the work. Observe the metal profile line executed.

Fig. 15. Side track performed after recording metal profiles online.

Fig. 16. View of the finished work.

Fig. 17. Track with the finished work.

6 Conclusions

This article aims to identify possible causes of erosion that occurred in the Federal District, addressing the problem of misuse of the local soil, as well as the consequences of erosion for the people who daily use the connecting portion of the satellite cities of Ceilândia-Wansbeck to several points, but also data such as: characterization of soil, stability of slopes, and checking the interference of water networks registration and sewage of the area in question generating important data for future studies on the problem of risk and consequences arising from erosion. Several actions were carried out in the area in order to lessen the damage to the population and to contain erosion on site. Traffic was diverted on site at various times with the goal to preserve the area in recovery. In terms of construction, the design with a view to recovery of the area, it adopted a system composed of crimped metal piles and bridge/slab for track stretch that broke. As for the reconstruction of embankments adopted the technique of soil nailing, shotcrete and wall in gabião. As works were executed with the start of the bridge on 26.03.2014, 10.21.2014 and end and, according to companies responsible. On the monitoring of the erosion process, the progress of the work performed, studies and analyzes of various materials, performance testing, verification of existing documents, has the following main conclusions:

(a) The main cause of the incident is related to the occurrence of infiltration of water coming from a pipe Environmental Sanitation Company of the Federal District (CEBS), which already had a degree of commitment to integrity, as noted in one of the site visits. This fact was boosted by the impact of the rainy season;

(b) Confome consulting the technical register base CEBS, has sewage networks nearby confirming the incident that potentiated the erosion process;

(c) It was possible to verify the occurrence of erosion also type "piping" in which this phenomenon starts by percolation of water through the soil pores. When this percolation achieves a preferred path, the flow velocity increases, coming to erode this preferred path, forming a "tube" in the soil mass;
(d) grading tests indicate it is clayey-silty soil with dispersant, the silty-clay soil, when executed without dispersant. It is noteworthy that are characteristic soils of the Federal District and, in general, are susceptible to erosion process;
(e) analyzing maps and the region is possible to see great influence of urban occupation in the vicinity of erosion and poor land use and occupation that can cause various problems. Among them soil impoverishment increasing the incidence of erosion;
(f) In terms of stability it appears that the situation in question was about to break with safety factor close to 1.0. Then with the partial saturation of the soil mass been causing the decrease of the safety factor and the consequent disruption of the system;
(g) The cracks in the soil generated by the advance of disruptions directly influenced the evolution of erosion. This was mainly observed before the onset of total collapse. This fact indicated for appropriate action that should be taken in order to avoid the proportions that the process has taken;
(h) As a result of the process, there are many disorders in the function of prohibiting means for several days, both in the period when the slope breaking in period of corrective measures on the floor;
(i) it is deemed to be irregular occupations and unplanned accelerate soil degradation and the environment in many regions, with the consequent growth of invasions and population occur in a disorderly manner, without previous studies, without planning of environmental impacts, education, health and without infrastructure works. It is observed that, to waterproof the ground for the construction of buildings, water precipitation does not seep into the ground, causing the erosion. Given the above, and need to raise awareness of the use and preservation of natural resources.
(j) This work shows the importance of preventive measures instead of corrective to the erosive processes in order to prevent damage to people and the environment. also points up on cuidadados that should be taken by public agencies, since, for the case in question, had already been observed the presence of erosive process in the region. The fact of the break quite raised the costs of the work, in addition to the long period of inconvenience to the population.

Acknowledgments. I thank God that blesses me constantly, the Catholic University of Brasilia (UCB), and Professor. Dr. Rideci Farias, for all the support and guidance.

References

Bertoni, J., Lombardi Neto, F.: Conservação do solo. 4.ed., 355 p. Ícone, São Paulo (1999)
http://www.diariodopoder.com.br/estados/distrito-federal/erosao-que-atingiu-asfalto-da-elmo-serejo-esta-contida/. 25 fev de 2014
https://www.google.com.br/search?q=avenida+elmo+serejo&source=lnms&tbm=isch&sa=X2Ferosao-na-elmo-serejo-esta-completamente-contida%252F%3B1000%3B664
Tecnologia Para Prevenção E Controle De Erosão Em Vias Urbanas. http://www.labogef.iesa.ufg.br/links/simposio_erosao/articles/T015.pdf. Acesso 28 março 2014
http://g1.globo.com/distritofederal/noticia/2015/03/defesa

Performance of Shallow Foundation Overlaying Cavernous Limestone

Ahmed M. El-Tohamy[1,2(✉)]

[1] Faculty of Engineering, Beny Sufe University, Beni Suef, Egypt
ahmed.eltohamy@egypt.nbad.com
[2] Faculty of Engineering, Umm Alqura University,
Mecca, Kingdom of Saudi Arabia

Abstract. New Minia City, Egypt is one of the most successful and highly growing new urban communities. Construction projects in the city were faced by a foundation soil that can be described as soft, highly fractured and jointed limestone which when exposed to dissolution by acidic rainfall or pipes leakage creates shrink holes and caves. The present numerical analysis study aims at investigating the effect of limestone cavity with varying dimensions and occurrence depth on the performance of shallow foundations. It was concluded that there is no clear information relating positions, shape, dimensions and level of caves. It is recommended and furtherly intended to conduct a detailed field study in order to supply complete information about existing and predicted caves. Mode of failure of limestone soil underneath a strip footing is different from the traditional modes of shear failure. There exists a critical cave width beyond which caves existence underneath shallow strip footing results in reducing bearing capacity of the foundation. Caves with roof level deeper than or equal to 9 times footing width with a cave to footing width ratio less or equal to 3.5 do not result in reducing the bearing capacity of strip footing compared to no-cave case. For practical design purpose charts for reduction factors of ultimate bearing capacity and bearing capacity factors were presented.

1 Introduction

In the last thirty years, urban development has been one of the high priorities of the Egyptian government. Through the last century, the extension of existing Egyptian cities was directed towards agricultural land. As a result, the Egyptian government specified new communities outside the Nile valley for urban development. At present time New Minia city is one of the most successful and highly growing new cities in Egypt. The city is located at the east bank of river Nile in front of old Minia city 250 km south Cairo, 200 m above sea level and 110 m above Nile level. Early construction stages of the city started at 1986. Construction projects in the city were faced by a foundation soil that can be described as soft, highly fractured and jointed limestone which when exposed to dissolution by acidic rainfall creates shrink holes and caves in the heterogeneous bedrock. Four years later the city planners wanted an assessment of any geo-environmental problems that may face construction, Hassan (2002). National Egyptian Association of Nuclear Power (NEANP) conducted a

H. Shehata and Y. Rashed (eds.), *Numerical Analysis of Nonlinear Coupled Problems*,
Sustainable Civil Infrastructures, DOI 10.1007/978-3-319-61905-7_26

detailed geo-physical, and chemical studies. Results of this study were presented in a report at Hussan (1990). According to this report, it was concluded that limestone soil at New Minia City (NMC) is non-homogenous. It consists of a thin layer of fractured soil of abbreviated limestone and sand with varying depths. This layer is followed by limestone layers of yellowish white to pure white colored limestone. The strength of limestone at (NMC) may be characterized as ranging from very low to moderate strength. Soil compressive and tensile strength, modulus of elasticity, unit weight, cohesion and angle of friction are illustrated in Table 1.

Table 1. Properties of limestone at NMC

	Strength Range* 10^3 (Comp.) (kN/m^2)	Strength (Comp.) (kN/m^2)	Strength (Tension) (kN/m^2)	Modulus of elasticity (GPa)	Unit weight (γ) (kN/m^3)	Cohesion (C) (kN/m^2)	Angle of friction φ^o
Very low	1.25–5.0	3000	300	12.07	16.7	480	38
Low	5.0–12.50	9000	730	16.39	20.4	1250	55
Moderate	12.5–50.0	30000	2233	36.26	33.4	5000	58

Fig. 1. Relatively large caves at a plateau at the entrance of (NMC).

Voids and cavities have been known to be a common feature of limestone and dolomite formation all over The World, Baus (1983) and Wang and Badic (1985). This is also valid at (NMC) as indicated by (NEANP) report, Fig. 1 illustrates relatively large caves at a plateau at the entrance of (NMC). The report indicated that the caves detected in the study are presented in maps which were not included with the available copy of the report. As indicated by Tan (2005), the process of voids and caves formation in limestone is attributed to chemical dissolution process formed by water moving through limestone. Figure 2 illustrates dissolution cycle of limestone by raining water. Carbon dioxide from the atmosphere is fixed or converted in the soil in

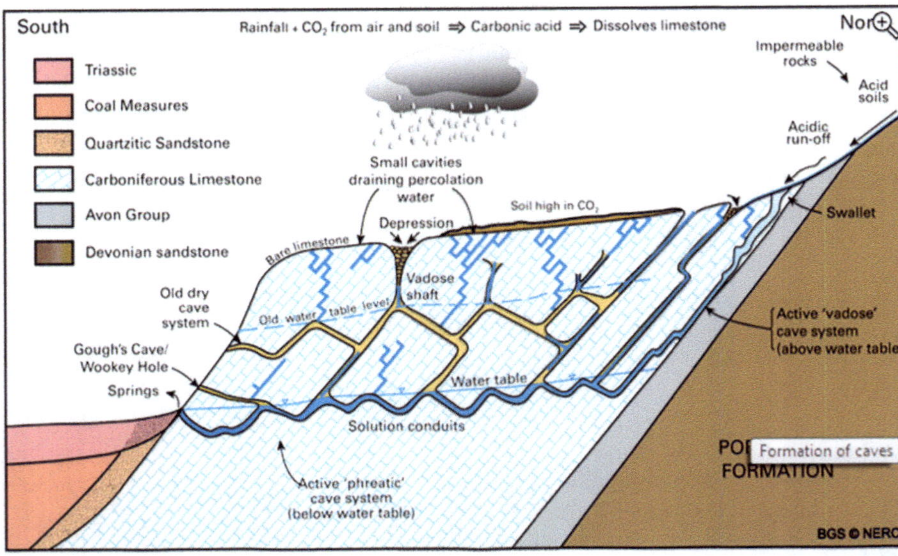

Fig. 2. Dissolution cycle of limestone by raining water, British Geological Survey (2016)

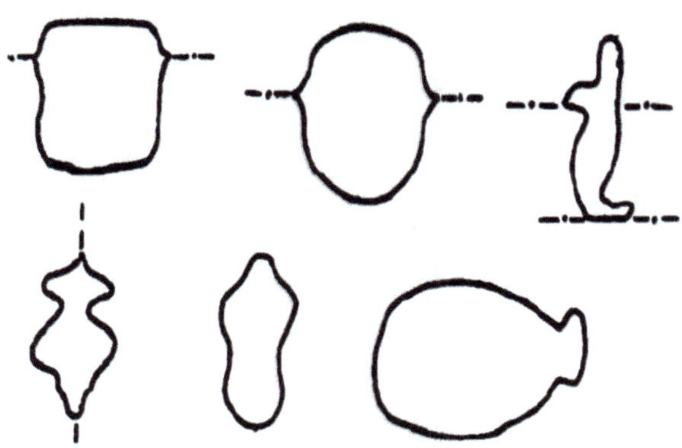

Fig. 3. Different shapes of limestone caves, Tan (2005)

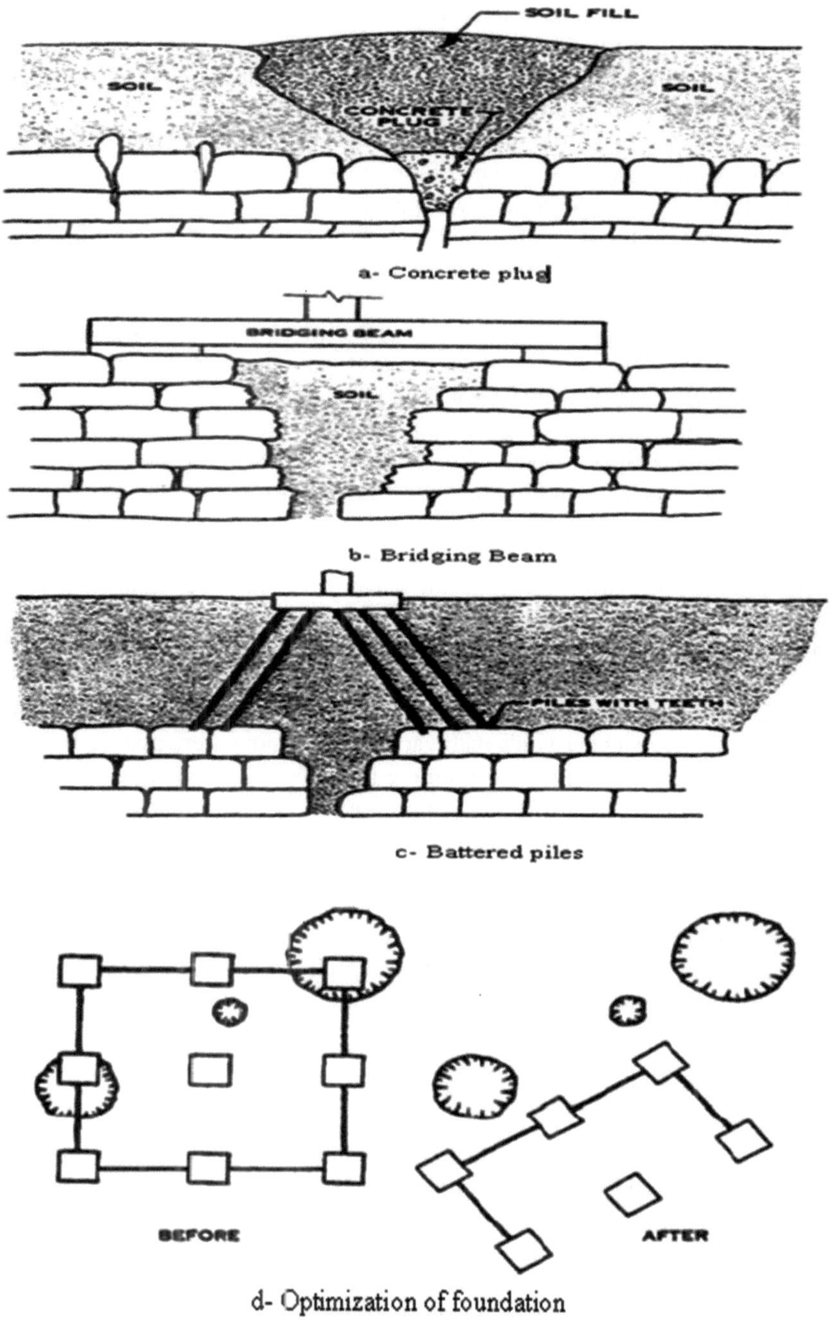

Fig. 4. Suggested alternatives for founding above caves, Sowers (1996).

an aqueous state and combined with rain water to form carbonic acid, which dissolves carbonate rocks. As also indicated by Tan, water flow through path ways in limestone such as bedding planes, joints and faults occurs under gravity force in dowry words direction. The water flow results in enlargement of path ways allowing for more water to flow which increases the dissolution rate. Canakci (2007), indicated that the most significant factor that may affect failure of limestone caves of Gazantep City, in Turkey is the saturation of limestone from sewer pipes and surface water which caused considerable reduction in the strength of soft limestone. Sowers (1996), specified a rate of 5–200 mm/1000 years for limestone dissolution. However, the dissolution rate is dependent on the characteristics of limestone. Muhammad and Beng (2002), depending on results of an experimental study specified higher dissolution rates of 134–369 mm/1000 years. Caves shape is also an important factor in specifying reduction of bearing capacity of shallow foundations. Tan (2005), after Wilford (1964) illustrated different limestone caves at East Malaysia as illustrated in Fig. 3.

Among different shapes caves with flat roofs are the most unstable and critical caves. Design of structures on limestone must account for both of pre-existing cavities as well as cavities that can develop during the life time of the structure, Azam (1988). Structural solutions for cavities existence in foundation soil may include high slump concrete grouting of cave, bridging the cave by reinforced earth fill, bridging, Abdeltuab (2013), foundation beam, battered piles, concrete plugging of cave and optimization of foundations layout to avoid caves, Fig. 4 as cited after Sowers (1996), illustrates some of these solutions. Baus (1983), studied a numerical model to investigate the effect of voids existence underneath a strip footing of varying width in clay soil. Circular and rectangular voids with varying dimensions were assumed. The study concluded that bearing capacity of the footing is reduced as the void becomes closer to foundation level. The effect of void vanishes when the ratio of cave roof depth to footing width reaches a critical value that increases with the increase of width of cave. Agaiby and Jones (1996) conducted a two dimensional finite element study to investigate the effect of rectangular void with varying depths and dimensions in clay soil. The study concluded that when the bedrock exists at relatively shallow depth smaller than 6 times footing width it affects bearing capacity of the footing. Void shape and location were indicated to affect the bearing capacity of footing.

2 Research Scope

As can be concluded from the introduction presented, urban development at (NMC) is faced by the risk of existence and/or formation of limestone voids and cavities. Formation of such cavities may be assisted by probable heavy rain falling and leakages in water and sewage networks. This motivated the researcher to conduct the present numerical analysis study to investigate the effect of limestone cavity with varying dimensions and occurrence depth on the performance of shallow foundations. The results of this research may gain their practical importance by providing an estimation of critical roof of caves depth and cave width present in (NMC) limestone of intermediate geotechnical properties. Reduction of shallow foundation bearing capacity may also be concluded from the results of present research.

3 Finite Element Model Setup

3.1 Model Geometry and Material Properties

The numerical analysis of this research was carried out using plaxis 2d (Ver.7.2), which is two dimensional finite element software for soil and foundation structures. The model dimension was chosen as 21 * 14 times footing width (Bf) for width and depth, respectively. The side of the model was 10 times footing width so as not to affect the results. Mohr column model was used for modeling low strength limestone layer with properties and shear strength parameters according to (NEANP) report, ($\gamma = 20.4$ kN/m^3, E = 1.207 E + 06 KN/m^2, Possion's ratio $\nu = 0.25$, C = 480 kN/m^2 and $\varphi = 38°$). The strip footing was medaled by linear elastic model with material properties (stiffness EA = 5.0E + 05 kN/m and EI = 8500 kN/m^2/m). Figure 5 illustrates plaxis two dimensional model showing the strip footing overlaying a cavity for the case of cave roof depth (Dc) to footing width (Bf) ratio (Dc/Bf = 3.0) and cave width (Bc) to (Bf) ratio (Bc/Bf = 2.0).

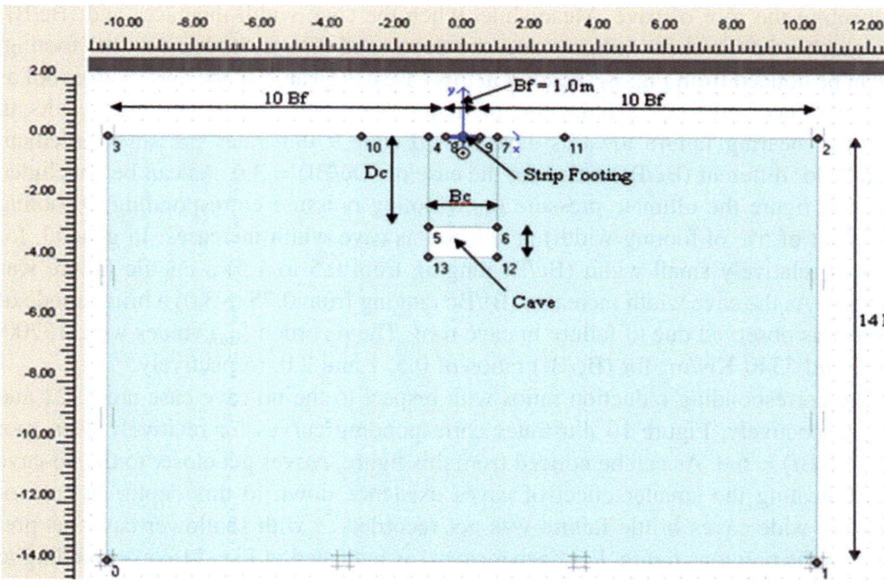

Fig. 5. Plaxis numerical model (Bc/Bf = 2.0 and Dc/Bf = 4.0)

3.2 Analysis Program

In order to investigate the effect of cave existence underneath shallow foundation, the following analysis program was designed. (Dc/Bf) ratios adopted in this program ranged from 3.0 to 10.0 with step of 1.0. At each cave roof level, caves with varying width were considered. The adopted (Bc/Bf) ratios ranged from 0.5 to 3.5 with step of 0.25. A constant footing width of 1.0 m was assigned. As being the most critical case, a flat roof cave with rectangular section was chosen. A constant cave depth of 1.0 m was adopted. The middle of the footing was aligned with the middle of the cave. For different cases, the first loading phase was assigned for generating initial stresses in the model due to the excavation of cave prior to loading the strip footing. In the second loading phase, the footing was loaded till reaching failure in soil body. The ultimate load on strip footing was considered as the load that produces a footing settlement of 5% of footing width.

4 Results and Discussion

Figures 6 and 7 illustrate incremental shear shadings for no-cave and existing cave case of ((Dc/Bf) = 3.0 and (Bc/Bf = 2.0)), respectively. As can be seen from Fig. 6 for no-cave case the model of failure under strip footing is a local shear failure. On the other hand, as indicated in Fig. 7 for of a strip footing overlaying a cave shearing failure surface is different from the well known failure modes of shallow foundation (General, Local and Punching) modes of failure. In this case, the shear failure surface is at first inclined from footing edge towards the sides of cave and then it is directed downwards till meeting the roof of cave. Meanwhile, when the cave width increases and (Bc/Bf) ratio reaches 3.0 no shear failure planes are developed directly underneath the footing. As can be noticed from Fig. 8, shearing in this case has an arch shape above cave roof as cave existence produces a reduction in material strength the matter that results in directing shearing failure towards cave roof. Figure 9 illustrates pressure settlement diagram for different (Bc/Bf) ratios for the case of (Dc/Bf) = 3.0. As can be concluded from this figure the ultimate pressure (q_u) (footing pressure corresponding to footing settlement of 5% of footing width) is reduced as cave width increases. In general, for caves of relatively small width (Bc/Bf ranging from 0.5 to 1.5) a ductile failure was recorded. As the cave width increases (Bf/Bc ranging from 0.75 to 3.0) a brittle mode of failure was observed due to failure in cave roof. The recorded (q_u) values were 13700, 11800 and 5340 KN/m^2 for (Bc/Bf) ratios of 0.5, 1 and 2.0, respectively.

The corresponding reduction ratios with respect to the no-cave case are 9, 22 and 65%, respectively. Figure 10 illustrates corresponding curves for relatively deep roof with (Dc/Bf) = 6.0. As can be noticed from this figure, curves get closer to the no-cave case indicating the smaller effect of caves existence down to this depth. In case of relatively wide caves brittle failure was not recorded as with shallower caves as presented in the previous figure. For deeper caves, as indicated in Fig. 11 corresponding to (Dc/Bf) = 10, caves with (Bc/Bf) up to 3.5 seem not to have any effect on footing pressure and settlement, as curves corresponding to different (Bc/Bf) coincide with no-cave curve. Critical cave width may be defined as the maximum cave width (existing at certain depth) up to which no reduction in bearing capacity of shallow foundation occurs due to cave existence.

Figure 12 illustrates the relation between critical (Bc/Bf) ratios against (Dc/Bf) ratio. As can be noticed from this figure critical (Bc/Bf) ratio increases with the increase of roof of cave depth up to (Dc/Bf) up to 9.0 and remains constant thereafter. The recorded critical (Bc/Bf) are 0.25, 0.5, 0.75, 1.25 2.25, 3 and 3.5 for (Dc/Bf) ratio ranging from 3.0 to 9.0. This indicates that existence of caves under shallow foundation at a depth of equal to or more than 9.0 times foundation depth does not affect the

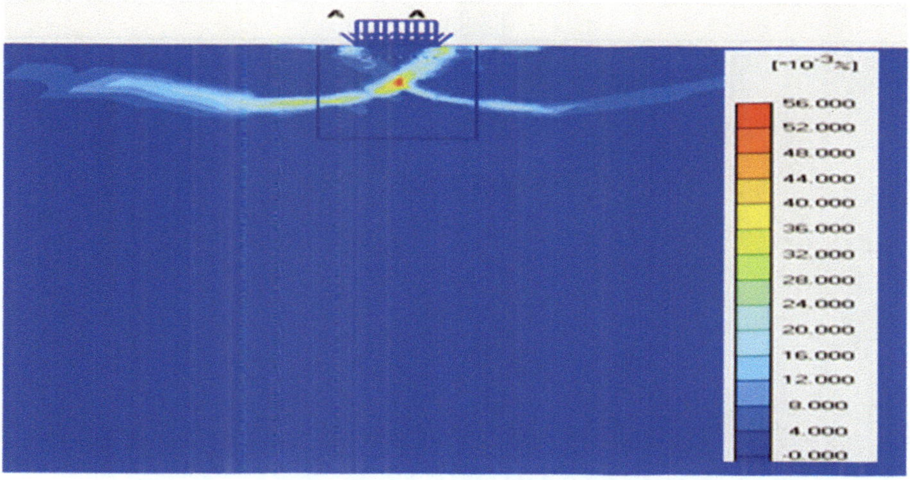

Fig. 6. Incremental shear shading for no-cave case

Fig. 7. Plaxis numerical model (Bc/Bf = 3.0 and Dc/Bf = 2.0)

bearing capacity of foundation regardless of cave width up to 3.0 times footing width. Azam (1991), indicated a (Bc/Bf) critical ratio of 6.0 for rectangular voids under lagging shallow foundation over clay soil. For practical design purpose of shallow strip footings on limestone soil with an underground cave.

Figure 13 presents reduction ratios of ultimate bearing capacity (q_u) with respect to no-cave case for different (Bc/Bf) and (Dc/Bf) ratios. As can be indicated from this figure, caves of a relatively narrow width slightly affect bearing capacity of shallow foundation for different cave roof depths. The recorded reduction ratios $q_{u(cave)}/q_{u\ (no\text{-}cave)}$ ranged

Fig. 8. Plaxis numerical model (Bc/Bf = 3.0 and Dc/Bf = 3.0)

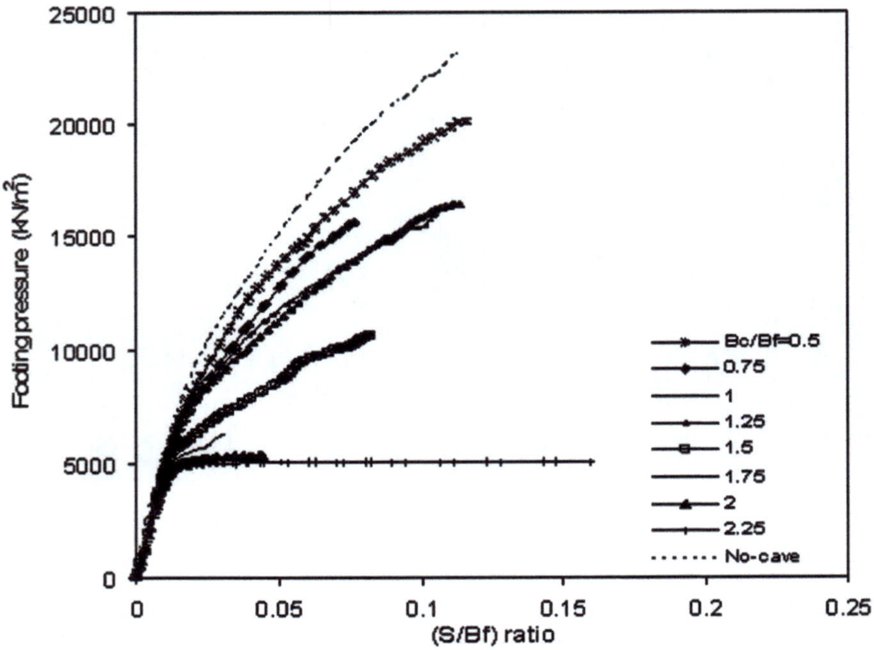

Fig. 9. Footing pressure against settlement for the case of (Dc/Bf) = 3.0. for different (Bc/Bf) ratios

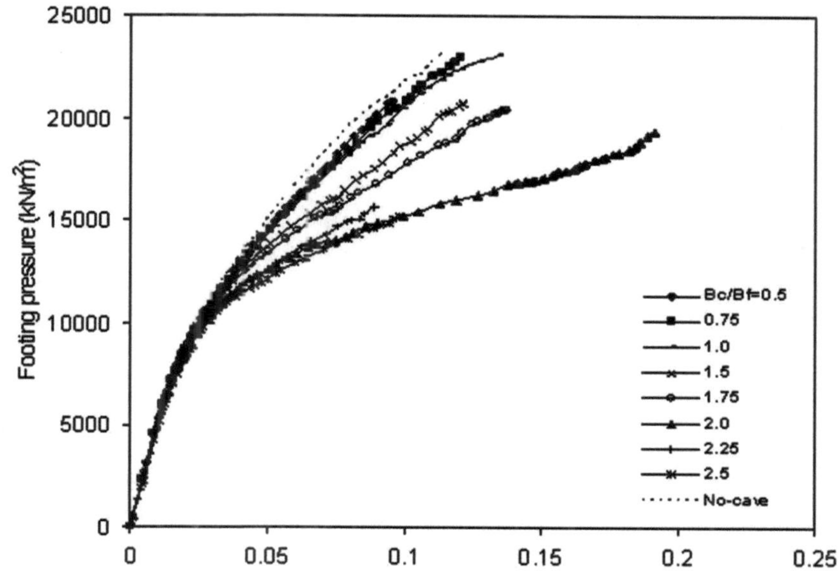

Fig. 10. Footing pressure against settlement for (Dc/Bf) = 6.0 and different (Bc/Bf) ratios

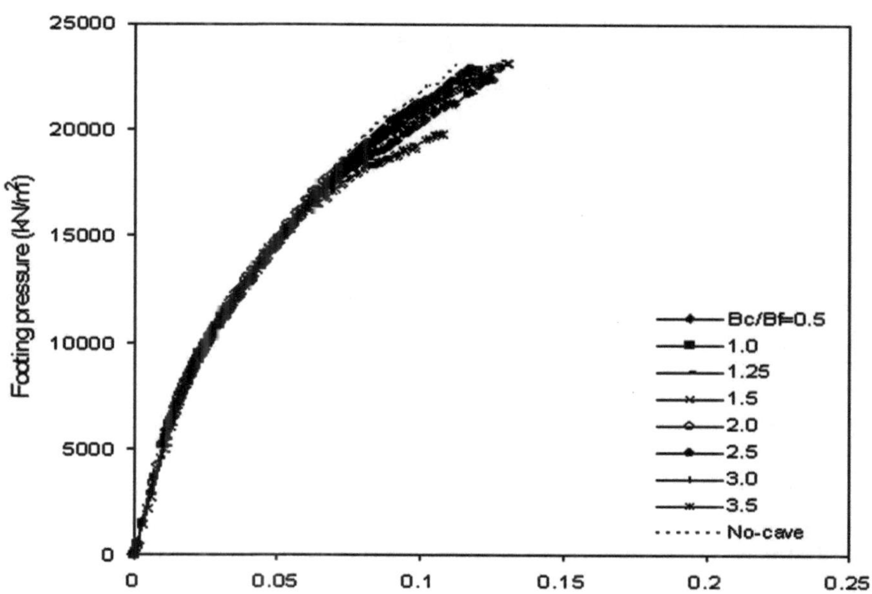

Fig. 11. Footing pressure against settlement for (Dc/Bf) = 10.0 and different (Bc/Bf) ratios

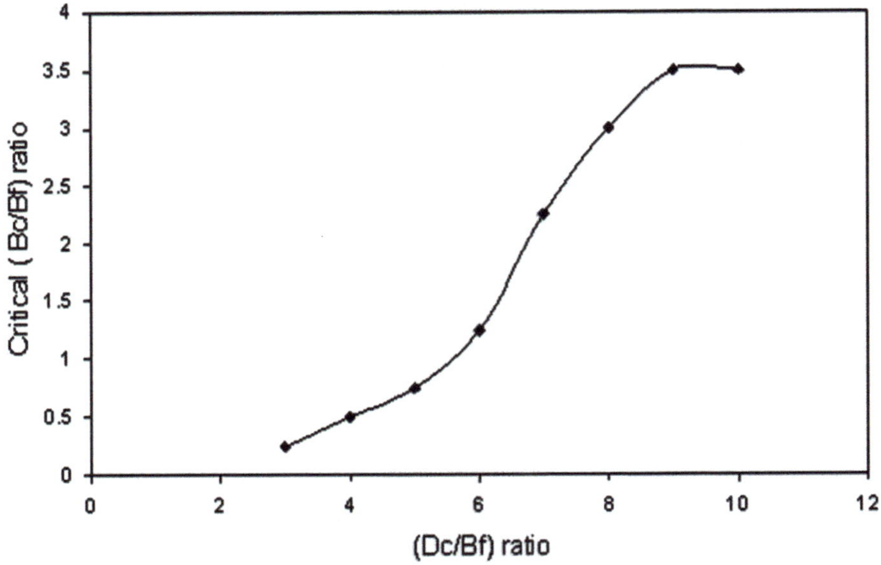

Fig. 12. Critical (Bc/Bf) ratios for different (Dc/Bf) ratios

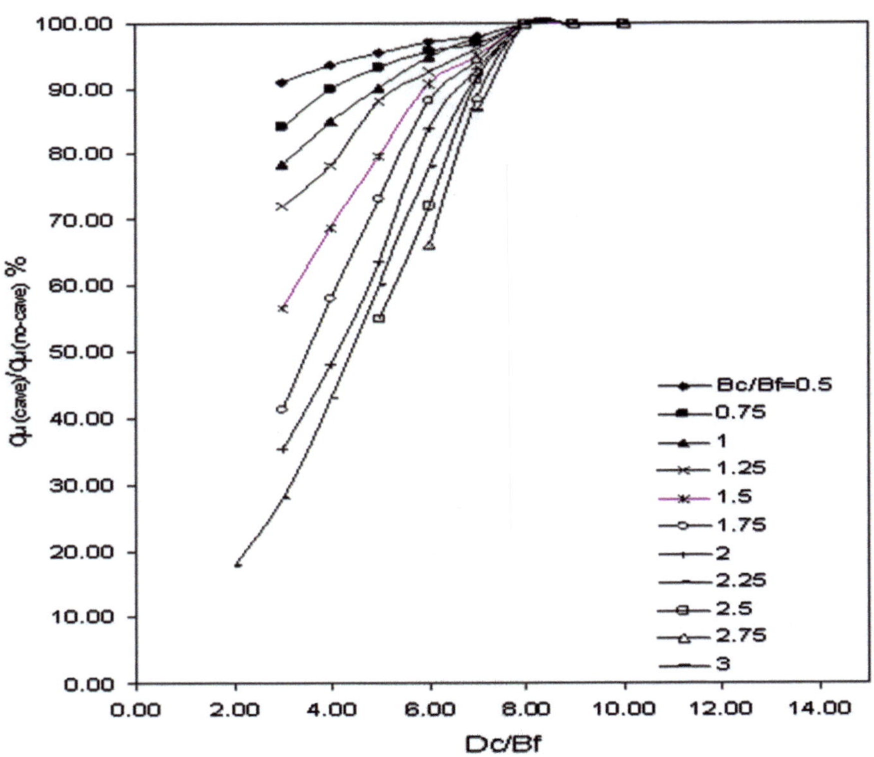

Fig. 13. Reduction ratio of bearing capacity against (Dc/Bf) ratio.

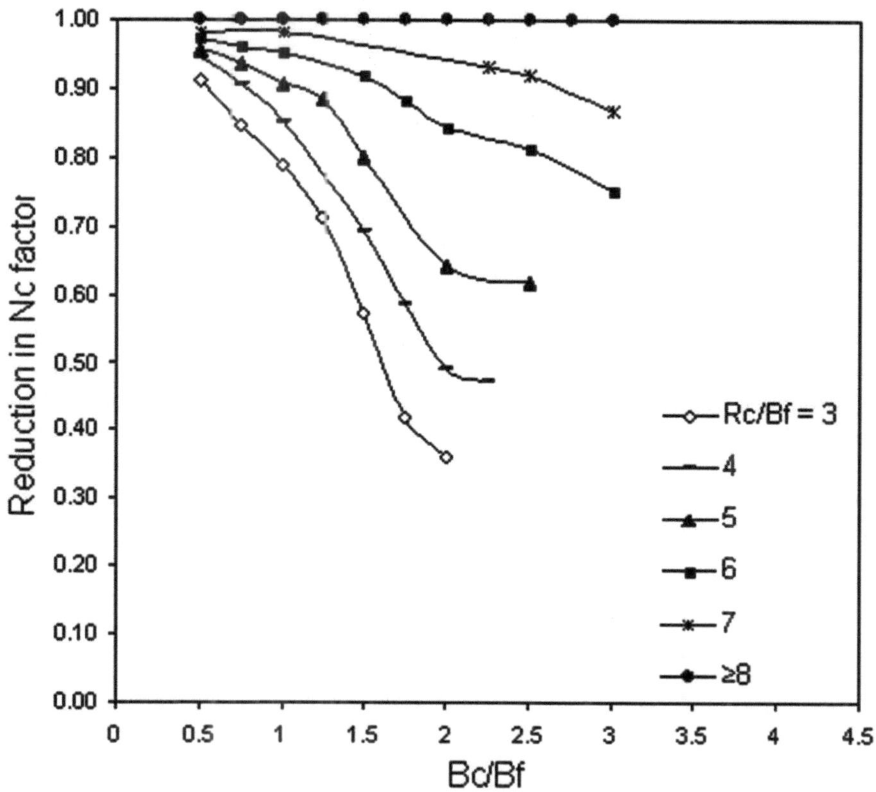

Fig. 14. Reduction in Nc factor against (Bc/Bf) for different (Rc/Bf) ratios

from 9% to 3% for (Dc/Bf) ratios ranging from 3 to 6, respectively. The inclination of curves becomes steeper as corresponding to higher cave width. For (Bc/Bf) = 1 the corresponding reduction ratios were 22% and 5% for (Dc/Bf) ratios ranging from 3 to 6, respectively. The cases of caves with (Bc/Bf) > 2.5 at relatively shallow depths were not analyzed as their model collapsed under their own weight of covering limestone soil of low strength. This was indicated by the uncompleted curves corresponding to (Bc/Bf) of 2.5, 2.75 and 3.0 as appearing in Fig. 13.

Murthy (2007), presented Terzaghi bearing capacity equation as:

$$q_u = C\,N_c + \gamma\,Df\,N_q + 0.5\gamma\,Bf\,N_\gamma$$

Where: (γ) = Unit weight of soil, (Df) = depth of foundation level, (Bf) = strip footing width and Nc, Nq, N γ are bearing capacity factors and they are a function of angle of friction of soil and may be stated as:

$$Nc = (Nq - 1) * \cot\varphi$$

$$Nq = \frac{a_\theta^2}{2\cos^2(45 + \frac{\varphi}{2})}$$

$$a_\theta = e^{\eta \tan \varphi}, \ \eta = (0.75\pi - \varphi/2)$$

$$N\gamma = \frac{1}{2}\tan\varphi(\frac{kp\gamma}{\cos^2\varphi} - 1)$$

$kp\gamma$ = passive earth pressure coefficient.

As a more critical case (and from practical point of view for foundation executed at NMC) no overburden pressure was assumed in the present study so the second term may be omitted. Depending on back substitution in bearing capacity equation with (q_u) corresponding to different studied cases and comparing the computed bearing capacity factors Nc and Nγ to those corresponding to no-cave case, reduction factors of bearing capacity factors may be computed. Figures 14 and 15 present reduction factors for Nc and Nγ, respectively for different (Bc/Bf) and (Dc/Bf) ratios.

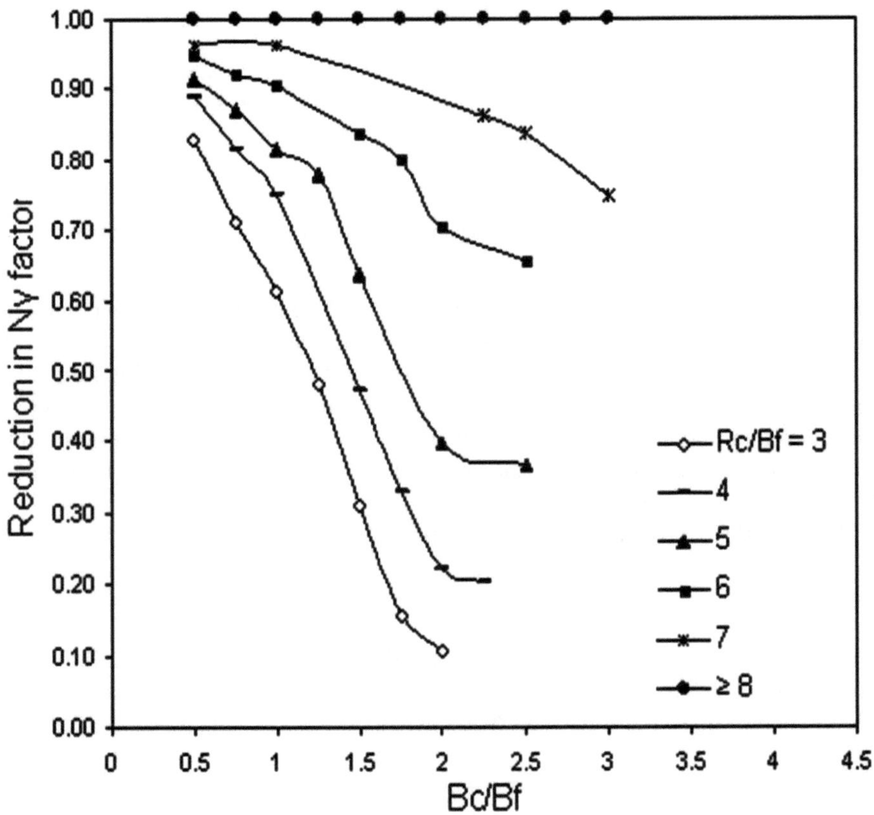

Fig. 15. Reduction in Nγ factor against (Bc/Bf) for different (Rc/Bf) ratios

5 Conclusions

Limestone voids and cavities of (NMC) are produced due to the dissolution of soil by heavy rainfall and leakage of water and sewage networks. Hence, they present an obstruction to urban development in the growing city as they may reduce bearing capacity of shallow foundation. There is no clear information relating positions, shape, dimensions and level of caves. It is recommended and further intended to conduct a detailed field study in order to supply complete information about existing and predicted caves of (NMC). The present study aims at indicating the effect of existence of caves on bearing capacity of shallow foundations. It was concluded that:

- Mode of failure of limestone soil underneath a strip footing is different from the traditional modes of shear failure.
- There exists a critical cave width (with respect to footing width corresponding to a certain depth of cave roof) beyond which caves existence underneath shallow strip footing results in reducing the bearing capacity of the foundation.
- Caves with roof level deeper than or equal to 9 with cave to footing width ratio less or equal to 3.5 do not result in reducing the bearing capacity of strip footing.
- For practical design purpose charts for reduction factors of ultimate bearing capacity and bearing capacity factors were presented.

References

Abdeltuab, S.: Karst limestone foundation geotechnical problems, detection and treatment: case studies from egypt and saudi arabia. Int. J. Sci. Eng. Res. **4**(5), 376–387 (2013). ISSN 2229-5518

Agaiby, S.W., Jones, C.J.: Design of reinforced fill system to support footing overlaying cavities. Geotext. Geomembr. **14**, 22–57 (1996)

Azam, G.: Performance of Strip Footing on Stratified Soil Deposit with Void thesis Presented to Pennsylvania Stat University, at university Park, Pa., on-partial fulfillment of requirements of the degree of Master of Science (1988)

Baus, R.L.: Bearing capacity of strip footing above voids. J. Geotech. Eng. **109**(1), 1–14 (1983)

British Geological Survey, (2016). http://www.bgs.ac.ukNERC

Canakei, H.: Collapse of caves of shallow depth in gazantep city center, turkey: a case study. Environ. Geol. **53**, 915–922 (2007)

Muhammad, R.F., Beng, Y.E.: Estimating limestone dissolution rates in the kinta and lengong valleys using micro-errosion meter, a preliminary study. In: Geological Society of Malaysia Annual Geological Conference (2002)

Hassan, A.: Engineering geomorphology assessment of new minia city, Egypt using GIS. Texas A&M University, News and Information Bulletin (2002)

Hussan, H.A., et al.: Final report on examining suitability of new minia city. Egyptian Association of Nuclear Power (1990) (in Arabic)

Murthy, V.N.S.: Advanced Foundation Engineering, Satish Kumar Jain (BS). CBS Publishers & Distributors PVT. Limited, Bangalore (2007)

Sowers, G.F.: Buildings on shrink holes, design and construction of foundations in karst terain. In: Geological Conference, p. 26. ASCE, New York (1996)
Tan, S.M.: Karstic features of Kualalampur limestone. IEM Bull. **5** (2005)
Wang, M.C., Badic, A.: Effect of underground void on foundation stability. Journal of Geotechnical Engineering **111**(8), 1008–1019 (1985). ASCE

Numerical Investigations on Lateral Load Response of Fin Piles

K.V. Babu[1,2(✉)] and B.V.S. Viswanadham[1]

[1] Department of Civil Engineering,
Indian Institute of Technology, Bombay, India
kv.babu@larsentoubro.com
[2] L & T Hydrocarbon Engineering, Mumbai, India

Abstract. Wind energy is one of the most cost effective and renewable energy options for power generation. With this aim, monopiles are widely used to support offshore and onshore wind turbines. Unlike onshore, offshore foundations are subjected to large environmental loads from wind and wave forces, which are in the order of 20 to 30% of gravity loads. These loads act at significant height on pile top with respect to seabed level, thereby causing an eccentricity. To cater for the large amount of lateral loads, provision of fins to monopiles is one of the viable options to enhance their lateral load carrying capacity. In this paper, some of the ongoing numerical model studies on the lateral load response of regular piles (pile without fins) and fin piles in sand are presented. In the present study, three dimensional finite element analyses were performed on regular piles as well as fin piles. Analyses were performed in sand with different relative densities, namely 40%, 55% and 85%. Pile material is mild steel for all analyses. Pile section is modeled as a linear elastic material and soil is modeled using Mohr-Coulomb constitutive model with non-associated flow rule. Regular and fin piles having four and eight fins are considered during the analyses. Influence of sand relative density, fins orientation and their position on lateral load response of fin piles is highlighted. The analyses and interpretation of results shown that, fins placed near the top portion of the pile are more effective than fins placed at bottom portion. Among various fins orientation, star fin piles offer more lateral load compared to straight and diagonal fin piles. Further, it was also noticed that sand relative density influences the lateral load carrying capacity of fin piles significantly.

1 Introduction

Energy demand is increasing rapidly day to day. Among various sources of energy generation, wind is one of the cleanest sources for generation. Monopile foundations are the most common foundations for offshore wind turbines. Unlike conventional piles, monopiles generally have large diameters ranging from 1 to 6 m (Lensy and Wiemann 2006) and these foundations are generally suitable for water depth up to 35 m (Doherty and Gavin 2012). Large diameter of piles is required due to presence of weak soil and less over burden soil pressure near the pile top portion. Improvement in the pile capacity can be achieved by providing fins near the top portion of the monopiles. Few investigators have been performed field tests, 1 g, Ng as well as numerical modeling on fin

H. Shehata and Y. Rashed (eds.), *Numerical Analysis of Nonlinear Coupled Problems*, Sustainable Civil Infrastructures, DOI 10.1007/978-3-319-61905-7_27

piles and it was reported that lateral load carrying capacity of the monopiles was increased significantly when fins were introduced. Duhrkop and Garbe (2008) studied the behavior of bulged piles through 1-g model studies. Bienen et al. (2012) performed series of centrifuge tests on winged piles subjected to monotonic and cyclic loading in medium dense sand. Lutenegger (2012) performed field tests on plain piles and piles with fins placed at top and bottom of piles located in sandy silty soils. Application of these piles was to support solar panels. From the field tests, it was reported that piles with fins provide higher lateral load capacity compared to plain piles. Nasr (2013) performed 1 g model and numerical model studies on rectangular and triangular fin piles in loose and dense sands. And it was reported that, for resisting lateral loads, rectangular fins are more effective compared to triangular shape fins. Peng et al. (2010) performed numerical studies in medium dense with various fin lengths and suggested optimum fin dimensions in medium dense sand. In all of the above studies it was observed that fins are effective in improving the lateral load carrying capacity of the piles.

From the above studies, it was revealed that performance of the fins to resist the lateral load is mainly depending on sand relative density, fin orientation and its position. It is evident that very limited information is available related to fin pile behavior with the above parameters. And further, increase in lateral load was not quantified with respect to fins orientation, size and its position. In view of the above, present study focuses on lateral load response of fin piles in sand with different relative densities to evaluate the influence of relative density, fins orientation and fins position in sandy soils. Figure 1 show the typical straight fin pile elevation and plan details.

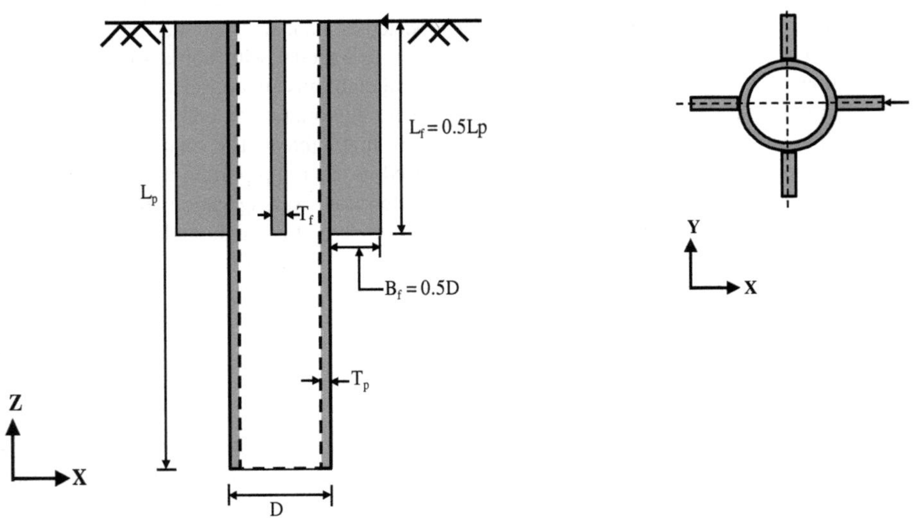

Fig. 1. Elevation and plan view of the straight fin piles

2 Numerical Model Description

The three dimensional finite element models were developed in Abaqus (Dessault systems 2012) to study the lateral load response of regular and fin piles in sandy soils. Finite element mesh type is mainly depend on loading type, number of piles to be modeled and the interface condition between pile and soil. During the analyses, fin thickness was kept constant, which is equal to pile wall thickness. Embedment length of pile (10 m) was kept constant in all the analyses. Lateral load was applied at the ground surface, where the top of the pile and ground surface coincide with each other. Figure 2 shows a schematic view of the three dimensional finite element mesh discretization of the soil and fin pile system. Using symmetry, only half of the soil and pile section in the direction of lateral load is considered for analyzing the problem. Soil and pile was modeled as a three dimensional continuum eight noded reduced integration elements (C3D8R). All boundaries of the soil are supported by roller supports, which represent the zero displacement perpendicular to the boundaries. The bottom portion of the soil is restrained in the X, Y, and Z direction. There is one plane of symmetry that allows sliding in X and Z direction as shown in Fig. 2. The accuracy finite element analyses are mainly depends on the number of elements employed in the numerical model. In order to establish approximate number of elements, mesh sensitivity analyses were performed varying from coarser to finer mesh. Series of analyses was performed with coarser mesh having 7200 elements to finer mesh which consists of 26242 elements. In the present study, typical finite element mesh consists of approximately 21784 elements, 27105 nodes and interface elements generated by Abaqus was in the order of 2432 elements.

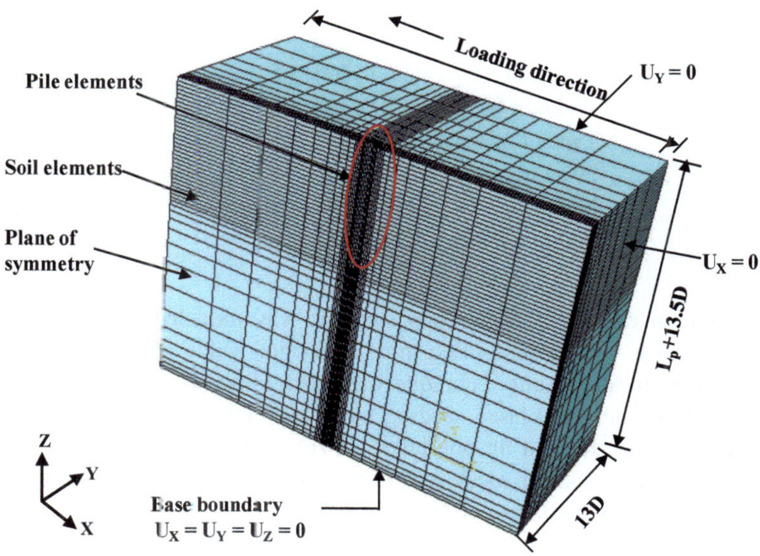

Fig. 2. Three dimensional view of the finite element mesh

Soil properties used in finite element analysis are given in Table 1. Dilation angle considered for sand during analyses was nearly one third of the friction angle (Karthigeyan et al. 2007). Fin and pile properties used in finite element analysis are given in Table 2. Pile soil interaction is surface based and therefore matching of nodes is not required in Abaqus. Coulomb friction model was used to describe the interaction along the contact surfaces between piles, fins and soil. The Coulomb friction model relates the maximum allowable shear stresses across an interface to the contact pressure between the contacting bodies. The interaction between soil and fin pile was modeled by defining tangential and regular contact behavior in the finite element model. Pile surface was modeled as a master surface while soil surface was modeled as slave surface.

Table 1. Properties of soil used in finite element analysis

Parameter	Symbol	Dense sand	Medium sand	Loose sand
Young's modulus	E[MPa]	47.5	27	20
Poisson's ratio	ν	0.3	0.3	0.3
Friction angle	φ[°]	38	36	30
Dilation angle	ψ[°]	12	10	6
Unit weight of soil	γ [kN/m^3]	16.2	15.3	14.9

Table 2. Properties of fin piles used in finite element analyses

Parameter	Symbol	Value
Pile outer diameter	D [m]	1.2
Pile wall thickness	T_p [m]	0.075
Fin wall thickness	T_f [m]	0.075
Length of pile	L_p [m]	10
Length of fin	L_f [m]	5
Width of fin	B_f [m]	0.6
Type of material	-[a]	Mild steel
Poisson's ratio	ν	0.15
Modulus of elasticity	E [GPa]	200

-[a]: Not relevant

The tangential contact between two surfaces was modeled by defining friction coefficient, which is equivalent to two third of friction value of sand. In the vertical direction uniform mesh seeding technique was used. However, in the lateral and longitudinal direction one way biased seeding mesh technique was used. From Fig. 2, it can be noticed that at soil and fin pile interface zone, finer mesh density was adopted and towards boundary coarser mesh was used. This type of meshing provides better accuracy over uniform mesh to get better results with optimum computational time. Initially, regular pile analyses were performed and then subsequently followed by fin pile analyses. Loads shown in the present study corresponds to full piles section. Lateral load carrying capacity of piles was calculated based on the reaction forces developed at nodal points corresponding to the given displacement. Ultimate lateral load carrying capacity of the pile (H_{ult}) in the present study is considered as the pile deflection that corresponds to 10% pile diameter.

3 Infleunce of Sand Relative Density

In cohesionless soils, one of the main factors influencing the lateral resistance of pile is its relative density. In the present study, to evaluate the influence of relative density on lateral resistance of fin piles with different fin configurations, three density indices are considered ranging from loose (40%), medium dense (55%) and dense state (85%) conditions. Figures 3a, b, c and d show the normalized lateral load carrying capacity of regular and fin pile with various density indices having different fin orientations. It can be observed that lateral load carrying capacity of regular pile in dense sand is more than twice compared to piles in loose sand as shown Fig. 3a. Figures 3b, c and d show the normalized lateral load response of fin piles in loose, medium dense and dense sand for various fin configurations. It can be noticed from the figures (Figs. 3b, c and d) that for piles in dense sand carry more than 50% of the load compared to loose sand irrespective of its fin orientation. Increase in lateral load in the case of regular pile as well as fin piles in dense sand was mainly due to development of high confining stresses around the fins and piles.

4 Influence of Fin Orientation

Nasr (2013) considered rectangular and triangular shape fins, while Duhrkop et al. (2010) and Bienen et al. (2012) also considered two fins with rectangular shape for short piles. Two fins were considered and are placed perpendicular to the loading direction. On the other hand, Peng et al. (2010) considered four numbers of fins and all are rectangular in shape only. In practice, lateral load may act in any direction; hence it is always encouraged to consider more than two number of fins to resist lateral loads. In the present study, four and eight numbers of fins were considered to study the behavior of pile under lateral loading. Figures 4, 5 and 6 show the variation of normalized lateral load deflections for different fins orientation in loose, medium dense and dense sand for L_f/L_p and B_f/D ratio 0.50. It can be observed from the figure that, in the case of $L_f/L_p = 0.50$, star fin configuration exhibits marginally higher lateral load capacity compared to straight fin piles irrespective of sand relative density. On the other hand, diagonal fin piles exhibit significantly lower resistance compared to straight and fin piles. The main reason to offer more load resistance of star fin piles compared to straight and diagonal fin piles was mainly due to increase in flexural rigidity of pile as well as mobilization of more frictional stress along the fins surface.

Figure 7 shows the normalised lateral deflection of fin piles along pile length for various fin orientations in medium dense sand for L_f/L_p ratio 0.25. It is evident from the figure that fin orientation is not all significant in resisting deflections for a given design load at lower L_f/L_p ratios. From this it can be concluded that fin orientation is significant when the fin length is more than $0.25L_p$. Further, it can be noted that even though star fin configuration exhibits marginal improvement in lateral load carry

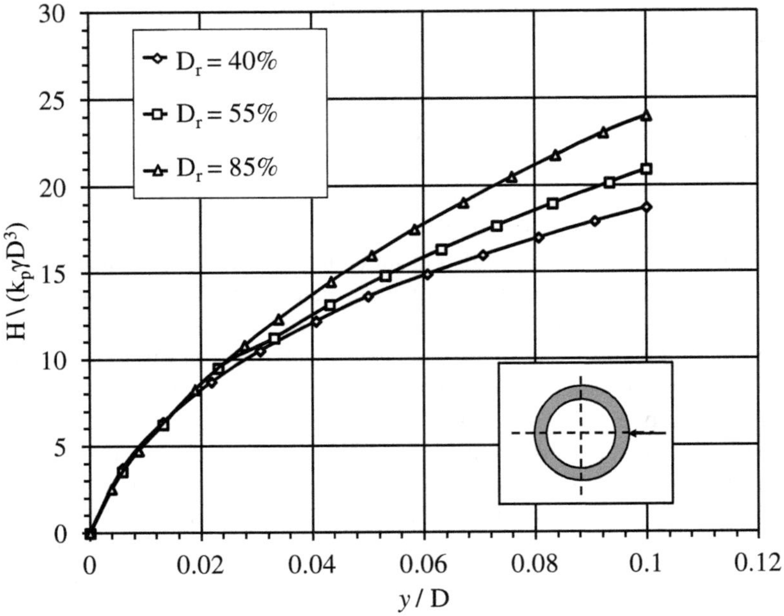

a) Regular pile

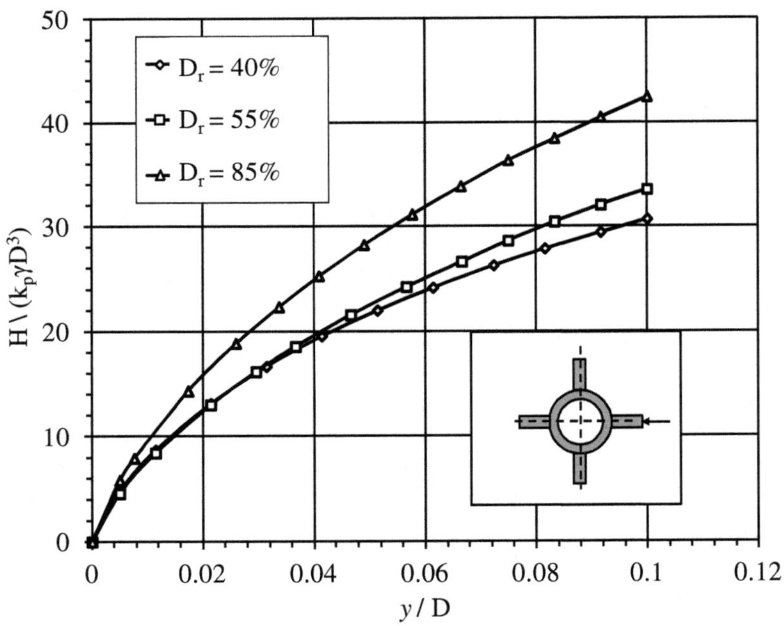

b) Straight fin piles

Fig. 3. Normalized lateral load deflection curves for regular and fin piles, (a) Regular pile, (b) Straight fin piles, (c) Diagonal fin piles, (d) Star fin piles

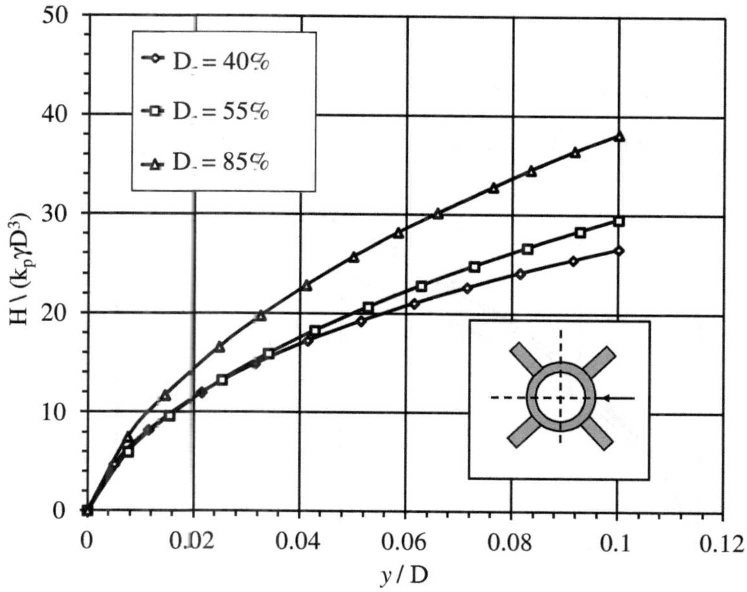

c) Diagonal fin piles

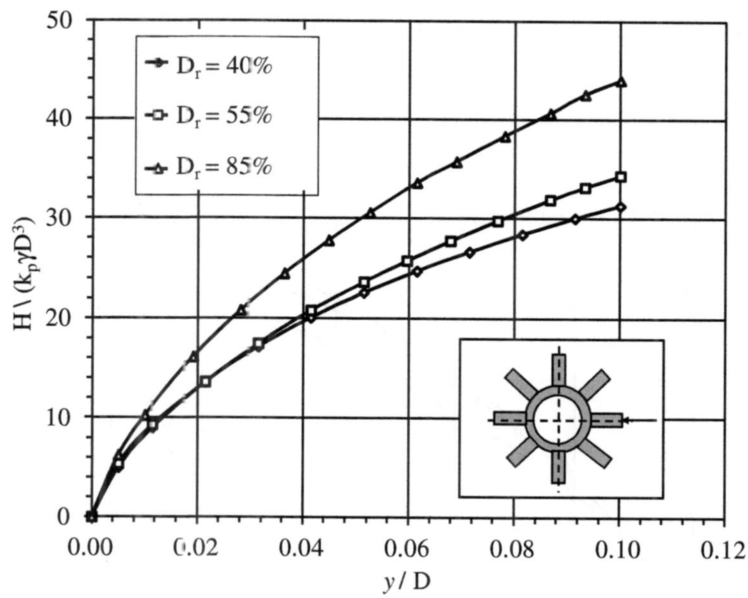

d) Star fin piles

Fig. 3. (continued)

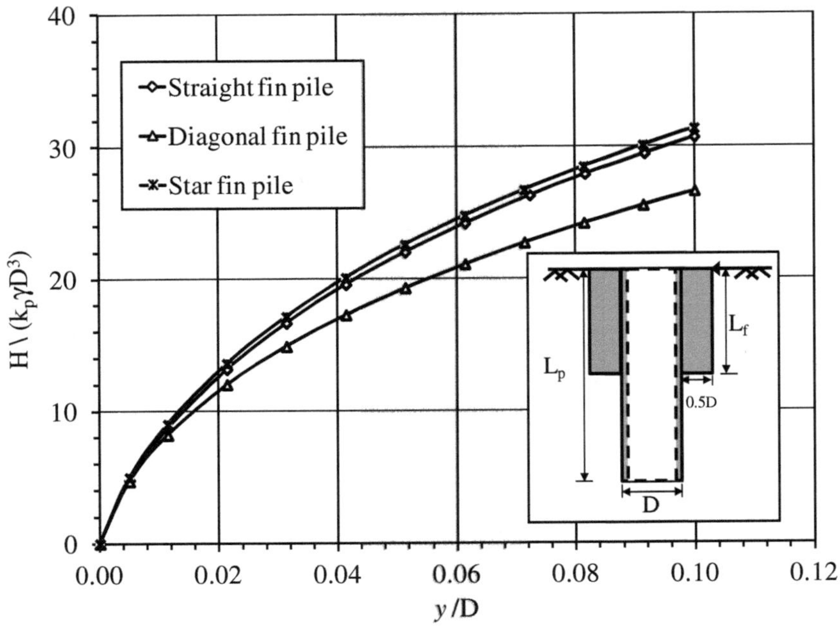

Fig. 4. Normalized lateral load deflection curves for fin piles in loose sand

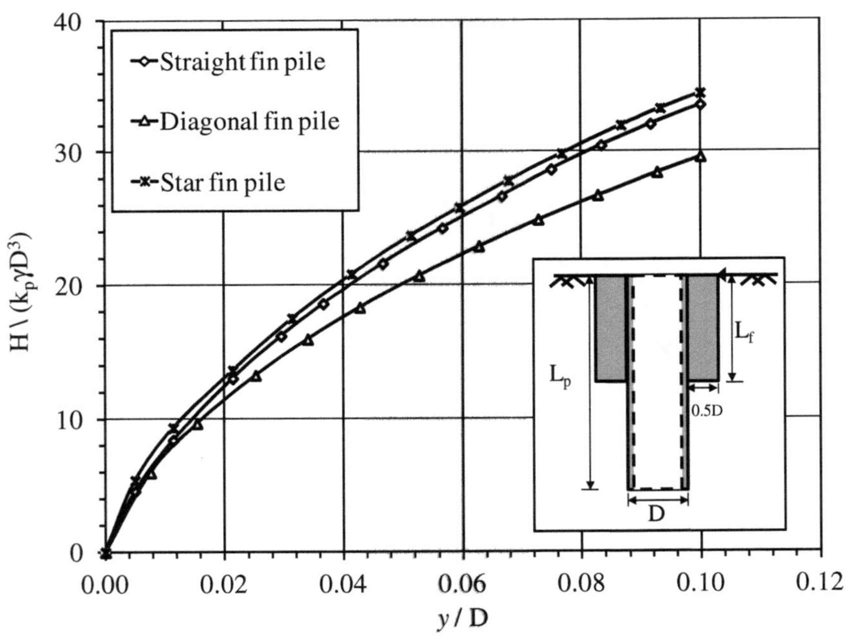

Fig. 5. Normalized lateral load deflection curves for fin piles in medium dense sand

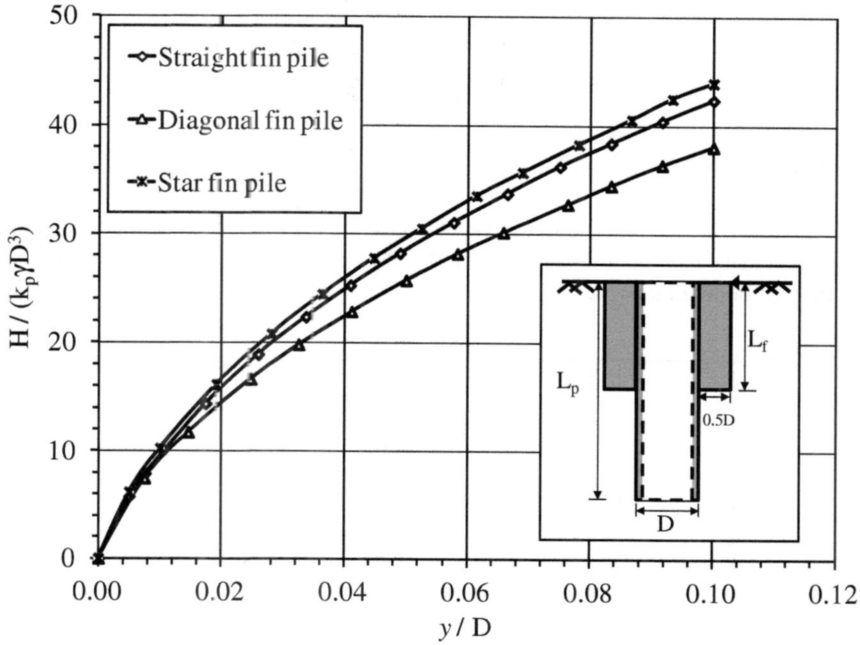

Fig. 6. Normalized lateral load deflection curves for fin piles in dense sand

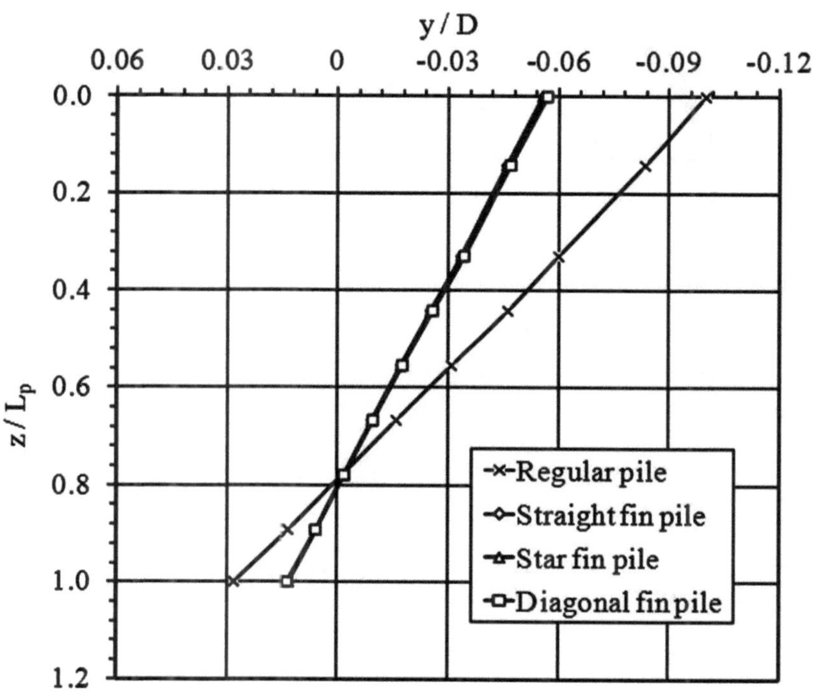

Fig. 7. Normalised lateral deflections along piles depth in medium dense sand for $L_f/L_p = 0.25$

capacity compared to straight fin piles for L_f/L_p ratio is 0.5. However, considering material requirement as well as installation efforts, straight fin piles are more preferable over star fin configurations.

5 Influence of Fin Position

From the above discussions, it could be understood that, fins increases the lateral load carrying capacity of regular pile significantly. And placing of fins along the pile length even though beneficial, which will help to improve the load carrying capacity of pile.

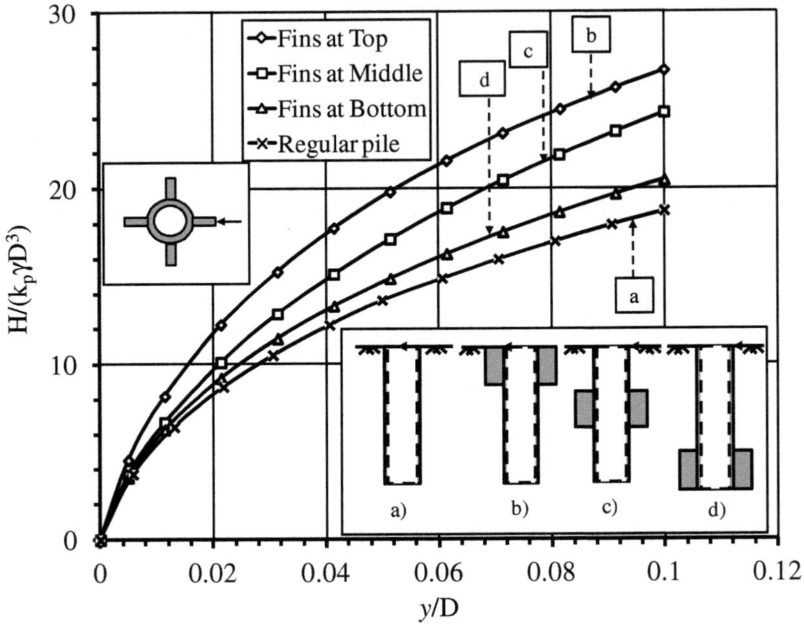

Fig. 8. Normalized lateral load deflection curves of straight fin piles in loose sand

Table 3. Influence of fin position for straight fin piles in loose sand

Description	Lateral load at deflection of 0.05D (kN)	Lateral load at deflection of 0.1D (kN)	Percentage increase in fin efficiency at deflection of 0.05D	Percentage increase in fin efficiency at deflection of 0.1D
Regular pile	1015	1445	–	–
Fins at top	1370	2060	34	42
Fins at middle	1169	1875	15	30
Fins a bottom	1025	1580	1	9

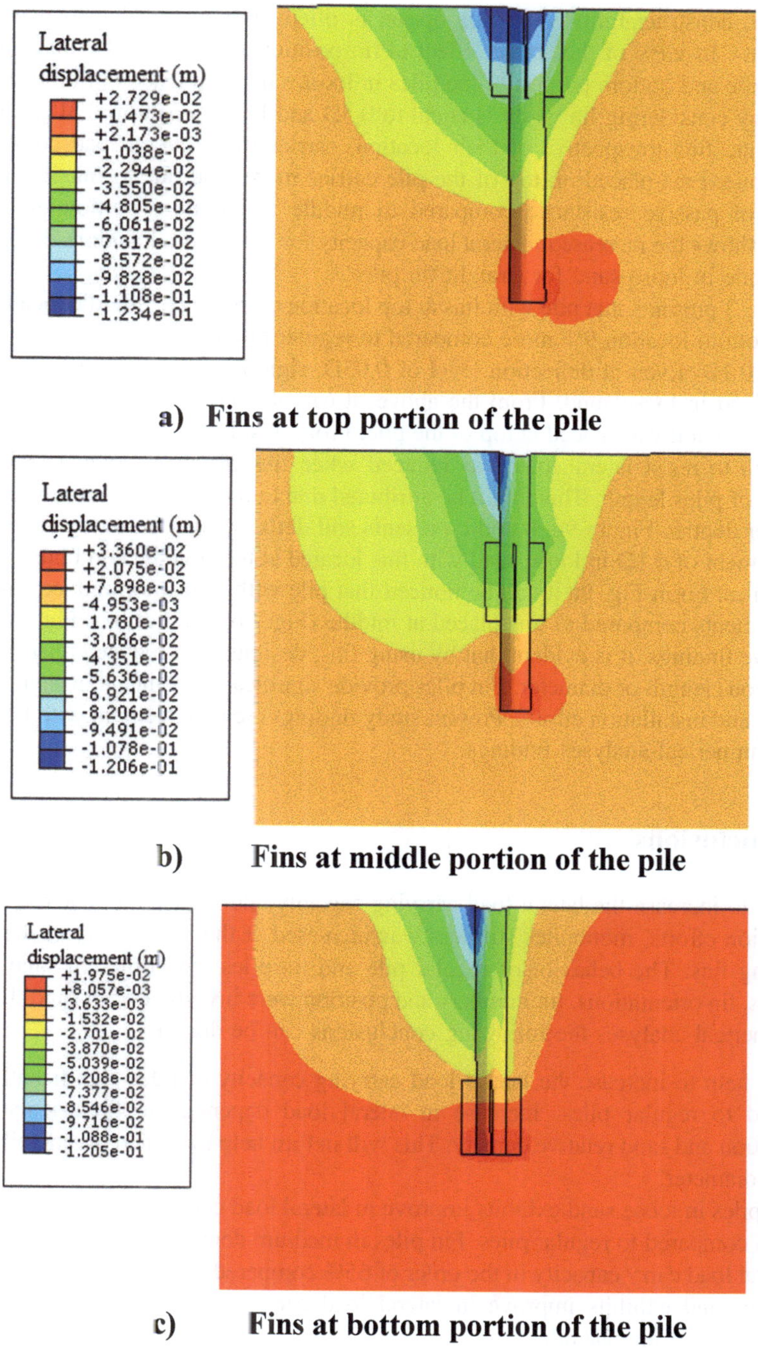

Fig. 9. Deformations of soil due to straight fin pile displacement of 0.1D in loose sand, (a) Fins at top portion of the pile, (b) Fins at middle portion of the pile, Fins at bottom portion of the pile

However, constructability view this option is often discouraged due to pile driving constraints. In view of the above, effect of fin position was studied by placing fin on top, middle and bottom portion of the piles in loose sand. Influence of fin position was studied by considering fin width is equal to 0.5D and length equal to $0.3L_p$. Figures 8 shows that, fins irrespective of their location, carries more lateral load compared to regular pile. Fins placed on top of the pile carries more load due to maximum mobilization of passive resistance compared to middle and bottom portion of the pile. Table 3 shows the increase in lateral load capacity for various fin locations compared to regular pile in loose sand for straight fin piles.

Table 3 presents that pile with fins at top location carries 42%, middle location 30% and at bottom location 9% more compared to regular pile in loose sand at a deflection level of 0.1D. Even at deflection level of 0.05D, significant increase in fin efficiency was noticed in loose sand. From the above, it can also be noticed that fins are more effective when it was placed at top of the pile. From this, it could be noticed that, effect of the fin to resist lateral load was reduced when fins placed at middle and bottom location of piles length. This might be attributed due to reduction in soil displacements at deeper depths. Figure 9a, b and c presents soil deformation contours due to fin pile displacement of 0.1D in loose sand with fins located at top, middle and bottom portion of the piles. Form Fig. 9a it can be noticed that pile with fins at top induces more soil displacements compared to fins placed at middle (Fig. 9b) and bottom (Fig. 9c). From the above findings, it is evident that by using fins, designer can reduce either monopile foundations length or diameter. Fin piles provide significant advantage in saving of pile material and installation efforts. Present study findings are in agreement with Peng et al. (2010) numerical analyses findings.

6 Conclusions

In order to increase the lateral load carrying capacity and reduce the material cost and installation efforts, monopiles stiffness was increased at the top portion of the pile by providing fins. The behavior of regular pile and fin piles with different sand relative densities, fin orientations, fin numbers and position were investigated in sand. Based on the numerical analyses the following conclusions can be drawn.

- Fins help to increase the lateral load carrying capacity of piles significantly, compared to regular piles. Increase in lateral load depends on, fin orientation, fin position and sand relative density. This will in turn help to reduce overall pile length and diameter.
- Fin piles in loose sand exhibits improve in lateral load carry capacity in the order of 60% compared to regular piles. Fin piles in medium dense sand exhibits improve in lateral load carry capacity in the order of 65% compared to regular piles. Fin piles in dense sand exhibits improve in lateral load carry capacity in the order of 75% compared to regular piles.
- At higher fin length, star fin piles carry more lateral load and is followed by straight and diagonal fin piles. This is due to increase in stiffness of the pile as well as more contact area, which in turn helps to mobilize more frictional stress between fins and

soil compared to straight and diagonal fin piles. Fins orientation is significant when L_f/L_p ratio is more than 0.25.

- Fins placed near the pile top provide more resistance than those placed near the pile bottom. Piles with fins placed at top in loose sand carries 42%, middle location 30% and at bottom location 9% more compared to regular pile.

References

Bienen, B., Duhrkop, J., Grabe, J., Randolph, M.F., White, D.J.: Response of piles with wings to monotonic and cyclic lateral loading in sand. J. Geotech. Geoenviron. Eng. ASCE **138**(3), 364–375 (2012)

Systems, D.: Abaqus Analysis User's Manual. Simula Corp, Providence (2012)

Doherty, P., Gavin, K.: Laterally loaded monopile design for offshore wind farms. Proc. Inst. Civil Eng. Energy **165**(1), 7–17 (2012)

Duhrkope, J., Grabe, J.: Laterally loaded piles with bulge. J. Offshore Mech. Arctic Eng. **130**(4), 1–5 (2008)

Duhrkope, J., Grabe, J., Bienen, B., White, D.J., Randolph, M.F.: Centrifuge experiments on laterally loaded piles with wings. In: Springman, S., Laue, J., Seward, L. (eds.) Proceedings of the 7th International Conference on Physical Modeling in Geotechnics (ICPMG), Switzerland, vol. 2, pp. 919–924. Taylor & Francis group (2010)

Karthigeyan, S., Ramakrishna, V.V.G.S.T., Rajagopal, K.: Numerical investigations of the effect of vertical load on the lateral response of piles. J. Geotech. Geoenviron. Eng. ASCE **133**(5), 512–521 (2007)

Lensy, K., Wiemann, J.: Finite element modeling of large diameter monopiles for offshore wind energy converters. In: Degroot, D.J., DeJong, J.T., Forst, D., Setunge, L.G. (eds.) Proceedings of Geocongress: Geotechnical Engineering in the Information Technology Age, pp. 1–6. ASCE, Reston, VA (2006)

Lutenegger, A.J.: Tension tests on driven piles for supporting of solar panel arrays. In: Rollins, K., Zekkos, D. (eds.) Geo Congress, State of the art and practice in Geotechnical Engineering, Oakland, California. ASCE, GSP226, pp. 305–314 (2012)

Nasr Ahmed, M.A.: Experimental and theoretical studies of laterally loaded finned piles in sand. Can. Geotech. J. **51**(4), 381–393 (2013)

Peng, J.R., Rouainia, M., Clarke, B.G.: Finite element analysis of laterally loaded fin piles. Comput. Struct. **88**(21), 1239–1247 (2010)

Experiences with Tip Post Grouted Drilled Shafts in China

Zhihui Wan(✉) and Guoliang Dai

Key of Laboratory for RC and PRC Structure of Education Ministry,
School of Civil Engineering, Southeast University, Nanjing, China
seuwanzhihui@163.com, daigl@seu.edu.cn

Abstract. It is known that the tip post-grouting technology can enhance the bearing capacity of drilled shaft and can reach the purpose of optimizing shaft length, and reducing cost. In this work, the key points of design and construction technology for tip post grouting in Chinese practices are introduced primarily. Furthermore, the mechanism of improvement for tip post-grouting drilled shafts is analyzed on the basis of the strengthening mechanism of tip post grouting, which indicates that the tip resistance of the grouted shaft is mobilized before performing vertical load due to the preloading effect of soil beneath shaft tip arising from tip grouting. Consequently, the asynchrony of the side resistance and tip resistance of pile are improved. Finally, two cases are selected to further study the influence of post grouting on the bearing characteristics of drilled shafts.

1 Introduction

Drilled shaft foundations in China have been widely used due to their ability to resist the large load and due to consideration of scour for highway bridges. However, soil relaxation beneath the shaft tip due to drilling process, which brings great influence to the tip resistance, and debris remaining after cleanout will further reduce the tip resistance (Sliwinski and Fleming 1984, Safaqah et al. 2007). Additionally, the soil stress release and disturbance surround the shaft due to drilling will also reduce the side friction (Sliwinski and Philpot 1980, Dapp et al. 2006, Safaqah et al. 2007). It can be seen that the construction process has a negative impact on the bearing capacity of drilled shafts. On the other hand, the end bearing and the side friction cannot develop simultaneously. The side friction is fully mobilized at displacements between 0.5 and 1.0% of the shaft diameter (*D*), but that end bearing fully develops at displacements of 10–15% *D* (Bruce 1986, Hirayama 1990, Mullins et al. 2000). Therefore, the end bearing is not fully mobilized within the service displacement limits, and the end bearing component is unavailable to the useful shaft capacity.

To effectively mitigate these issues, tip post-grouted technology has been used widely in China, and has been proven to be an effective method. In gravel layer, the bearing capacity of pile can increase by up to 40%; in sand layer, the bearing capacity can increase by 20%–35%; in clayey layer, the bearing capacity can increase by about 15% (Zhang 2009). However, these grouting effects vary greatly in practice. In this paper, the key points of the construction technology for tip post grouting in China are

H. Shehata and Y. Rashed (eds.), *Numerical Analysis of Nonlinear Coupled Problems*,
Sustainable Civil Infrastructures, DOI 10.1007/978-3-319-61905-7_28

expounded, the improvement mechanisms for tip resistance and side friction of drilled shafts due to the pressure grouting at the shaft tip are studied, and combined with two field case histories, the influence of one-way grouting and cyclic grouting on the bearing performance of shaft are studied, and the time effect of tip post-grouting drilled shaft is studied by static load test.

2 Overview of Chinese Practices for Tip Post Grouted Drilled Shafts

Tip post-grouting technology was first applied in Beijing in the 1980s, and the grouting method was used to strengthen the gravel stratum. Post-grouting technology, from the development of closed grouting to open grouting, from one-way grouting (the straight pipe) to cyclic grouting (U-shaped pipe), has now developed into a more perfect construction technology. The major bridge engineering has been widely used in China, such as Runyang Bridge, Donghai Bridge, Sutong Bridge and Hangzhou Bay Bridge and so on. The application of this technology has achieved remarkable economic and social benefits. At present, it has been incorporated into "Technical Code for Building Pile Foundation" (JGJ94-2008) and "Code for Design of Ground Base and Foundation of Highway Bridges and Culverts" (JTJD63-2007).

Tip post-grouting process entails installation of grouting system during the fabrication of reinforcement cage. After the concrete in the shaft has cured and gained sufficient strength, cement grout can be injected to the region around the shaft tip under high pressure. The pressure grout splits and permeates the surrounding soils and compresses any debris left by the drilling process, making the end bearing resistance stiffer and stronger. Therefore, tip post-grouting technology can increase total shaft capacity and decrease settlement to reach a purpose of optimizing shaft length, saving investment and reducing cost.

Tip grouting device is generally divided into two broad categories: straight pipe method (one-way grouting method), or U-shaped pipe method (cyclic grouting method). One-way grouting method usually consists of grouting pipes with a rubber sleeve wrapped underneath. As shown in Fig. 1(a), the straight pipe is a segment of pipe with two screw ends. The grout delivery holes, 60 mm intervals vertically and 8 mm diameter at 90° intervals circumferentially, are wrapped with rubber membrane in rows to act as one-way valves. The number of grouting pipes depends on the shaft diameter (D), generally two pipes for D < 1200 mm, three for 1200 mm < D < 2500 mm and four for D > 2500 mm. Cyclic grouting method typically consists of one grouting pipe, one return pipe and tip-manchette. As shown in Fig. 1(b), grouting pipe and return pipe are connected by two 90° elbows. The number of U-type pipes typically depends on the shaft diameter (Dapp and Mullins 2002, Dapp et al. 2006).

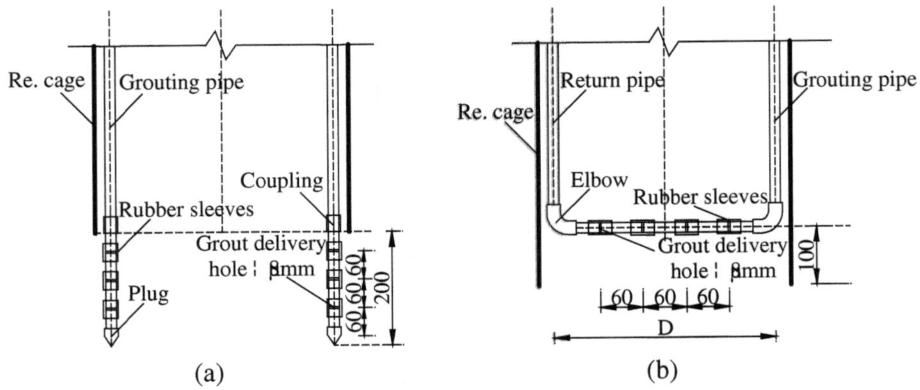

Fig. 1. Grout distributor arrangements (Unit: mm): (a) Straight (b) U-shaped

3 The Key Points of Design and Construction Technology for Tip Post Grouting

Tip pressure grouting process includes grouting pipes preset in the shaft, after the concrete in the shaft has cured, water is pumped through the grout delivery system, and the concrete cover on the tip-manchette is broken off by the water pressure, and then cement grout can be injected to the region around the shaft tip. The grouting take and the grouting pressure are both controlled during grouting process. Therefore, the grouting take and grouting pressure are the key parameters.

3.1 Water Injection Test

Water injection test is an indispensable important process before grouting. It can be proven the groutability of shaft tip by water injection test before tip pressure grouting. Water injection test, which can determines water/cement ratio, grouting amount and grouting pressure and so on, is one of the important basis for selection the grouting parameters. Additionally, water injection test is also responsible for proving and dredging grouting pipes, it can improve the special effect of the groutability of the shaft tip.

In general case, this test requires a certain time and amount of water injection. Water injection is generally controlled at about 0.6 m^3; open-plug pressure is generally less than 8 MPa. If a grouting pipe is injected water, the other grouting pipe overflows, it indicates that is connected. If single grouting pipe is injected constantly water, it also shows that water pressure test has been successful.

3.2 Grout Concentration

The rheological properties of grout with different concentrations were different. Dilute grout (with a water/cement ratio of about 0.8:1), which could be conveniently transported and had a strong penetrability, was used. Medium-concentration grout (with a water/cement ratio of about 0.6:1), which could fill or compress the surrounding soils, was used to reinforce the core portion of predetermination range. Dense grout (with a water/cement ratio of about 0.4:1), was utilized to reduce the pressure loss before curing. During the grouting process, the initial concentration of grouting depended on water injection test. The grout was generally changed from a dilute to a dense concentration.

The water/cement ratio of the grout should be determined depending on the permeability and saturation of the soil. It should be 0.5–0.7 for saturated soil and 0.7–0.9 for unsaturated soil (with a water/cement ratio of 0.5–0.6 for loose detritus soil and gravel sand). Water reducing agents generally were added to increase the pumpability of grout mixtures by increasing their fluidity, therefore, dense grout generally mixed with water-reducing admixture. The grout should be added an accelerator if the groundwater is flowing.

3.3 Determination of Grouting Amount

The amount of grout directly determines the grouting effect. There are many factors affecting the actual grouting amount for a shaft, such as the soil layer properties at the shaft tip and shaft side, shaft geometry, mud content, debris remaining after cleanout, and increment of a shaft's bearing capacity. Generally, the larger shaft diameter is, the higher a shaft's bearing capacity is. Therefore, the grouting amount is correspondingly greater. The grouting amount can be estimated using the following equation (Dai et al. 2011):

$$G_c = \alpha_p D \tag{1}$$

where G_c = grouting amount of single shaft (t); α_p = empirical coefficients, the valid ranges are given in Table 1; and D = shaft diameter.

Table 1. Empirical coefficients of grouting amount α_p

Bearing layer	Clayey soil, silt	Silty sand	Fine sand	Medium sand	Coarse sand	Gravel sand	Detritus soil
Reasonable range	2.1-2.5	2.5-3.2	2.4-2.7	2.3-2.7	3.1-3.8	3.1-3.8	2.3-2.8

3.4 Control of Grouting Pressure

The diffusion radius of grouting is closely related to the grouting pressure. Therefore, high pressure grouting can be used to carry out the shaft tip grouting, which splits the soil and helps to improve the groutability. In addition, high pressure grouting also helps to squeeze out excess water in the grouting, so that the strength of the grout stone can be enhanced. However, once the grouting pressure exceeds the shaft weight and the side resistance, it is possible to lift. To avoid heaving, the grouting pressure should be less than the sum of the shaft weight and the side resistance, divided by the shaft area, the pressure value at this time can be used as the grouting control pressure. As given below:

$$p_c = \frac{G + \pi D \sum q_{sik} l_i}{A_p} \tag{2}$$

where p_c = grouting control pressure (Pa); G = shaft weight (N); D = shaft diameter; q_{sik} = average side friction in layer i; l_i = shaft depth in layer i; and A_p = shaft area.

The working pressure of grouting should be based on the premise of little disturbance of the soil. The pressure is affected by many factors including shaft geometry, the soil layer properties at the shaft tip and length of the grouting pipes. The grouting working pressure can be estimated using the following equation:

$$p_g = p_w + \xi_r \sum \gamma_i l_i \tag{3}$$

where p_g = grouting working pressure (Pa); p_w = hydrostatic pressure at shaft tip (Pa); γ_i, l_i = effective unit weight and depth of soil in layer i above the grouting point, respectively; and ξ_r = empirical coefficient for grout resistance, it is related to geological conditions, saturation, density, grout consistency and length of the grouting pipes and so on. It should be 1.0–1.5 for soft soil; 1.5–2.0 for saturated clay, silt, silty sand; 2.0–4.0 for unsaturated clay, silt, silty sand; and 1.2–3.0 for medium and coarse sand, gravel.

It should be noted that the grouting pressure above is that at the shaft tip. However, the grouting pressure is measured at the ground surface by the grout pump. The actual pressure at the shaft tip is significantly smaller than the pressure at the outlet of the grout pump, because there are many factors affecting the actual grouting pressure, such as the line losses, grout viscosity and geometry of the grouting pipes.

Based on the described above, it is seen that due to grouting process is affected by a variety of complex factors, the grouting amount and grouting pressure should be determined according to field tests.

4 Analysis on the Mechanism of Improvement for Tip Post-grouting Drilled Shaft

4.1 The Strengthening Mechanism of Tip Post Grouting

With the extensive application of post grouting technology, many researchers around the world have investigated strengthening mechanisms for tip post-grouting drilled shaft. At present, it is considered that the strengthening mechanisms of tip post grouting on the shaft's capacity are as follows: improvement of the soil beneath shaft tip (Bruce 1986, Mullins and Dapp 2001); the pre-stressing effect of tip with side shear stress reversal along the shaft (Mullins et al. 2006, Ruiz and Pando 2009); increased tip area (Ruiz and Pando 2009, Wan et al. 2016a, b); and the upward grout penetration along the shaft side (Dai et al. 2006, Huang and Gong 2006, Wan et al. 2016a, b). However, owing to associated uncertainties, the study of strengthening mechanism is not in-depth. Therefore, it is necessary to further study the improvement mechanism of post grouted shaft.

The ground improvement of the soil beneath the tip is the main the strengthening mechanisms of post grouted shaft. The grout pressure permeates, compacts and splits the soil and any debris beneath the shaft tip and the surrounding soil to effectively strengthen them, thereby making it stronger and stiffer. A spherical grout bulb is formed by the spherical diffusion, or the soil is densified by spherical grout bulb due to tip pressure grouting, and the formation of grout bulb is believed to effectively increase the tip area. The boundary conditions at shaft-soil interface can be improved due to the penetration of the grout along the shaft, which the shaft diameter is increased and the resistance and roughness of the shaft side shear interface are strengthened. Additionally, the soil around the shaft is compacted due to the penetration upward a certain distance of the grout along the shaft side, resulting in the increase of horizontal effective stress and hence improvement of side friction. Mechanism of grouting and loading process of post grouted shaft is shown in Fig. 2.

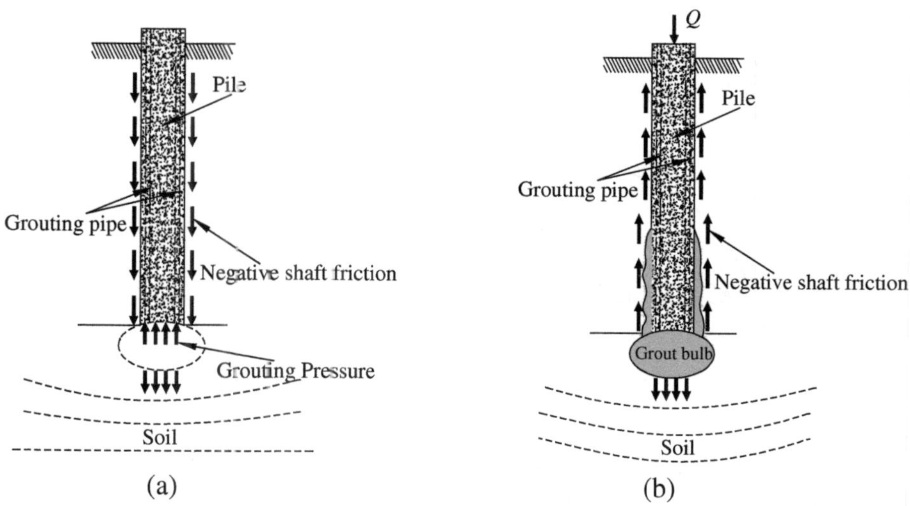

Fig. 2. Mechanism of shaft tip grouting: (a) Grouting process (b) Loading process

The grout pressure causes a bidirectional force at the shaft tip during the grouting injection process. Owing to the pre-compress effect of the grout, the grout pressure compresses the soil beneath the shaft tip and the surrounding soils. Meanwhile, the grout pressure also acts on the shaft tip, the shaft moves upward, thus the negative side friction is mobilized, as is shown in Fig. 2(a). After the grouting injection, the grout pressure at the shaft tip is gradually dissipated, and it cannot be dissipated to form a residual stress below the shaft tip. When the top of shaft is loaded with the vertical load and its downward movements tend to reverse the side friction of shaft-soil interface from negative to positive. In the mean time, the soil below the shaft tip and surrounding soils pre-complete a part of deformation, and also makes the tip resistance participate in the action in advance, as is shown in Fig. 2(b). The asynchrony of tip resistance and the side friction of shaft are effectively improved by tip post grouting, which has a significant influence on the bearing characteristics of the shaft.

4.2 The Mechanism of Improvement for Tip Post-grouting Drilled Shafts

According to the strengthening mechanism analysis of tip post grouting, the improvement mechanisms whereby tip post grouting affects tip resistance and side friction are given in Figs. 3 and 4, respectively.

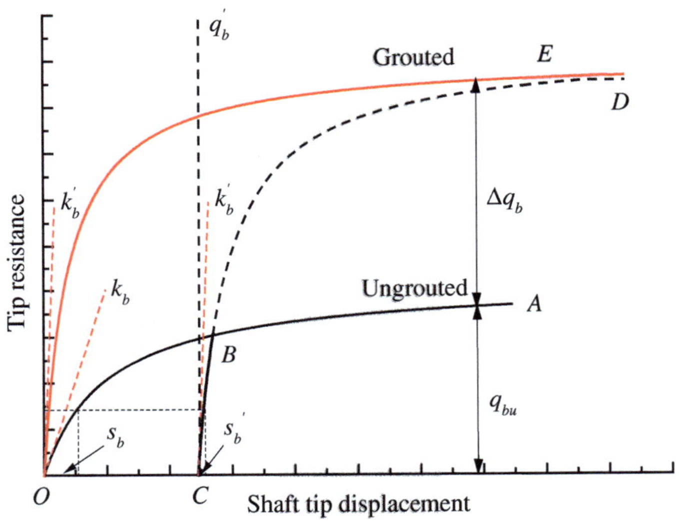

Fig. 3. Mechanism of shaft tip resistance improvement

It can be seen from Fig. 3 that curve OA represents the tip response for the shaft without tip pressure grouting, while curve OB represents tip response during the grouting injection. After the grouting injection, the grout pressure at the shaft tip is gradually dissipated, and line BC represents the complete dissipation of the grout pressure at the shaft tip. When the shaft is loaded with the structural load, the load is transferred to the shaft tip, and the tip response of the grouted shaft can be represented by curve BC (OE). The grouted shaft has less tip displacement compared with the ungrouted shaft under the same tip resistance, which indicated that end bearing capacity can be developed within smaller displacement after tip pressure grouting. Due to the grout compacts and splits the soil at the shaft tip and mixes with the soil to effectively strengthen them, making it stronger and stiffer. Additionally, the initial stiffness of the soil below the shaft tip increases up to k_b' and tip resistance of the grouted shaft is increased by Δq_b compared with the ungrouted shaft, and then the bearing performance of the shaft tip is improved.

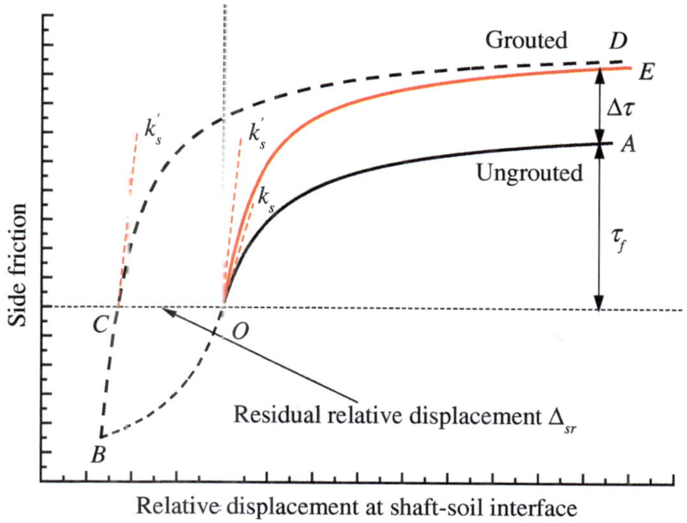

Fig. 4. Mechanism of shaft side friction improvement

It can be seen from Fig. 4 that the load transfer curve for ungrouted shaft is represented by the curve OA. The negative side friction is mobilized during the grouting injection due to the preloading of the shaft tip is shown by the curve OB. Owing to dissipation of the grout pressure, the complete dissipation of the grout pressure produces a residual displacement between shaft-soil interface. When the top of shaft is loaded with the vertical load, a part of load is offset by the prestressing and negative side friction at shaft-soil interface gradually transformed into positive side friction. The load transfer curve for the grouted shaft can be represented by curve CD (OE). The boundary conditions at shaft-soil interface can be improved due to the migration of the grout along the shaft, which the shaft diameter is increased and the strength and stiffness of the surrounding soils are improved. Similarly, the initial stiffness of the soil around the shaft increases from k_s to k_s' and side friction of the grouted shaft is increased by $\Delta\tau$ within the height of grout penetration. Therefore, tip pressure grouting has a significant influence on the bearing performance of the shaft.

Based on the mechanisms analyzed above, it can be found that the tip resistance is mobilized within smaller displacement by preloading the soil below the shaft tip due to tip pressure grouting. The strain incompatibility between tip resistance and the side friction of shaft is improved. Therefore, tip post grouting has a significant influence on the bearing performance of both tip resistance and side friction of the shaft. Total shaft capacity improvement and settlement decrease can be realized by tip post grouting.

5 Selected Case Histories From China

Case 1

The influence of one-way grouting and cyclic grouting of the shaft tip on the bearing performance of shaft are investigated using a field case study conducted in Taizhou, China. Two test shafts were one-way grouting and cyclic grouting, respectively. Both of the drilled shafts had diameter of 1.5 m and were 72 m in length. Shaft TS1 was tipped in silty clay soil while the other shaft TS2 was tipped in gravel with clayey soil. The detailed soil profiles and properties can be found in Wan and Dai (2016a, b). Both of two shafts were grouted after its static load test. The grouting pressures are 5.21 MPa and 3.5 MPa, respectively, and the grouting amounts are 4.8t and 5.0t, respectively. Load tests were conducted on both shafts in accordance with the Chinese slow maintained load procedure (it is similar to the ASTM standard loading procedure). The displacements at the top of both shafts were measured using dial gauges and stress gages were installed at the shafts. Curves of the tip resistance and displacement for both shafts are illustrated in Fig. 5. The load and displacement curves for both shafts are shown in Fig. 6.

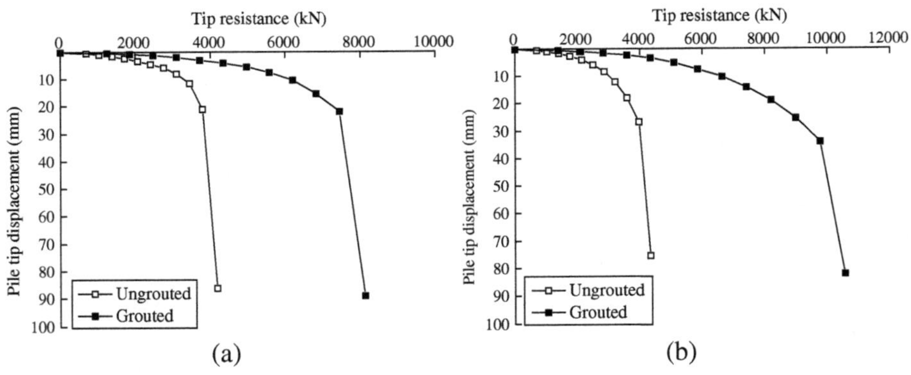

Fig. 5. Curves of the tip resistance and displacement of the shafts TS1 and TS2: (a) TS1 (b) TS2

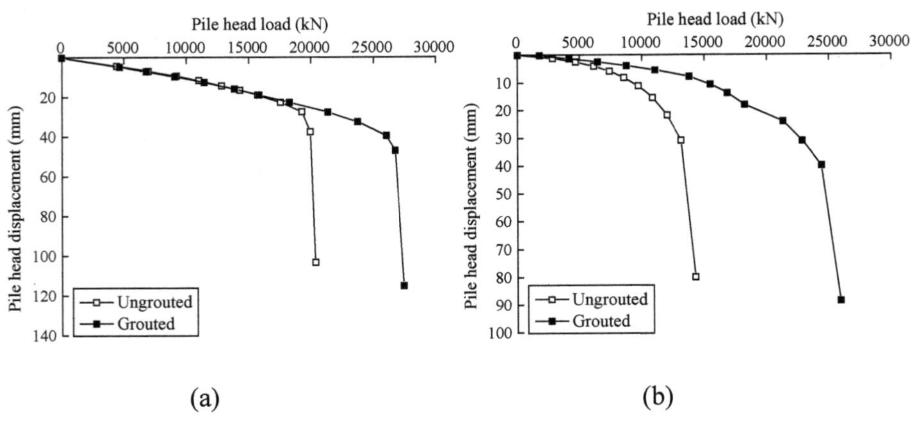

Fig. 6. Curves of the load and displacement of the shafts TS1 and TS2: (a) TS1 (b) TS2

Figure 5 shows that the change of tip resistance-displacement curves before grouting for both shafts is similar to that of after grouting. The tip displacement after grouting is less than that of before grouting under the same tip resistance, which indicated that end bearing capacity can be mobilized within smaller displacement after grouting. As mentioned earlier, the initial stiffness and tip resistance of the grouted shaft are increased compared with the ungrouted shaft, and then the bearing performance of the shaft tip is obviously improved. Figure 6 shows that the shaft head load is fully supported by the side friction under low loading level before the tip load is mobilized. The ultimate load for TS1 and TS2 after grouting is significantly larger than that of before grouting, and the increment are 34.09% and 85.52%, respectively.

Based on the results analyzed above, it can be found that the effect of tip grouting is significant. The end bearing capacity for TS1 and TS2 after grouting is significantly larger than that of before grouting, and the increment are 96.31% and 146.36%, respectively. It also reflects that the effect of U-shaped pipe grouting is more significant than that of straight pipe grouting.

Case 2

The grout is mixed with the soil below the shaft tip after tip pressure grouting. Since strength of mixture increases with time passing by, performance of total shaft is raised accordingly, which reveals time effect of post grouting. In order to confirm the time effect of post grouting, the time effect of tip post-grouting drilled shaft is studied by static load test. The test site was located in Yinchuan, China. The shaft had a diameter of 1.2 m and was 70 m in length. Shaft was tipped in fine sand. O-cell static load test was conducted on the shaft before and after the grouting, and the long-term bearing performance of test shaft was studied. The description of the grouting process can be found in Gong and Wan (2016a, b).

The displacement at the top of the shaft was measured using dial gauges and the shaft was instrumented with strain gauges along the shaft length. The elevations of strain gauges along the shaft are also shown in Fig. 7. Load test was conducted on the shaft after 28 days of base grouting, and two times of O-cell loading test were carried out at different times. After 10 months, the first long-term loading test was carried out and the second long-term loading test was carried out after 17 months.

Curves of the tip resistance and displacement for the shaft are illustrated in Fig. 8. This shows that the tip load is gradually mobilized with increasing load. The end bearing capacity after grouting is 21.07% larger than that of before grouting. The end bearing capacity at later test stage after grouting are 104.84% and 106.71% respectively larger than that of before grouting due to strength of mixture increases with time. Test shaft results at different times are given in Table 2, which shows that, the load at the shaft tip accounts for 13.11% and 13.09% of the ultimate load before grouting and after grouting, respectively. Therefore, the shaft head load is mostly supported by the side friction. As previously described, side friction of the grouted shaft is increased due to the penetration of the grout along the shaft, resulting in ratio of tip to total capacity after grouting less than before grouting. Ratio of mobilized tip load to shaft head load of the first and second long-term loading after grouting is increased, but their increasing range is gradually smaller. Therefore, tip resistance tends to be stable with time passing by.

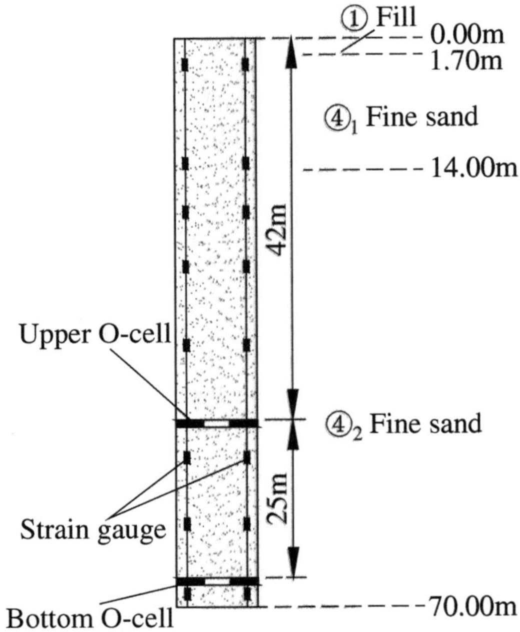

Fig. 7. Soil profile of test shaft and shaft instrumentation

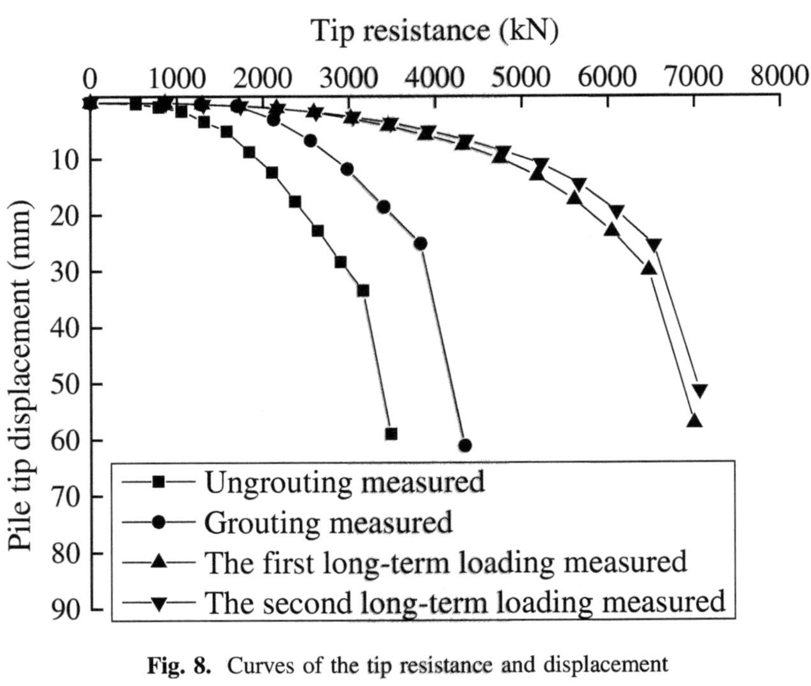

Fig. 8. Curves of the tip resistance and displacement

According to the relevant rules (Gong et al. 2002), the results of O-cell static load test shall be converted to the equivalent load-displacement curve. The equivalent load and displacement curves for the shaft are shown in Fig. 9. It can be found from Fig. 6 that there is a clear inflection point in the *Q-s* curves before grouting, the *Q-s* curves before grouting is gentler than that of before grouting, and the total capacity is 21.23% larger than that of before grouting. The shaft head load is fully supported by the side friction under less than 15000kN, which shows that the load has not been transferred to the shaft tip and tip grouting is not mobilized. The *Q-s* curves of before and after

Table 2. Test shaft results at different times

Test time	Shaft head		Shaft tip		Ratio of tip to total capacity (%)
	Ultimate load (kN)	Displacement (mm)	Ultimate load (kN)	Displacement (mm)	
Before grouting	24115	59.03	3161	33.50	13.11
After grouting	29235	58.85	3827	25.19	13.09
The first loading after grouting	32435	69.72	6475	30.00	19.96
The second loading after grouting	32435	65.14	6534	25.42	20.14

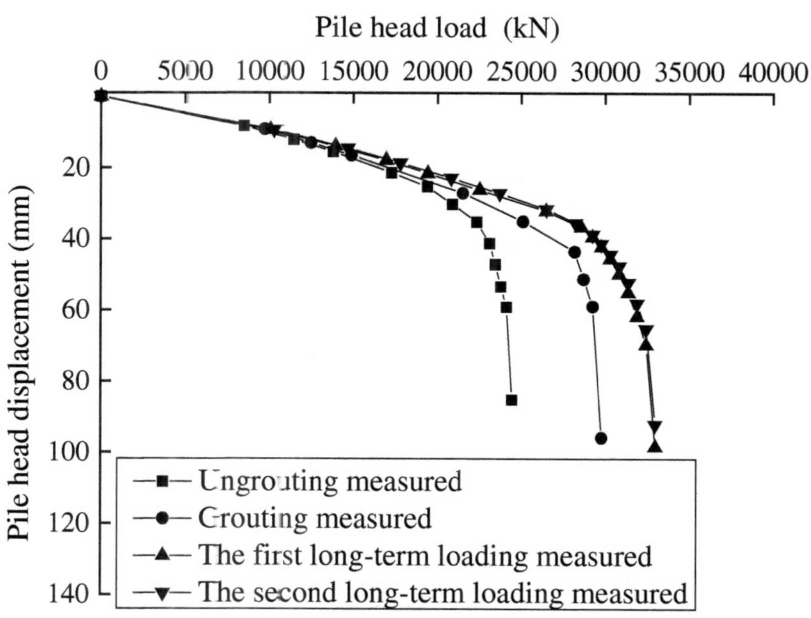

Fig. 9. Curves of the load and displacement

grouting are separated under more than 15000kN, and the total shaft capacity after grouting is obviously higher than before grouting at the same displacement condition, which indicates that tip grouting is mobilized.

The total capacity of the first long-term loading is higher than that of after grouting, which indicates that strength of mixture formed by tip post grouting at the shaft tip and part of the shaft side increases with time passing by. It can also be seen from Table 2 that the total capacity of the first long-term loading is 10.95% higher than that of after grouting. It is further studied that the total capacity of the second long-term loading after grouting has not been improved, but its settlement has been reduced. Therefore, the total capacity at later test stage after grouting are obviously higher than that of before grouting, and the settlement of tip grouted shaft decreases with the increase of time and tends to a stable value under the same load.

6 Conclusions

In this paper, the key points of the construction technology for tip post grouting in Chinese practices are expounded. The improvement mechanisms for tip resistance and side friction of drilled shafts due to the pressure grouting at the shaft tip are studied, and two Chinese case histories are selected to further study the influence of post grouting on the bearing characteristics of drilled shafts. Some key findings are summarized in the following:

1. Post grouting technology is widely used in China. China's Practical experiences and technical key points are introduced. The grouting amount and grouting pressure are two important key construction parameters in post grouting technology, which should be determined by field tests.
2. The mechanism of improvement for tip post-grouting drilled shafts is analyzed on the basis of the strengthening mechanism of tip post grouting. It shows that the tip resistance of the grouted shaft is mobilized before application of vertical load due to the preloading effect of soil beneath shaft tip arising from tip grouting, and the asynchrony of the side resistance and tip resistance of pile are improved.
3. Tip post grouting obviously increases the total capacity of drilled shafts and the grouting effect is significantly obvious. It also reflects that the effect of U-shaped pipe grouting is more significant than that of straight pipe grouting.
4. Strength of mixture formed by tip post grouting at the shaft tip and part of the shaft side increases with the increase of time. The bearing characteristics of total shaft are significantly improved, which reveals time effect of post grouting.
5. The settlement of tip grouted shaft decreases with the increase of time and tends to a stable value under the same load. Therefore, the long-term bearing performance of tip post-grouted drilled shaft should be considered in the future design and construction.

Acknowledgments. This work was support by the Major State Basic Research Development Program of China (973 program) (No. 2013CB036304), and the science and technology project on transportation construction in Yinchuan, China (K2014K063-01-05).

References

Bruce, D.A.: Enhancing the performance of large diameter piles by grouting. Ground Engineering (1986)

Dai, G., Gong, W., Xue, G., et al.: Effect examination for a base post-grouting overlength drilling pile. Rock and Soil Mechanics. **5**, 35 (2006). (in Chinese)

Dai, G., Gong, W., Zhao, X., et al.: Static testing of pile-base post-grouting piles of the Suramadu bridge (2011). doi:10.1520/GTJ102926

Dapp, S.D., Mullins, G.: Pressure grouting drilled shaft tips: Full-scale research investigation for silty and shelly sands. Geotechn. Spec. Publ. (2002). doi:10.1061/40601(256)24

Dapp, S.D., Muchard, M., Brown, D.A.: Experiences with base grouted drilled shafts in the Southeastern United States. In: Proceedings of 10th International Conference on Piling and Deep Foundations (2006)

Gong, W., Dai, G., Jiang, Y., et al.: Theory and practice of self-balanced loading test for pile bearing capacity. Journal of building structures. **1**, 14 (2002). (in chinese)

Hirayama, H.: Load-settlement analysis for bored piles using hyperbolic transfer functions. J. Geotech. Eng. Div. (1990). doi:10.3208/sandf1972.30.55

Huang, S., Gong, W.: Study on bearing behavior of super long-large diameter piles after grouting. Chin. J. Geotech. Eng. **28**(1), 113–117 (2006). (in Chinese)

Mullins, G., Dapp, S.D., Lai, P.: Pressure-grouting drilled shaft tips in sand. Geotech. Spec. Publ. (2000). doi:10.1061/40511(288)1

Mullins, G., Dapp, S., Fredrerick, E., et al.: Pressure grouting drilled shaft tips-Phase I final report. Final Report Submitted Florida Department of Transportation (2001)

Mullins, G., Winters, D., Steven, D.: predicting end bearing capacity of post-grouted drilled shaft in cohesioniess soils. J. Geotech. Geoenviron. Eng. (2006). doi:10.1061/1090-0241(2006)132:4(478)

Ruiz, M.E., Pando, M.A.: Load transfer mechanisms of tip post-grouted drilled shafts in sand. Contemp. Topics Deep Found. (2009). doi:10.1061/41021(335)3

Safaqah, O., Bittner, R., Zhang, X.: Post-grouting of drilled shaft tips on the Sutong Bridge: a case history. Contemp. Issues Deep Found. (2007). doi:10.1061/40902(221)33

Sliwinski, Z.J., Flemming, W.G.K.: The integrity and performance of bored piles. In: Proceedings of the International Conference on Advances in Piling and Ground Treatment for Foundations (1984). doi:10.1680/padt.01855.0022

Sliwinski, Z.J., Philpot, T.A.: Conditions for effective end-bearing of bored cast in-situ piles. In: Proceedings ICE Conference on Recent Developments in the Design and Construction of Pile. doi:10.1680/raditdacop.00827.0009

Wan, Z., Dai, G., Gong, W., et al.: Research of load settlement relationship for post grouted based on load transfer function method on Yueqing Bay Bridge. Bridging the East and West, ASCE (2016a). doi:10.1061/9780784479810.017

Wan, Z., Dai, G., Gong, W., et al. Experimental study on bearing behaviors of tip grouting on pile tip resistance. In: Proceedings of the 14th China National Conference of Ground Improvement (2016b). (in Chinese)

Zhong-miao, Z.: Post-grouting technology of bored piles and its engineering application. China Architecture & Building Press, Beijing (2009). (in Chinese)

Analysis on Post-peak and Creep Mechanical Behavior of Highly-Weathered Rock

Yinghua Tan[1] and Qian Zhang[2(✉)]

[1] Geotechnical and Structural Engineering Research Center, Shandong University, Ji'nan 250061, Shandong, China
sdtanyh@sina.com

[2] Structure Health Monitorng and Control Institute, Shijiazhuang Tiedao University, Shijiazhuang, China
bowanghoul@sina.com

Abstract. Highly-weathered rock mass's failure is usually caused by creeping with time and high pressure, and the rock deformation process under the two conditions was analyzed in this study respectively. Firstly, In order to further learn the rheological characteristics of soft fractured rock mass, the creep test of highly-weathered breccia with different water content was processed, and the strain-time curves under different water conditions and the corresponding variation of creep curves with different stress levels and moisture states were obtained. The curves showed that influence of water content on the rheological characteristics of rock mass was significant and softening critical load (long-term strength) and softening critical depth of soft rock were greatly reduced due to the effect of water on rock structure. Simultaneously, in-depth analysis on stress-strain curve especially after peak strength was carried out. According to piecewise function and hyperbolic function, the rule that strength parameters of rock mass after peak strength changed with the softening parameter was established. Then, the axial strain was taken as the softening parameter and the cohesion and internal friction angle were taken as the strength parameter to give post-peak stress-strain formula providing the fundamental reference for the follow-up tests and relevant engineering.

Keywords: Rheological characteristics · Highly-weathered rock · Creep deformation · Post-peak

1 Introduction

The main research object of rock mechanics is the mechanical properties of rock, and the study of which is conducive to preventing or decreasing the related natural calamities such as rock landslides, foundation instability, cracking of dam foundation and rockbursts of jambs, and so on. However, with the deepening of the knowledge about rock medium and its engineering characteristics, the elastic and elasto-plastic theories are found to often have obvious defects in the description and processing for time effect and rheological characteristics of rock material, and the time effect phenomena only can be explained from the rheological point. In many cases, the instability and failure of rock engineering always lag for some time after the completion of

H. Shehata and Y. Rashed (eds.), *Numerical Analysis of Nonlinear Coupled Problems*, Sustainable Civil Infrastructures, DOI 10.1007/978-3-319-61905-7_29

excavation or whole project. For the tunnel project, the surrounding rock presents a stable state after tunnel excavation; the deformation of surrounding rock has constant development over time, and after some time later, the tunnel may be unstable or destroyed, i.e., the deformation of tunnel surrounding rock is closely related to the time factor. With the time goes by, the significant creep phenomena can be produced in the strata of low-strength rock, expansive mudstone, soft interlayer, faultage fracture zone and fissured rock under the influence of concentrated stress caused by self-respect, tectonic stress and excavation. In addition, the existing researches show that the influence of water content on the rheological characteristics of rock mass is also significant. The softening critical load (long-term strength) and softening critical depth of soft rock are greatly reduced due to the effect of water on rock structure and mechanical parameters [1, 2].

Many academics have done a lot of theoretical explorations and experimental researches for rock rheological characteristics with increasingly abundant achievements since 30s of the 20th century [3–5]. Not only because the deformation and failure of rock material are considerably affected by different types and sizes of micro-defects inside rock, but also the rheological constitutive models describing various kinds of rock deformation and failure are still imperfect, so many challenges still would have to be overcome in theoretical study of rock rheological constitutive model. The rheological (creep) test of rock, by contrast, is the basic work of rock rheological research providing indispensable experimental basis for the establishment of rock mechanical model and promoting the study of rock rheological theory. The long-term strength characteristics of soft fractured surrounding rock with different water contents are discussed in this paper, and the creep test of highly-weathered breccia with uniaxial compression is processed to obtain the strain-time curves under different water conditions and the corresponding variation of creep curves with different stress levels and moisture states providing fundamental reference for the follow-up tests.

2 Uniaxial Creep Test of Highly-Weathered Breccia

2.1 Testing Instrument

The testing instrument adopted in this paper is low-intensity uniaxial creepmeter developed by Shandong University, which is divided into four parts of leveling system, loading system, testing system and buffer protection system with advantages of high precision, simple operation and reliable results (shown in Fig. 1).

2.2 Experimental Condition

The breccia is a kind of sedimentary clastic rock cemented by angular gravel with size of greater than 2 mm. The detrital components of highly-weathered breccia are mainly formed by detritus as well as a small amount of mineral fragments with interstitial materials of sand, silt, clay and chemical precipitation. The rock samples are directly sampled form the tunnel excavation surface in order to make the test results have better representativeness, moreover, the rock samples are rough machined into blocks with

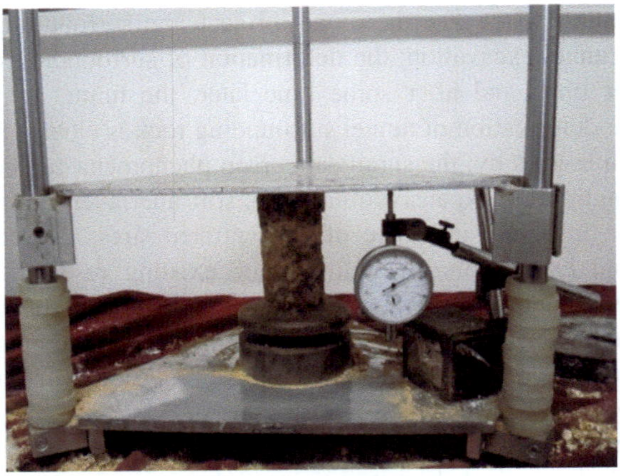

Fig. 1. Uniaxial creepmeter

horizon, azimuth and stress direction marked for the convenience of transportation and drilling in laboratory. The multilayer food wrappers are used to wrap the surfaces of samples to keep the natural water contents of specimens as far as possible and avoid the weathering effect of outside atmospheric environment on samples; the rock samples are bound by laminated papers and deposited in the well-set wooden cases filled with sawdust and leatheroid for purpose of reducing possible damage in transit, then the wooden cases are transported to the laboratory of Geotechnical Engineering Center.

The samples should be processed into specimens in time for the efficient preparation of rock specimens. The method for preparation of rock specimens in this test is that the indoor small quarrying drill is adopted to take cylinders from rock samples, and the cylinders are cut and grinded into the required specimens with size of about 10 cm in height and about 5 cm in diameter by rock cutting machine. Meanwhile, in order to ensure the consistency of specimens, the specimens of same group are densely drilled from one rock. After the drill of specimens, both end faces should be smoothed carefully with planeness less than 0.05 mm and be perpendicular to the axis of specimen with deviation less than 0.25°. Then, the specimens having defects of cracks, beddings and stripes on surface are rejected, and the rest ones are chosen by ultrasonic testing to select the specimens with good uniformity and consistency for standby application. The standard specimens are divided into three groups with different water contents for creep test, each of which includes three specimens. The specimens of 1-1, 1-2 and 1-3 are adopted for the creep test with minimum water content; the specimens of 2-1, 2-2 and 2-3 are adopted for the creep test with medium water content; specimens of 3-1, 3-2 and 3-3 are adopted for the creep test with maximum water content. The sizes of each specimen are measured by vernier caliper; the two mutually perpendicular diameters respectively corresponding to three sections of both ends and middle are measured and averaged; the lengths are measured by the symmetrical four points on the center and surrounding of specimen with averaging later.

The water content, also known as moisture content, is the ratio of lost water mass to dry mass of specimen when the rock specimen is dried to the constant, which can be expressed by percentage:

$$w = \frac{m_w - m_d}{m_d} \times 100\%$$

In which, m_w is the mass of specimen under water-bearing state, m_d is the mass of specimen after drying, w is the water content.

The water content is measured by oven-drying method based on the regulations of "Code for rock tests of hydroelectric and water conservancy engineering". The water contents of nine rock specimens used in the test are shown in Table 1. The multi-stage loading method adopted in the compression creep test of rock mass is that the different stresses are successively applied on the same specimen from low to high, and when the creep deformation of specimen is stable under a certain stress level, the next level of greater stress is applied until the specimen failures. The typical creep curves can be obtained after the procession of test results by multi-stage loading method. All the results are shown in Table 1. In the test, the loading procedures are processed for five times, and each loading weight is 2.5 kg with observation every 4 h.

Table 1. The list of masses, sizes and water contents of rock specimens

Specimen numbers	Mass/g	Diameter/cm	Length/cm	Water content
1-1	353	4.97	9.87	0.65%
1-2	360	5.02	10.15	0.65%
1-3	347	4.91	9.76	0.65%
2-1	350	4.98	9.84	1.88%
2-2	346	5.08	10.40	1.88%
2-3	352	4.96	9.94	1.88%
3-1	365	5.07	10.32	4.03%
3-2	368	5.02	10.24	4.03%
3-3	354	4.95	9.90	4.03%

3 Creep Deformation and Creep Speed

It can be seen from Table 1 that the water contents of specimens are divided into three groups of 0.65%, 1.88% and 4.03%, and each group can obtain corresponding deformation data of specimens with time, which are shown in Fig. 2. The curves of creep strain-time show that in the hierarchical loading process, the rock axial strain typically experiences two stages of initial creep and stable creep after each loading with lower stress levels. The rock will experience the third creep stage with high stress level, namely accelerated creep stage. The strain value corresponding to the creep failure increases with the increase of water content. With the increase of water content under same loading level, the creep deformation gradually becomes larger and the time of creep deformation reaching stable is gradually longer. In the whole loading process, the speed of reaching destructive creep stage is quickened with the increase of water content, and the corresponding yield strength of creep failure is also diminished. The instantaneous strain characteristics is produced in rock mass at the new added pressure moment before creep failure, i.e., the instantaneous deformation is proportional to stress level, and the instantaneous strain by creep loading under some load level gradually increases with the increase of water content.

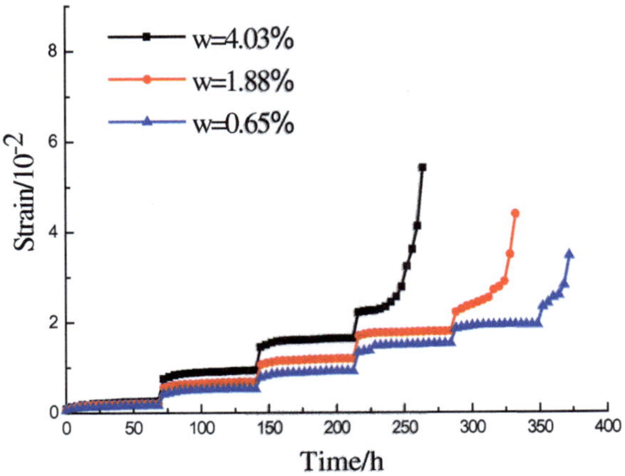

Fig. 2. Curves of creep strain-time

In the last loading process, the strain data are used to work out the corresponding creep rate, which is shown in Fig. 3. The corresponding variation of axial creep strain rate also has relevant three stages of initial creep rate, stable creep rate and accelerated creep rate in time sequence. The creep rate quickly decays to a constant with the growth of time at the first stage; the creep rate basically remains unchanged (the value is zero

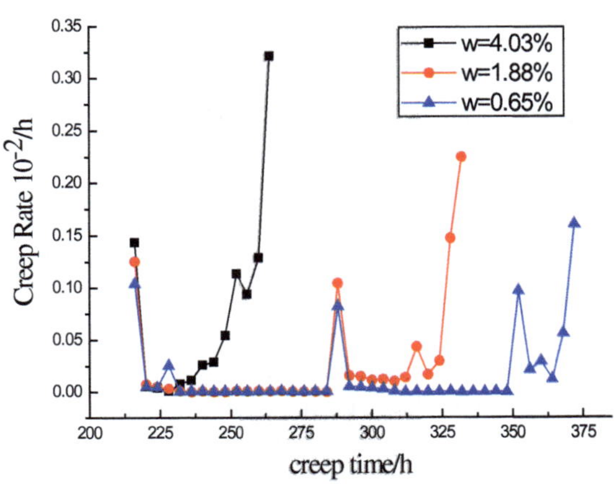

Fig. 3. Curve of creep rate-time

or constant) with the growth of time at the second stage, and the creep strain tends to a steady value (i.e. ultimate creep strain) or is proportional to a certain constant; the creep rate does not remain stable at a limit value but increases quickly until the rock failures at the third stage.

4 Post-peak Stress-Strain Analysis

It is generally believed that post peak rock mass has two stages of strain softening and plastic flow [6–10]. The stress value (point 1) is the peak strength σ_{1p}, and the corresponding strain was ε_{1p}. In the curve section of strain softening (corresponding to the AB section of Fig. 4), the stress decreased with the increase of the strain, and strain softening occurred. When the confining pressure became higher, the final macro crack and the maximum principal stress direction formed a certain angle. Caused by the axial stress, failure surface of the specimen formed at this period, and the mechanical properties of stress decreased sharply, but the strain continued to increase. The stress value of point C was called the residual strength σ_{1r}, the corresponding residual strain value was called ε_{1r}. In the plastic flow segment (corresponding to BC section of Fig. 4), the stress remained basically unchanged ($\sigma = \sigma_{1r}$), while the strain still continued to increase with time. At this time, the specimen had been completely destroyed and the strength of samples was gradually stabilized with the increase of plastic deformation.

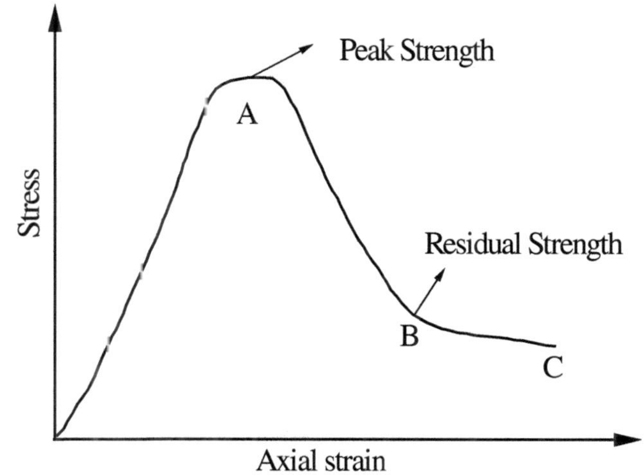

Fig. 4. Typical stress-strain curve

In this study, based on Mohr-Coulomb strength criterion, the stress-strain formula formed of hyperbolic functions was proposed to express the rules of each post-peak strength parameters of the rock mass (only in the condition that the post-peak cohesion and internal friction angle changed with same trend), the relationship of strength parameters and strain softening parameters was established, whose expression was:

$$\xi(\gamma) = \gamma_p(\xi_p - \xi_r)/\gamma + \xi_r, \gamma > \gamma_p \tag{1}$$

In which, γ was strain softening parameter, γ_p was the value of γ at peak, $\xi(\gamma)$ was the strength parameter, ξ_p was the value at peak, ξ_r was the value of the residual strength parameter. The relationship between the shear parameters cohesion and internal friction angle and axial strain was established as follows:

$$c(\varepsilon) = \begin{cases} c_p, 0<\varepsilon<\varepsilon_p \\ c(\varepsilon) = \varepsilon_p(c_p - c_r)/\varepsilon + c_r, \varepsilon > \varepsilon_p \end{cases} \tag{2}$$

$$\varphi(\varepsilon) = \begin{cases} \varphi_p, 0<\varepsilon<\varepsilon_p \\ \varphi(\varepsilon) = \varepsilon_p(\varphi_p - \varphi_r)/\varepsilon + \varphi_r, \varepsilon > \varepsilon_p \end{cases} \tag{3}$$

In which, ε_p, c_p and φ_p were axial strain, cohesion and internal friction angle at the peak, c_r and φ_r were cohesion and internal friction angle at the beginning of the residual phase, the corresponding evolution curve was shown in Fig. 5:

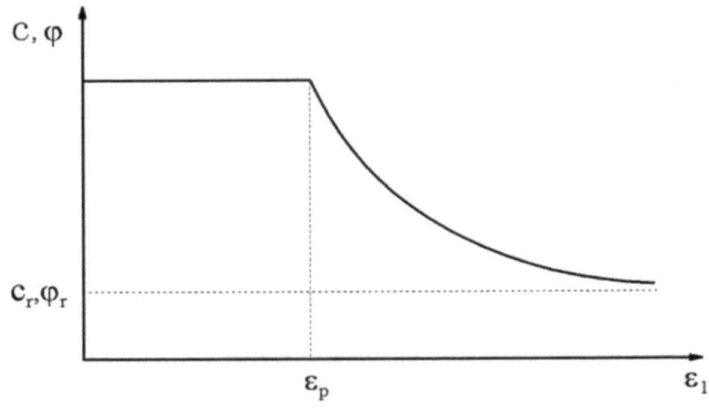

Fig. 5. Evolutional curve of shear strength parameters

Since the Mohr-Coulomb strength criterion was expressed as

$$\sigma_1 = \frac{1 + \sin\varphi(\varepsilon)}{1 - \sin\varphi(\varepsilon)}\sigma_3 + \frac{2c(\varepsilon)\cos\varphi(\varepsilon)}{1 - \sin\varphi(\varepsilon)} \tag{4}$$

Take (2) – (3) into Eq. (4), the rock stress - strain relationship could be obtained.

5 Conclusion

The instability of rock mass is controlled by creep characteristics of rock. The creep test of highly-weathered breccia with different water content is processed to obtain the strain - time curves under different water conditions and the corresponding variation of creep curves with different stress levels and moisture states, and all the results show the following laws:

(1) The creep failure is produced inside rock with low strain; the strain of creep failure increases and the corresponding yield strength increases with the increase of water content, which can explain that the softening critical load (long-term strength) and softening critical depth of soft rock are greatly reduced due to the effect of water on rock structure and mechanical parameters.
(2) With the increase of water content under same loading level, the creep deformation gradually becomes larger, the time of creep deformation reaching stable is gradually longer, and the instantaneous strain by creep loading gradually increases.
(3) In the whole loading process, with the increase of water content, the initial instantaneous strain gradually is increased and the speed of reaching destructive creep stage is quickened; the creep strain caused by creep effect (i.e. the total strain minus initial strain) increases with the increase of loading level.
(4) According to piecewise function and hyperbolic function, the rules that strength parameters of rock mass after peak strength changed with the the softening parameter was established. Then the axial strain was taken as the softening parameter and the cohesion and internal friction angle were taken as the strength parameter to give post-peak stress - strain formula. All that could provide the fundamental reference for the follow-up tests and relevant engineering.

Acknowledgements. This work was supported by the National Natural Science Foundation of China (51609138). Great appreciation goes to the editorial board and the reviewers of this paper.

References

1. Yan-qing, W.: Groundwater flow and geological hazards. J. Undergr. Space **19**(4), 303–310 (1999)
2. Li, P., Liu, J.: Experimental and theoretical studies on the effects of water content on shear creep behavior of weak structural plane of sandstone. J. Hydrogeol. Eng. Geol. **6**, 49–53 (2009)
3. Sun, J., Hu, Y.Y.: Time-dependent effects on the tensile strength of saturated granite at Three Gorges Project in China. Int. J. Rock Mech. Min. Sci. **34**, 381–384 (1997)
4. Maranini, E., Brignoli, M.: Creep behavior of a weak rock: experimental characterization. Int. J. Rock Mech. Min. Sci. **36**(l), 127–138 (1999)
5. Li, Y.-S., Xia, C.-C.: Time-dependent tests on intact rocks in uniaxial compression. Int. J. Rock Mech. Min. Sci. Geomech. **37**, 467–475 (2000)

6. You, M., Hua, A.: Triaxial confining depressure test rock sample. Chin. J. Rock Mech. Eng. **17**(1), 24–29 (1998)
7. Wang, X.: Characteristics of post-peak deformations of rock in uniaxial compression based on gradient-dependent plasticity. Chin. J. Rock Mech. Eng. **23**(Supp. 1), 4292–4295 (2004)
8. Shengqi, Y., Weiya, X.U., Chengdong, S.: Study on the deformation failure and energy properties of rock specimen in uniaxial compression. Acta Mech. Solida Sin. **27**(2), 213–216 (2006)
9. Zhang, C.-H., Zhao, Q., Huang, L., et al.: Post-peak strain softening mechanical of rock considering confining pressure effect. Rock Soil Mech. **31**(Supp. 2), 193–197 (2010)
10. Wang, S.-L., Wang, W., Wu, Z.-J.: Study of relationship between evolution of post-peak strength parameters and stress-strain curves of geomaterials. Chin. J. Rock Mech. Eng. **29** (8), 1524–1529 (2010)

Author Index

H. Shehata and Y. Rashed (eds.), *Numerical Analysis of Nonlinear Coupled Problems*,
Sustainable Civil Infrastructures, DOI 10.1007/978-3-319-61905-7

MIX
Papier aus verantwortungsvollen Quellen
Paper from responsible sources
FSC® C105338

If you have any concerns about our products,
you can contact us on
ProductSafety@springernature.com

In case Publisher is established outside the EU,
the EU authorized representative is:
Springer Nature Customer Service Center GmbH
Europaplatz 3, 69115 Heidelberg, Germany

Printed by Libri Plureos GmbH
in Hamburg, Germany